BREWING SCIENCE, TECHNOLOGY AND PRINT, 1700–1880

Science and Culture in the Nineteenth Century

Series Editor: Bernard Lightman

Titles in this Series

1 Styles of Reasoning in the British Life Sciences:
Shared Assumptions, 1820–1858
James Elwick

2 Recreating Newton: Newtonian Biography and the Making of
Nineteenth-Century History of Science
Rebekah Higgitt

3 The Transit of Venus Enterprise in Victorian Britain
Jessica Ratcliff

4 Science and Eccentricity: Collecting, Writing and Performing Science for
Early Nineteenth-Century Audiences
Victoria Carroll

5 Typhoid in Uppingham: Analysis of a Victorian Town and School in Crisis,
1875–1877
Nigel Richardson

6 Medicine and Modernism: A Biography of Sir Henry Head
L. S. Jacyna

7 Domesticating Electricity: Expertise, Uncertainty and Gender, 1880–1914
Graeme Gooday

8 James Watt, Chemist: Understanding the Origins of the Steam Age
David Philip Miller

9 Natural History Societies and Civic Culture in Victorian Scotland
Diarmid A. Finnegan

10 Communities of Science in Nineteenth-Century Ireland
Juliana Adelman

11 Regionalizing Science: Placing Knowledges in Victorian England
Simon Naylor

12 The Science of History in Victorian Britain: Making the Past Speak
Ian Hesketh

13 Communicating Physics: The Production, Circulation and Appropriation of Ganot's Textbooks in France and England, 1851–1887
Josep Simon

14 The British Arboretum: Trees, Science and Culture in the Nineteenth Century
Paul A. Elliott, Charles Watkins and Stephen Daniels

15 Vision, Science and Literature, 1870–1920: Ocular Horizons
Martin Willis

16 Popular Exhibitions, Science and Showmanship, 1840–1910
Joe Kember, John Plunkett and Jill A. Sullivan (eds)

17 Free Will and the Human Sciences in Britain, 1870–1910
Roger Smith

18 The Making of British Anthropology, 1813–1871
Efram Sera-Shriar

Forthcoming Titles

Science and Societies in Frankfurt am Main
Ayako Sakurai

The Making of Modern Anthrax, 1875–1920: Uniting Local, National and Global Histories of Disease
James F. Stark

The Medical Trade Catalogue in Britain, 1870–1914
Claire L. Jones

Uncommon Contexts: Encounters between Science and Literature, 1800–1914
Ben Marsden, Hazel Hutchison and Ralph O' Connor (eds)

The Age of Scientific Naturalism: Tyndall and his Contemporaries
Bernard Lightman and Michael S. Reidy (eds)

Astronomy in India, 1784–1876
Joydeep Sen

BREWING SCIENCE, TECHNOLOGY AND PRINT, 1700–1880

BY

James Sumner

Published by the University of Pittsburgh Press, Pittsburgh, Pa., 15260

Printed on acid-free paper
10 9 8 7 6 5 4 3 2 1

Cataloging-in-Publication is available from the British Library

ISBN 13: 978-0-8229-8166-4 Hardback: 978-1-84893-423-8
ISBN 10: 0-8229-8166-1

CONTENTS

For Leucha

ACKNOWLEDGEMENTS

Authorial self-positioning should never be taken at face value, as this book aims to make clear, and one of the most problematic conventions of print is the myth of the sole author. Every publication is to some extent a collaborative effort, and this one owes a great deal to a great many people.

The text is descended distantly from a PhD thesis completed in the Division of History and Philosophy of Science, University of Leeds, and was largely composed at the Centre for the History of Science, Technology and Medicine, University of Manchester. It has benefited significantly from the advice and support of many staff and graduate students including, at Leeds, Graeme Gooday (supervisor of the PhD), Jon Topham (who first focused my attention on the power of print) and John Christie; and at Manchester, John Pickstone (for general guidance on the fundamental interconnectedness of all things) and the members of the History of the Physical Sciences and Technology Reading Group, who generously read and commented on several drafts, and whose convenor, Jeff Hughes, has done more than anyone to sharpen my standards concerning authoritative source coverage.

Many important questions were opened, and many others resolved, through talk or email exchanges with colleagues elsewhere, including Will Ashworth, Alan Gall, Jan Golinski, Frank James, Anita McConnell, the late John Millburn, David Philip Miller, Alison Morrison-Low, James Nicholls, Richard Noakes, Jonathan Reinarz and Anna Simmons. Parts of chapters 2 and 4 derive from research conducted for articles in *Ambix* and the *British Journal for the History of Science*, and thus benefited from the editors' and referees' guidance. Chapter 6 was largely inspired by the great Jack Morrell, who kindly lent me his notes on the 1804–6 Scottish Excise experiments, and was notably tolerant when I returned them just under a decade later.

As the draft began to resolve itself into something like a monograph with a head and a tail, Simon Schaffer offered useful advice on publication routes, and Mick Worboys and Aileen Fyfe were, in their different ways, instrumental in affirming the point that it would be better for everyone were the thing completed rather than otherwise. An especially crucial contribution came from Sophie Forgan, who was able to explain to me what the book is about (see p. 1) at a time when I had very little idea myself.

Research along the way relied heavily on efforts by the staff of the various archives identified on p. 259, and also of Special Collections at the Brotherton Library, University of Leeds, home to many of the print volumes cited. I received particular help from Don Marshall and Eleanor Possart in Special Collections at Oxford Brookes University, home of the National Brewing Library; Mary Kelleher of the RDS Library, Dublin; Nichola Court, formerly of the Royal Society Library and Archives; Phoebe Fox-Bekerman at the Royal Society of Arts Archives; and, alas, more people at the London Metropolitan Archives than I can account for.

The project was also sustained by a less formal support network of friends, colleagues and helpful citizens. To list only the most readily defined contributions, Michael Stuart Williams equipped the corporate authorial persona with the ability to read Latin, and performed useful research on 'Mr Allum'. Dave Pritchard and Isabel Stainsby provided a base for some of the research expeditions, confirming the marked tendency in the history of chemistry for Mancunian ideas to prosper best through assistance from Glasgow. My thanks also to my parents, who have been putting up with this business for rather longer than is usually necessary; and to all who have helped to keep me sane or something like it.

The book's emergence also benefited from the efficiency of Bernie Lightman as series editor, and to two referees whose extensive and thoughtful comments prompted significant and useful revisions. At a later stage, Ray Anderson and Martyn Cornell both read through the whole manuscript, rescuing it from several imbecilities of detail. For an exemplary production experience at Pickering & Chatto, I must thank Ruth Ireland, as commissioning editor, Frances Lubbe, for her scrupulous copy-editing, and Eleanor Hooker, by whose efforts you were able to find this book in the first place. My thanks also to the unsung graphic designers, whose effect on the credibility of any print text is as crucial as it is subtle.

The images appear through the efforts of those colleagues at the British Museum, University of Liverpool Library, National Library of Scotland, Science and Society Picture Library and Wellcome Library who deal daily with the unfathomable business of digital-era permissions clearance.

These people, and numerous others, made this book; but the fiction of the solitary author must be preserved, if only in the interests of keeping the catalogues comprehensible. Those errors and inanities that survive in the text are, as such, my responsibility alone.

A note concerning the dedication. I briefly considered following the example of William Littell Tizard, one of my pet brewer-authors, who showed his customary enterprise in dedicating his debut work of 1843 to every single one of his readers individually (p. 207). I am certainly much obliged to you for investigating this volume, and hope, as did Tizard, that you will find its contents both credible and useful. The prime dedication, however, can only be to Leucha Veneer, who has experienced several years of its gestation at alarmingly close quarters, and who, while I have frequently driven her to distraction, has consistently driven me only to important destinations lacking in adequate rail links. My heartfelt thanks to her.

LIST OF FIGURES

PRINCIPAL DRAMATIS PERSONAE

Accum, Friedrich (1769–1838). Chemical analyst, retailer and lecturer, 'London's pet chemist' in the 1810s. His sensational *Treatise on Adulterations* (1820), together with a brewing treatise compiled from second-hand knowledge, earned him the hostility of the trade.

Banks, Joseph (1743–1820). De-facto Perpetual President of the Royal Society from 1778; professional octopus, wire-puller and relentless inserter of fingers into pies. Orchestrated philosophical involvement in the Excise project around 1800, and probably responsible, via William Brande, for safeguarding the alcoholometric interpretation of drink.

Baverstock, James (1741–1815). Hampshire brewer who adopted wort hydrometry in the 1760s and promoted it privately to other brewers, gaining support from Henry Thrale. Later engaged in a short pamphlet war with John Richardson over priority in the invention, but seems to have retreated from writing and consultancy as his core business became more successful. His son, **James Hinton Baverstock (1785–1837)** continued, and in places dissented from, his work.

Black, William (fl. 1830–44). Aberdeen brewer who moved to London and became increasingly involved in consultancy and authorship. First known English-language author to discuss diastase. Later wrote under the chemical patronage of George Birkbeck and Thomas Graham. Developed the doctrine of electrical influence on fermentation, still supported by Frank Faulkner in the 1880s.

Blake, George (fl. 1765–1817). Time-served practical brewer and consultant, and antagonist of John Long.

Booth, David (1766–1847). Briefly a brewer in Fife; chiefly a journalist and miscellaneous writer in London, associated with the Society for the Diffusion of Useful Knowledge. His *Art of Brewing* (1830) was unconventional for an SDUK production, containing anti-Excise polemic in the style of John Richardson.

Böttinger, Henry (1820–74). House chemist to Allsopp's of Burton from 1845, later becoming brewer; Giessen-trained, and a key link to Liebigian chemists in London.

Boys, Jeffrey (1646–1703). Jurist turned country landowner. Wrote a procedural guide to the brewing process which, for want of alternatives, dominated the fragmentary print literature of the early eighteenth century.

Brande, William Thomas (1788–1866). Royal Institution chemist who bid fair, in his youth, to become a defining figure in public chemistry. In the 1810s, chief upholder of the alcoholometric interpretation of drink in the English-language literature.

Brown, Horace Tabberer (1848–1925). Long-lived 'Master' of the first generation of professional brewing scientists. Brown's influential recollections of 1916 rewrote the history of science in the trade, favouring Burton and Pasteur, and sidelining the early presence of trained chemists in London breweries.

Combrune, Michael (fl. 1740–68; d. 1773). Hampstead brewer whose publications of 1758 and 1762 first combined the conventions of the brewing process narrative and the philosophical treatise. A zealous upholder of Boerhaavian chemistry, Combrune wrote chiefly to affirm his philosophical status, encouraged by Matthew Maty and Peter Shaw, and to promote the thermometer. Also offered himself, in the quietest and politest of terms, as a brewing consultant.

Davy, Humphry (1778–1829). Romantic visionary turned pillar of the London chemical establishment, happy to receive the profits of industrial consultancy but firmly dismissive of 'trade' as a source of knowledge. Nonetheless, as the most memorable English public chemist, a totem for chemically inclined brewer-authors into the 1840s.

Donovan, Michael (1791–1876). Chemist to the Dublin Apothecaries, and author of the brewing volume in Lardner's *Cabinet Cyclopædia*.

Ellis, William (*c.* 1690–1759). Hertfordshire farmer who increasingly made his living by writing on husbandry, in the process exposing himself to the perils of the trade writer's paradox. On the basis of youthful involvement (of doubtful extent) in a London brewery, developed as a sideline *The London and Country Brewer*, the first extensive and (somewhat) systematic text on the trade in English.

Faulkner, Frank (1843–1925). Brewery employee and 'frequenter of technical libraries' as a young man; became the main scientific and technical advisor to the *Brewers' Journal*, and later a full-time consulting brewery analyst. Argued that theoretical work by Pasteur, O'Sullivan and others was crucial to commercial success in a harshly competitive age. Supported electrical influence on fermentation long after most brewers had discarded it.

Graham, Charles (1836–1909). Agricultural and mining chemist who preached the value of incorporating 'chemical technology' and industrial applications into university teaching, ultimately gaining a secure niche as a specialist brewers' consultant.

Graham, Thomas (1805–69). Eminent disciple of Thomas Thomson who became, as Professor of Chemistry at UCL, a leading light of chemical London across the middle third of the nineteenth century. Engaged with brewers as, variously, Excise consultant, advisor to William Black and collaborator with Böttinger and Hofmann in dispelling the 1852 strychnine panic.

Ham, John (fl. 1802–37). Brewer, vinegar-maker and miscellaneous chemical industrialist, committed to Whiggish improvement projects, and responsible for pamphlets (sold chiefly to private brewers) on the economies to be gained from brewing with unmalted corn and potatoes.

Hofmann, August (von) (1818–92). Giessen-trained chemist, hand-picked by Justus von Liebig to lead the Royal College of Chemistry in London, where he and his successor, Edward Frankland, trained a generation of brewing chemists.

Jackson, Humphrey (bap. 1721, d. 1801). Manufacturing chemist and son-in-law to Benjamin Martin. Gained an FRS through his work on economically useful projects such as isinglass production, but his attempts to court brewers backfired drastically. Assailed by John Richardson as a mercenary quack with no grasp of brewhouse practicalities; embarked on disastrous venture with Henry Thrale; later established in brewhouse legend as a heresiarch who destroyed the traditional character of London porter.

Liebig, Justus (von) (1803–73). Hessian chemist, chief architect of the research laboratory approach to organic chemistry, and international chemical icon of the mid-nineteenth century. Influenced British brewing chemistry chiefly through his disciple August Hofmann, and through Hofmann's own disciples, but made a remarkable (and in some eyes embarrassing) cameo in the 1852 strychnine scare.

Long, John (d. 1807). Mysterious projector, almost certainly a non-brewer, who promoted a patent system of brewing based on attemperation around 1790.

Martin, Benjamin (1704–82). Instruments retailer, lecturer and author, whose pattern of lecturing and publication provided a model for others including his son-in-law, Humphrey Jackson. Also promoted the possibilities of wort hydrometry to James Baverstock and John Richardson before (on Richardson's doubtful testimony) abandoning the business as beyond his competence.

O'Sullivan, Cornelius (1841–1907). County Cork-born Hofmann disciple, house chemist to Bass from 1866 and part of the first generation of active brewers to publish regularly in chemical journals. A focus on the products of starch decomposition led him to a patent on a malt-sugar substitute in 1874.

Pasteur, Louis (1822–95). French chemist and germ theorist who worked extensively on fermentation from the 1860s, successfully promoting a living-cell interpretation in opposition to Liebig's supporters. From 1870, seized upon beer as a vehicle for the restoration of French national pride, patenting a fermentation process; the British brewing community, however, was more impressed by his use of the microscope as a diagnostic tool.

Richardson, John (1743–1815). Brewer and author. Single greatest influence on the character and content of the brewery treatise: his publications of 1777–1805 defined a philosophical yet practical stance, contrasting an enlightened trade with an ignorant Excise, which remained a model for brewer-authors to the mid-nineteenth century. Of undetermined background, appeared around 1775 as a brewing instructor, later taking on a brewhouse in Hull and promoting the brewers' saccharometer. His son, **Philetus Richardson (1776/7–1838)**, was head porter brewer at Whitbread's.

Shannon, Richard (fl. 1805). Metalware dealer and professed 'M. D.', known only through his patents and one publication, a thousand-page quarto doorstop, published by a subscription raised among leading contemporary brewers. The book, though presented as a comprehensive treatise, was largely compiled by hack borrowing as a vehicle for Shannon's patent specifications, and later condemned by Thomas Thomson as the single worst item in the brewery literature.

Shaw, Peter (1694–1763). Physician to two successive King Georges, natural philosopher and early promoter of the trades improvement agenda taken up in the 1750s by the Society of Arts. Main philosophical patron of Michael Combrune. Shaw's early writings on fermentation, aimed largely at distillers, led some nineteenth-century brewers to address him as the first brewing scientist.

Southby, Edmund Richard (1833–86). Medically trained analyst who became resident consultant to the *Brewers' Guardian*. Career, until his early death, strongly mirrored that of his rival Frank Faulkner at the *Brewers' Journal*, including authorship of an introductory treatise.

Squire, William Stevens (1834–1906). Giessen-trained chemist and distillery consultant who published on starch conversion. Gained the upper hand in a spat with Frank Faulkner over the discredited electrical influence theory, but a more general attack on 'rule-of-thumbists' in 1880s brewing fell flat.

Thomson, Thomas (1773–1852). University and consulting chemist of Edinburgh and Glasgow. Around 1806, experimented to address valuation questions for the Scottish Excise, becoming the first professing chemist to publish data verifiably collected in a working brewhouse. Later wrote the *Britannica* entry on 'Brewing' which strove to abolish the internal literature, refounding the art on professional chemical lines. This brought him little opposition from the trade, which acknowledged him as an influential chemical reformer who did not stoop to the adulteration scaremongering of some of his peers.

Thrale, Henry (1728–81). Leading porter brewer and patron to James Baverstock and (with obscure, but apparently wholeheartedly negative consequences) Humphrey Jackson. His long-suffering wife, **Hester Thrale (Piozzi) (1741–1821)** was among Jackson's principal antagonists.

Tizard, William Littell (fl. 1841–64). Brewers' engineer, perhaps the first to describe himself as such, whose promotion of his patent mechanical apparatus involved a large and repeatedly updated treatise on brewing, featuring an unparalleled survey of the prior literature.

Tryon, Thomas (1634–1703). Anabaptist vegetarian autodidact, whose unusual dietetic opinions led him to describe brewing processes in print at a time when few around him would have done any such thing.

Tuck, John (b. *c.* 1761, fl. 1820–2). Master bookbinder turned brewer-author, possibly a protégé of the leading porter-brewer John Elliot. Focused firmly on the practical brewer's perspective, Tuck provided a vivid rejoinder to the accusations of Friedrich Accum.

Ure, Andrew (1778–1857). Disputatious Glaswegian chemist, from 1830 a private consultant in London. More open to brewery literature than his rival Thomas Thomson, but earned some hostility in the 1840s for negative comments regarding the technological state of the trade.

Warington, Robert (1807–67). Among the earliest trained chemists to be taken on by a brewery, serving Truman's from around 1830 to 1839; later Chemical Operator to the Society of Apothecaries and a key founder of the Chemical Society.

Wigney, George Adolphus (1789–1872). Brighton brewer and author who attempted, apparently with little success, to remodel brewery instruction on the pattern of elementary chemical treatises; also much given to deep, if rather abstruse, commentary on the state of brewery knowledge.

Yworth, William (d. 1715). Alchemist and distiller whose published works included a brief account of the brewing process.

INTRODUCTION

> Though the art of brewing is undoubtedly a part of chemistry, and certainly depends upon fixed and invariable principles as well as every other branch of that science, these principles have never yet been thoroughly investigated. For want of a settled theory, therefore, the practice of this art is found to be precarious; and to succeed unaccountably with some, and misgive as unaccountably with others.[1]

This is a book about credibility. Its characters are the many researchers who, across the eighteenth and nineteenth centuries, tried to address beer-brewing in ways they called 'theoretical', 'philosophical' or 'scientific'. It is easy to imagine why such approaches were not always taken seriously. We are used to thinking of scientific investigators performing systematic experiments, searching for universal explanations in nature and communicating their findings on paper, usually by open publication. The craft and trade of beer-brewing conjures up an opposing set of ideas: down-to-earth artisans, mistrustful of theorists and tinkering; local customs of production, fiercely guarded from outsiders; skills passed down to a chosen few apprentices by hands-on experience. Nevertheless, a credible and coherent enterprise of brewing science existed by around 1880, the work both of theorists from outside the brewery, and of established brewers with theories of their own.

Credibility, for all investigators, meant showing that their claims were not merely valid, but actively *useful* to some relevant audience. There were many possible audiences: private brewers, commercial brewers, members of related trades such as malting, engineers, instrument-makers, scientific amateurs and professionals, and administrators of State. Various possible benefits could be claimed depending on the audience: production efficiency, uncontroversial regulation, general insights into the nature of matter or life. To make any such argument, however, required gaining a sympathetic hearing in the first place. Investigators usually relied on displays of past achievements or current professional context which suited their target audiences' pre-existing expectations: they thus needed to achieve a careful rhetorical balance of conservatism and innovation. By examining the techniques involved, we can better understand not only the specifics of the brewery case, but more generally how the knowledge and practices of scientific investigation and of industrial manufacture come to terms with each other – or, in some cases, don't.

Why Study Brewing?

Since the mid-twentieth century, the common view of beer in affluent alcohol-consuming societies has been the one supposedly formulated by Brendan Behan after a mammoth empirical study: it 'makes you drunk'.[2] Thus considered, it is not the most obvious topic for sober analysis. Indeed, some authors have cleverly put beer into historical context by showing just how seriously it was once taken.[3] Eighteenth-century beer did not necessarily make you drunk, but might be a staple foodstuff, a focus for community organization, a tool for governments to manage taxation and a source of private fortunes which launched political dynasties. In the nineteenth century, as anti-spirits campaigns gave way to full-blown teetotalism, all these associations fed into fierce debates over the moral character of beer. Brewing, then, was part of a web of economic and social connections broad enough and deep enough to matter to everyone. Much the same is true of other trades such as distilling, metalworking, textiles or ceramics; beer-brewing, however, offers a particularly productive case for exploring philosophy–trade interactions, for several reasons.

Unusually, brewing was both a manufacturing industry and a domestic art. In the eighteenth century, as the greatest commercial brewhouses became conspicuous public marvels of concentrated mechanized production, traditional 'house-brewing' on the estates of the landed gentry remained an embedded feature of economic and social life.[4] Brewing thus resembled agriculture in being constantly visible, and materially important, to the people most likely to be philosophically engaged. Allan Chapman, writing of an earlier period, offers a useful analogy:

> every clergyman, academic, and country gentleman would have received his income from land rentals, and such gentlemen would have watched the profitability of their farms, brew-houses, and dairies with the same assiduity as their modern-day colleagues might monitor the instincts of their stockbrokers.[5]

Joseph Banks, for instance – who did his best to intrude into almost every dimension of British natural science around the turn of the nineteenth century – kept a brewhouse at Spring Grove which issued around 80 hogsheads (over 4,000 gallons) of beer annually in the 1780s. In 1800, as London's largest brewers were scrambling to commission Boulton and Watt's steam engines, Matthew Boulton himself was computing how much he might save by brewing for his own household.[6]

Equally important was beer's role in the growth of the State. Seventeenth-century governments had identified beer and its raw materials as a convenient target for indirect taxation, levied at the point of production, and duties rose heavily from the beginning of the eighteenth century.[7] Efforts to minimize fraud and disputes created a legion of highly literate Excise officials who patrolled the breweries on foot and horseback, applying standardized measurement techniques, mathematical approximations and rigorous record-keeping.[8] For their own protection, therefore, brewers got used to quantifying the value of their product, and were easily persuaded of the usefulness of precision instruments.

It was also significant that the major steps of the brewing process – malting, mashing, hop-boiling and fermentation – all involved the subtle internal rearrangement of material substances: finding a common framework to explain such transformations had been a popular preoccupation for philosophers from the seventeenth century onwards. Particularly fascinating was the fermentation process which transmuted sugary materials into intoxicating spirit and large volumes of a gas (interpreted successively as fixed air, carbonic acid gas and carbon dioxide) which was also found in mineral waters, and associated with various effects on the health of the human body. Alchemical, chemical, pneumatic, medical and dietetic studies intertwined in the early consideration of beer in print, as we shall see.

Trade Knowledge and the Theory/Practice Division

Notwithstanding all these commonalities, the vast majority of brewery authors and innovators took it for granted that a firm dividing line existed between 'theory' and 'practice'. The line, of course, was rarely drawn in the same way twice, and those who drew it generally did so for particular strategic reasons of their own. My interpretation here draws on the picture presented in the edited collection *The Mindful Hand*, which shows how the bureaucratic elites of early modern Europe based their authority on a series of oppositions: 'scholar/artisan, science/technology, pure/applied and theory/practice'.[9] Status gaps across these dividing lines made it legitimate for physicians to regulate apothecaries, engineers to control shipwrights, and 'professionals' and 'gentlemen' in general to overrule the most skilled and organized 'tradesmen'. In making a show of building bridges across a status gap, such men were careful to affirm that the gap remained, nonetheless, and that the ground remained higher on their side. This was certainly the strategy adopted across the whole of my period of study by public men of science who sought to instruct legislators, householders and even brewers on brewery questions, without having brewed a drop of beer for themselves.

Yet *The Mindful Hand*'s manifesto implies a reflexivity which it does not quite bear out. In focusing on the iniquities of the self-appointed elites who privileged the authority of the mind, it says less about artisans' and tradesmen's ability to offer alternatives. In the literature of brewing, we find robust contributions from such figures as John Tuck, a bookbinder turned brewer active in the 1810s and 1820s. Tuck inverted the conventional hierarchy of the theory/practice divide, savaging chemical theorists who formulated rules through 'science instead of practical acquirement'.[10] The plain-speaking 'practical man', of course, was as much a strategically adopted persona as the man of letters; practical authors seldom rejected theory outright, but rather claimed an authority of their own over how it could best be used. My survey examines brewery-minded theorists and theory-minded brewers in parallel, exploring the opportunities and difficulties on both sides.

In consequence, this book shows only limited overlap with familiar narratives of the history of science. Closely studied figures such as Banks, Sir Humphry Davy and John Tyndall receive only walk-on parts; the chemists I discuss more fully – Humphrey Jackson, Fredrick Accum, William Thomas Brande – are marginal or under-examined, while most of the theoretical brewers are entirely unknown outside the history of brewing. Their obscurity is, in itself, revealing of how the disciplinary identity of science was being reworked. In the eighteenth century, much brewery writing consisted of what Ursula Klein has identified as 'experimental history', recording experiments to systematize knowledge about processes without necessarily trying to find underlying causes. At the time, this work could be of similar status to the 'experimental philosophy' which demanded to know why as well as how: the professionalizing scientists of the nineteenth century, however, wrote it off as workmanlike and trivial.[11]

Focusing on unsung practitioners has its dangers. James Secord has usefully warned of a tendency towards 'parochial antiquarianism' in history: painstaking contextual studies may ultimately just reconfirm the obvious point that ideas, in science as elsewhere, are a product of their local circumstances, without contributing anything to our wider understanding.[12] One rather literal but effective way to find universal relevance in an esoteric community, however, is to focus on its own interest in universals. All brewery authors who described their work as 'scientific' or 'philosophical' sought to reduce brewhouse actions to a set of rules which somehow transcended the traditional patterns of local and particular knowledge. Their proposed universals – chemical principles and instrumentally determined quantities – usually had pre-existing meanings created outside the brewhouse: they mattered to communities built around trade improvement, the reform of chemistry, instrumental precision or polite curiosity. My study is built around the connections between these communities, assessing in tandem what science brought to the brewery, and what the brewery brought to science.

The Paradoxes of Beer Theory

Several impressive histories of British brewing have been written based on the extensive records and correspondence which survive from many breweries, or on the more limited material evidence of brewhouse structures.[13] This book, by contrast, focuses on printed texts. I consider how the systematic circulation of written information (usually, though not always, published) helped, and occasionally frustrated, the establishment of a credible science of brewing. As it removed the direct interaction between author and audience, print also removed the conventional guarantees that went with it: readers have always had to address the possibility that what is printed is not what it claims to be.[14] Publications on commercial arts such as brewing were particularly suspect. In the period in question, as far as most readers were concerned, writers did not brew, and brewers certainly did not write: if a brewer's insights were really valuable, his safest path to profit was surely to keep quiet and apply them in his own brewhouse.

We might term this the *paradox of the trade writer*. Tradesmen who tried to live by publishing invited suspicions that they could not live by the trade itself, perhaps for want of skill, and that their professed revelations might be based on misunderstanding or downright fakery. Factors that would make other kinds of text more credible – detailed coverage, high production values, wide dissemination – might here be read as signals that the writer was spending too long away from the brewhouse, or, indeed, had never been in it: professional hack authors typically knew nothing of the brewery at first hand, yet had the best opportunities to shape its representation in the general literature.

Rivalry between authors meant that the paradox was often spelt out explicitly. William Yworth (d. 1715), an alchemical adept who wrote on brewing and distilling, complained in 1705 that much literature on trade processes was written by men who 'have not practically known the same; or else by those, who have on purpose concealed that, which in reality ought to have been discovered'. Yworth presumed his readers to be wearily familiar with what happened when writers who had nothing to write wrote it anyway: 'needless Prescriptions, confused Workings, long and tedious Prolixity of Words and Circumlocutions'.[15] In identifying these failings, of course, Yworth hinted that his own work was free of them.

The paradox could be resolved or diminished in various ways. Some authors announced that they were not trade writers at all, but private gentlemen brewing for their own households, their research inspired by curiosity rather than commercial concerns. This stance, however, largely disappeared across the course of the late eighteenth century, as large industrial brewhouses supplanted most home production. A more durable defence drew on the founding self-image of English natural philosophy, proclaiming that the moral obligation to reveal useful knowledge ranked above considerations of personal gain.[16] Alternatively, trade authors might argue that there was no profit in privatizing what they knew: some innovations, such as those intended to open up new markets or substitute costly materials, might plausibly work best for everyone if they were adopted universally.

The most popular approach among authors with commercial brewing backgrounds, however, was strategic partial revelation. These authors explained their publications as the result of a balance struck between open communication, to promote their names and establish their expertise, and private communication for profit. The approach was often hazardous: many theorists, and also some practical brewers, objected to such deliberate withholding as unacceptably mercenary. A more palatable variation was to argue that while the text held back nothing that could be revealed in print, there were important details of practical implementation that could only be conveyed through personal, on-site engagement. This strategy was widely used from the mid-nineteenth century by authors proclaiming new professional identities as brewers' engineers, analysts or consultants for hire – roles which allowed them to claim specialized expertise which did not rest on a vested interest in any particular brewery.

There was another tension facing the systematizing innovator, which we might call the *paradox of experiment*. Although brewing practices from time immemorial had undergone frequent and sometimes dramatic change as a result of trial-and-error investigation, the search for universal understandings required something more: systematic comparison of widely variant approaches, with failure as a routine consequence. In 1838, George Adolphus Wigney, a philosophically minded gentleman-proprietor, wrote ruefully of his early endeavours:

> Malted some tick beans for curiosity; they absorbed so much water in the cistern, as to cause the charge of nearly or quite double duty. Mixed therewith some barley malt and brewed one guile, obtained but little extract, that was nauseous and spoiled the whole. Burnt down a malthouse by snapping porter malt on an improper kiln; killed a horse by feeding him on the half burnt barley and bean malt ... created a vacuum in the copper, and the pressure of the atmosphere broke the bottom of the back asunder, and forced it into the copper with a tremendous explosion.[17]

Wigney's account is uniquely candid, and separated from its origins by twenty years of (relatively) successful practice. Typically, to be seen to depart so radically from established trade norms would demolish the innovator's credibility, if not his brewhouse. The problem was made worse if the proposals relied on results from small-scale test apparatus which bore little obvious resemblance to commercial equipment.

This second paradox melted away, of course, if the innovator could show the brewers that unfamiliar methods led to improvements on some criterion that *was* familiar. Innovators who could not quickly achieve this, however, faced being written off as confidence tricksters, dilettantes or just plain bad at brewing. Adopting the rhetorical pose of the experimental philosopher, as we will see, tended to do more harm than good in defusing these charges. The only reliable defence was to demonstrate a time-served brewery background and awareness of the needs and constraints of the trade.

Almost all the developments discussed in this book can be read as attempts to resolve these paradoxes, establishing theories of brewing as credible to some chosen audience. The range of possible audiences was great, as was the range of possible intentions. Some theorists, perhaps sincerely, claimed to seek only the satisfaction of extending human knowledge or improving a vital industry. Others were out for more tangible rewards: patents or retail monopolies, bookselling profits, consultancy fees. Competition for these meant competition for theoretical credibility, which influenced how those involved accounted for themselves and their work. Experimenters who were brewery insiders, for example, emphasized their career histories to damage rivals who were not. The pattern of success in this credibility-building largely determined the priorities and assumptions of what gradually emerged as a self-conscious community of scientific brewers.

Caveat Lector

In a study focusing on printed texts, it is important to remember that such texts do not transparently represent either the worlds they describe, or their authors' intentions.[18] For the reasons outlined above, many brewer-authors thought strategically about committing details to paper, and drew sharp lines between publication, private correspondence and private instruction by demonstration. Evidence of the three mechanisms survives unequally and in descending order. What was written, moreover, inevitably focused more on novel technologies and theories than on what was typical for the time; many ideas described at length were never taken up, whereas points of fundamental importance were routinely passed over as obvious (though they may not be so today).

It is also worth noting that texts often borrowed heavily from prior texts, and so do not reliably represent the views or experience of their named authors. The brewing and distilling directions of the abovementioned William Yworth, for instance, lack the cryptic, cosmological tone of his better-known alchemical writings, and were probably modelled on accounts found in husbandry manuals.[19] Such borrowing does not necessarily imply incoherence or imposture. Yworth was simultaneously a sincere alchemist and a practical pharmacist and distiller, and knew that publishing on trade processes required different rhetorical and linguistic conventions from publishing on alchemy; to draw from other sources in meeting these needs was conventional in early modern scholarship. Positions we take for granted – that copying content without acknowledgment is morally wrong, and that original contribution is the basis of authority – were starting to be articulated before the eighteenth century, but were not systematically applied until the nineteenth.[20]

The apparent period and context of authorship, therefore, can never be taken at face value. Hack authors could be sophisticated in cloaking borrowed texts to suit new markets. A bucolic gentleman-farmer's account of his practice, in a letter to the *Farmer's Magazine* in 1777, turns out to be a facetious reworking of an earnest philosophical account of 1764.[21] Directions for brewing first found in a 1737 anthology resurface as new in a stand-alone pamphlet of 1789, and again, largely unchanged, attributed to one 'Thomas Threale' in 1802.[22] Such practices were rarely challenged: a few knowledgeable authors bewailed the effects of misinformation due to the scissors-and-paste tendency, but were not immune to similar behaviour themselves. Alexander Morrice, author of a much-reprinted treatise of 1802, was careful to state that his brewing instructions were compiled directly from his own commercial practice; his guidance on malting, however, is almost verbatim from a work of the 1730s.[23] Prior borrowings often led the borrowers to underestimate the ages of their sources, so that a narrative could endure as 'current' for a century.[24]

Indeed, the very idea of 'current theory' is itself a creature of nineteenth-century professionalization and its emphasis on consensus, originality, hierarchy and progress. For much of my period, theoretical interpretations in the latest systematic treatments of chemistry could, and would, be challenged by knowledge based on older systems or none; a decades-old treatise might demand respect, and old theory might inspire new investigations. When, in the nineteenth century, professionalizing scientists rejected such tolerance, they became increasingly oblivious to literature which endured among commercial brewers. The converse was also true: commercial brewers in Britain stuck largely to English-language sources well into the period when serious chemists kept up with the French and German literature as a matter of course. This was as much a practical as a rhetorical divide, and it was bridged only when chemists schooled in the German analytical tradition were hired directly onto the staff of breweries, around the middle of the nineteenth century.

My approach also requires attention to how readers use texts, and where they go once they have left their authors' control.[25] Texts served readers in a variety of roles: as straightforward sources of information; as indicators of the author's expertise when deciding whether to pay for equipment or consultation; as rallying points for change or compositional models for further writing. The breadth of my period, and the lack of surviving publishers' records and editorial files, limit the possibilities here: I do not know how many copies of, for instance, Michael Combrune's *Theory and Practice of Brewing* (1762) were printed, let alone, in more than a handful of cases, where they ended up. But I have done my best to outline the range of responses articulated by his known readers, and the most influential incidents which brought his work to new audiences.

The Structure of this Book

Chapter 1 sets the scene by surveying the state of public and private communication on brewing methods in the early eighteenth century. In chapter 2, I focus on Michael Combrune, author of the first known brewery text modelled on a natural-philosophical treatise. Here, as throughout, I pay close attention to the work's reception: Combrune courted philosophical gentlemen, particularly within the Royal Society, but drew responses mainly from brewers seeking to improve their production, and from a few with wider interests in trade improvement. Chapter 3 addresses the culture of 'philosophical' brewery instructors which sprang up in Combrune's wake. The activities of Humphrey Jackson, chemist and druggist, illustrate particularly the challenges facing projectors who approached the brewery as outsiders: in the 1770s, Jackson's programme of private brewery tuition drew charges of long-winded quackery and fraudulent artifice, including the promotion of toxic adulteration. By contrast, Jackson's principal rival, John Richardson, established a more palatable platform for instruction, playing on his status as a brewery insider.

As I show in chapter 4, however, this 'practical' standpoint did not preclude theory: Richardson successfully promoted quantitative, gravimetric interpretations of beer strength by shrewdly framing concepts in terms of issues which brewers already considered economically important. Chapter 5 considers nineteenth-century repercussions as such projects collided with fears over toxic adulteration in beer, a long-rumbling public concern which grew ever stronger as industrialization separated drinkers from the production of their drink. The 'scientific' brewers who analysed beer into unfamiliar materials, and declared that it might be synthesized from others, refused to speak for traditional 'purity', and were the more suspect because the usual alleged adulterants were the stock-in-trade of their chemist colleagues.

Nineteenth-century governments increasingly relied on scientific authority to cut through taxation controversies, and in doing so often faced challenges based on the difference between theoretical and practical regimes. Chapter 6 addresses an attempt to overcome this gulf by literally bringing chemical professors into a working brewery. The exercise, though not successful on its own terms, gave one of those involved, Thomas Thomson, significant agency in rewriting the rules of acceptable systematic communication on beer. Finally, chapters 7 and 8 concentrate respectively on the two principal modes of publication on brewing in the later nineteenth century, treatises and periodicals. Treatises served a range of purposes, from the educational agenda of the 'useful knowledge' movement to the promotion of commercial consultancy or patent apparatus. By their nature, however, they did not function effectively for the communication of current research: growing interest in the practical implications of material published in scientific journals created a need that was filled, in the 1870s, by commercially oriented trade publications.

Readers familiar with the history of brewing may be surprised that I terminate my survey in 1880, at a point when, some have argued, the interaction between science and brewing was still in its infancy. Richard Unger dates the 'scientists' invasion of the brewery' to Louis Pasteur's microbiology of fermentation around 1870, while Harold Platt presents a 'scientific brewing' movement, founded on Pastorian analysis, refrigeration and carbonation, making its way via German-speaking immigrants into Chicago around the 1860s to 1890s.[26] In fact, these movements' claims that brewing had *only just now* embraced science or systematic theory were merely the latest in a long line: similar rhetoric had recurred in every generation since the 1750s.[27] The microbiological turn, crucial as it was to the content and institutions of late nineteenth-century brewing science, wrote off an extensive body of prior work in analytical chemistry, statics and other forms of investigation.[28] The infancy claim was pressed for the last time in the last decades of the nineteenth century; thereafter, the credibility of 'brewing science' was secure.

1 THE CURIOUS BREWER

> The late Curious *Simon Harcourt* ... thought the true Art of Brewing of such Importance, that it is said to [have] Cost him near twenty Pounds to have an old Daysman taught it by a *Welch* Brewer, and sure it was that this very Man exceeded all others in these Parts afterwards in the Brewing of that which he called his *October* beer. So likewise in *London* they lay such stress on this Art, that many have thought it worth their while to give one or two hundred Guineas with an Apprentice.
>
> [William Ellis][1]

On 17 December 1662, the philosophical gentlemen of the Royal Society of London had their minds on drink. Christopher Merret, the physician, presented his collected observations on how best to manage wine in storage; Robert Hooke, newly appointed as the Society's Curator of Experiments, described the products of distilling common water; and Sir Robert Moray, soldier turned courtier, announced that he was preparing a study of beer-brewing in his native Scotland, which he hoped could be extended to cover 'all sorts of ale and beer'. Abraham Hill, a wealthy young man who devoted much of his time to Society administration, suggested that a Mr Philips could help with this project, and might then receive a Fellowship.[2] No such man was elected, however, and it seems the project was never completed.

This episode illustrates how early modern beer-brewing straddled the division between what was accessible to the virtuosi, and what was not. On one hand, some elements of the art were easily examined. Several founders of the Royal Society owned large land estates with purpose-built brewhouses, and one, Sir Kenelm Digby (1603–65), recorded the brewing methods of various aristocratic acquaintances among his culinary 'receipts'.[3] Indeed, brewing was common knowledge at all levels of society. Kitchen brewing was a traditional duty of rural wives; in towns, there was often a communal brewhouse, on the same principle as the wash house, where householders could brew from their own materials. Families not brewing for themselves were often supplied by itinerant retailers, and most drinking houses were kept by 'brewing victuallers', both groups brewing their own supply on a small scale.[4]

On the other hand, the concept of brewing as a general art was already disappearing in London by the time of the Restoration. Commercial partnerships organized around larger-scale breweries – often referred to as *common brewers* because they brewed 'in common' for multiple houses – were steadily displacing the brewing victuallers across much of south-east England, and in some large towns elsewhere in the kingdom. Samuel Pepys, elected to the Royal Society in 1665, developed a taste for the ale served at the Rose and Crown in Mile End, supplied from a central brewhouse in Shoreditch.[5] Like other urban industries, common brewing tended to be monopolized by local guilds or fellowships, whose apprenticeship regulations confined their methods within the trade; as brewing disappeared behind the walls of the large common brewhouse, so it disappeared from the immediate gaze of the virtuosi.

For a philosophical and experimental programme truly to consider 'all sorts of ale and beer', at least one of two changes would need to take place. Philosophers would have to be admitted to the common brewhouse and its knowledge culture, or working brewers would have to be admitted to the meetings, correspondence or publications of the philosophers. Neither change was socially impossible; yet neither developed, to any noticeable degree, for almost a century following Moray's proposal. Non-commercial brewing, geographically diverse and rich in comparable methods, remained available for inspection. Yet there were few models for writing about brewing, precisely because the core process was presumed to be generally known and practically accessible.

Only in 1700 did a significantly comparative account appear in print. The *Directions on Brewing* of Jeffrey Boys, a landowner supervising his own estate brewhouse, presents the 'curious' approach of a leisured gentleman who, like the virtuosi, could afford to experiment. Boys's widely pirated work influenced numerous philosophical authors, in particular those concerned with agricultural improvement. We might expect that the systematizers would routinely have bolstered their accounts by careful study in such brewhouses as their social connections gave them access to. The key conclusion of this chapter is that they did not. With only a handful of exceptions, published accounts relied entirely on the authority of Boys's text and others already in print.

This does not mean that the authors considered stepping into a brewhouse to be beneath them; indeed, some, such as the agricultural compiler John Mortimer, made firm statements to the contrary. Conventionally, however, the systematic knowledge projects of these authors focused on synthesizing texts: they did not consider it feasible, or necessary, to incorporate an element of direct observation. By the 1730s, however, several had realized that the major common breweries had acquired advantageous methods which were not being communicated to outsiders, and which might offer strong opportunities for comparative investigation. This ultimately became possible, around mid-century, through a significant revision of philosophical attitudes to trade.

Beer and the Virtuosi

There must be writing about the practical arts – but not by practising artisans. This was the common manifesto of the men who shaped the institutionalization of philosophical enquiry in Britain, from Francis Bacon through Samuel Hartlib's circle of correspondents to the founders of the Royal Society.[6] The virtuosi who set out to write what they termed the 'histories of trades' (narrative accounts of industrial processes) upheld the convention that workmen were 'illiterate', whether or not this was strictly true. Trade masters, on the other hand, could evidently write, but their commercial interests were seen to compromise their testimony. Steven Shapin has memorably demonstrated how Robert Boyle and his supporters in the Royal Society built up an image of the disinterested gentleman, independent both intellectually and financially, as the only reliable revealer of natural knowledge.[7] In this picture, even open, honest and articulate tradesmen were poorly placed to account for their processes, precisely because they were familiar with them. Robert Hooke argued that the curious 'Stranger' was in the best position to instruct other newcomers, just as 'such as travel into other Countries' gave the clearest descriptions of their surroundings.[8]

The gentleman-philosopher, therefore, must go among the wealth-seekers – apothecaries, engine-makers, gardeners, brewers – to take over and universalize their stock of useful knowledge and the processes they used to make that knowledge. Boyle (who in practice found engaging with artisans rather uncomfortable and time-consuming) boasted the particular merits of learning from the most 'illiterate persons'. More systematic workers who developed theories of their own, he argued, did so from observations purely internal to the trade, which would distract the philosopher from his inquiry into universal causes.[9] Artisan masters must also have an ingrained preference for the structures and boundaries of their arts as they stood: this conflicted with the philosopher's role of centralizing understanding, finding new applications, demonstrating that some practices were superior to others and accounting for products and processes in terms of wider natural laws.[10]

We must remember, however, that the gentleman/artisan contrast was chiefly a rhetorical device to shore up the fledgling Royal Society's public credentials. If the vision of an ingenuous philosopher exploring exotic industrial lands was just about representative of Robert Boyle, it certainly misrepresented others more closely involved in the history of trades project, erasing their commercial obligations and opportunities. William Petty was a clothier's son, Abraham Hill a landowner raised as a merchant, while John Beale, son of a gentleman farmer, profited from the cider-apple grafting initiatives he promoted in his letters to Hartlib; physicians such as Christopher Merret, whatever their professional status, ultimately relied on income from patients.[11] Such men's investigative expeditions did not range systematically across the useful arts: they usually began in their own commercial backyards, and from there followed such native guides and guarantees of safe passage as their connections allowed. Trade histories on paper might arise through overlap between philosophical and commercial identities (for example, among physicians and apothecaries), or through chance collisions of interest.

The uneven, contingent spread of opportunity is well illustrated by the fact that the virtuosi engaged much more successfully with various other drinks than with beer. In 1669, Merret and Walter Charleton published a trade history of wine, *The Mysterie of Vintners*, which directly incorporated process accounts first found in wine-dealers' manuscripts: such narratives were often more systematic than Boylean dogma suggested.[12] Raw materials sometimes provided a path for philosophical engagement, as with apples and cider-making: John Worlidge (*d.* 1693), whose agricultural treatises were the most widely read of his generation, followed John Beale in making a particular study of cider, which he dedicated to Elias Ashmole. For spirits, with their long-standing medical associations, investigative and manufacturing roles could publicly coexist. The Dutch alchemist William Yworth, who dedicated a treatise to Boyle and was later patronized by Isaac Newton, combined distilling, drug manufacture and chemical instruction, and published books joining alchemical philosophy to distillery process narratives.[13]

If the early trades historians did not address beer in similarly systematic fashion, it was not for want of interest, as Robert Moray's abovementioned proposals demonstrate; Moray himself got as far as delivering a short account of Scottish malt-making, which was published in the *Philosophical Transactions*.[14] In suggesting a disclosure of brewery knowledge as worth a Fellowship, Abraham Hill may have been thinking of the precedent set by Beale, who was then on the verge of election after corresponding extensively with the Society on cider apples. Beale's letters also touched occasionally on beer, referring to one region of Dorset whose waters, by common assent of the locals, produced drink stronger by one half than other nearby sources. When this letter was read to the Society, several members testified to similar observations made elsewhere, and both Charleton and Timothy Clarke claimed to speak with authority on the preferences of London's common brewers.[15]

These virtuosi intended their trades histories to be the first stage in a comparative process: collate credible testimonies of practice; identify which techniques are optimal, either in particular circumstances or (preferably) universally; export these techniques as appropriate. This model made sense where survival did not depend on market competition, as among landowners producing drink for consumption on their own estates: on this basis, Beale had promoted Herefordshire cider apple-growing as a 'Pattern for All England' by distributing grafts from his own orchards.[16] Into the eighteenth century, the Royal Society's improvement agenda focused increasingly on programmatic cataloguing, experiment, comparison and systematization of materials and techniques, intended to operate nationally and to serve the national interest.[17]

Commercial brewers, however, proved distinctly reluctant to contribute to such comparative projects. It is easy to understand why. In a competitive market, any public judgment was undesirable: if the producer's practices or products were judged superior, outsiders might be inspired to copy their advantages, if inferior, to explore alternatives to beat them. This wariness, we should note, did not mean that brewers were averse to systematic knowledge projects in general. Charles Cox, brewer and Member of Parliament for the brewery fiefdom of Southwark, was patron to the mathematical and experimental lecturer John Harris, compiler of the industrial dictionary *Lexicon Technicum* (editions from 1704).[18] Cox's fellow MP and brewer John Cholmley also subscribed to the *Lexicon*, as did two members of the brewing Hind family. If Cox directed Harris's attention to brewing problems, however, no hint of the result was published, in the *Lexicon* or anywhere else. Perhaps process accounts circulated, as they did among the wine dealers, in manuscript in the beer trade. If so, throughout this period, they stayed there.

Describing the Art of Brewing

A number of accounts of non-commercial practice were circulating, sometimes in print, by 1700. What these texts signified to their readers is often hard to determine across four centuries of changing assumptions about background knowledge. Modern authors readily see the need to cater for the non-brewers in their audiences, opening with brief explanations of the apparatus, the process and the arcane terminology of the field.[19] They often specifically address the popular canard that beer is 'made from hops', politely pointing out that the main ingredient is malt, which provides the sweet flavourings, body, colour and, by fermentation, spirit, whereas hops, the feathery flower-cones of a climbing plant, are added in small quantities for bitterness, aroma and preservation.[20] In the era of widespread domestic brewing, such information was deemed common knowledge and rarely recorded: the convention that the reader might literally have no idea how to brew emerged only around 1790, when publishers spotted a market for cheap pamphlets promoting the revival of small-scale brewing among urban families.[21] Even in 1850, *Household Words* thought it unlikely that readers would be 'such Cockneys as not to know that malt is barley, steeped in water, laid out on a floor, let be there till it is just about to sprout, and then dried on a kiln'.[22]

Most early literature, then, was compiled as an adjunct to extensive orally and practically acquired knowledge. Literate men and women at various social levels kept manuscript 'receipts' of their brewing as personal reminders, or occasionally to instruct others, much like recipes for food or medicines.[23] Sometimes, these found their way into print, as when Kenelm Digby's culinary collection, with its accounts of '*Scotch* Ale from My Lady *Holmbey*' and other household recipes, was published after his death.[24] Seldom exceeding a couple of pages in length, such 'receipts' were often little more than lists of quantities for the various ingredients, the entire method being taken as understood by the reader.

More explicit and comparative accounts appeared in topographic surveys, a strain of literature where the key rhetorical technique was to write always for an imagined outsider. The cleric William Harrison (1535–93) attempted a general definition of the brewing process in the 'Description of England' which accompanied Raphael Holinshed's *Chronicles* of 1577. Somewhat similar was a French source, *Maison Rustique* of 1569–70, available in an English translation by Richard Surflet from 1600, which straddled the topographic and instruction-miscellany approaches as it attempted to compare rural practices across France and, sometimes, other parts of Europe. By the standards of their period, these works are uncommonly rich in concrete statements about what was then widely considered obvious: for the twenty-first century non-brewer, they offer the best available historically situated introduction to the art.

'Our drinke', Harrison begins, is 'made of barleie, water and hops, sodden and mingled together, by the industrie of our bruers, in a certeine exact proportion'. Before it can be used for beer, however, barley must undergo malting, for which both Harrison and *Maison Rustique* provide stepwise accounts. The barley is steeped in water for one or more days, then left in heaps for a time – Harrison suggests three weeks – to partially 'sprout', or germinate. This process is then stopped by careful drying, over straw or wood fires, on a 'kill [kiln] couered with haire cloth', after which the malt is ground in a mill. Both accounts say comparatively little about hops, or of the water to be used.[25] Harrison next describes the brewing process 'once in a moneth practised by my wife & hir maid seruants'. The method may, indeed, be that of Marion Harrison, a Flemish refugee from the Pale of Calais, although we cannot be sure: the 'Description' borrowed freely from many sources, and the account may be entirely second-hand. In any case, in keeping with the topographic agenda, most of the text is written at a level general enough to be common to most small-scale domestic brewing across Britain and Ireland. *Maison Rustique*, by contrast, focuses on 'the manner of making beere at *Paris*' – probably commercially, from the quantities involved – with notes on the chief distinguishing features of German, English and other practices. The two narratives differ significantly in their details, but give a clear sense that certain core features were regarded as universal.

The first stage is to infuse the ground malt in water which has been heated to near boiling, producing a sweet-tasting extract. This infusion takes place in a separate vessel from the boiler used to heat the water, which is needed for the next stage: the liquid extract is drawn off and boiled vigorously with hops, a process which takes up to two hours in the 'Description of England' account and twelve hours in *Maison Rustique*. Both narratives address the fact that the malt retains some sweetness after all the liquid has been drawn off. The English procedure is the more elaborate: the residual sweetness is captured by a second and a third application of hot water, the additional extracts being drawn off separately and boiled with hops like the first. *Maison Rustique* concludes by outlining several operations which must have been common to both processes, yet which Harrison, for all his explicitude, skips over entirely. The hop-impregnated extract is poured into shallow vessels and allowed to cool. It is then mixed with yeast, prepared from earlier batches of fermenting beer, and soon begins to ferment itself. While still fermenting, it is transferred to casks, and the froth of fermentation works out through the bungholes. The casks are then bunged up tightly, and the beer laid down to mature and clarify.

Modern brewers encountering these sixteenth-century narratives often remark on how closely they map onto surviving practice. The apparatus and sequence of operations are immediately recognizable, as is much of the terminology. Harrison calls the infusion mixture of malt in water a 'mash': the term *mashing* for this operation, and *mash-tun* for the vessel used, became universal in published sources in English soon after 1700. The sweet liquid extract made by mashing is 'woort', later regularized as *wort*. Surflet, translating from his French source, calls the vessel for heating water and hop-boiling a 'copper': any such vessel is still a copper today, whatever materials it is made of. In private brewhouses, and at the lesser end of commercial production, operations around the mash-tun and copper continued very much as these sources indicate until well into the nineteenth century.[26]

The topographers' aim was usually to describe, not to prescribe. There are occasional glimmerings of attempts to frame natural laws in Harrison's work: the preservative quality of the drink 'is alwaie determined after the quantitie of the hops, so that being well hopped it lasteth longer'. There is also comparative judgment on the choice of water source:

> although the waters that run by chalke or cledgie [clay] soiles be good, and next vnto the Thames water which is the most excellent, yet the water that standeth in either of these is the best for vs that dwell in the countrie.[27]

More typically, though, Harrison states that his directions should not be taken as an authority for imitation: in particular, he is 'scarce a good malster' himself. What drives him is the desire to correct 'forreine writers', who have published badly misinformed narratives. When such authors wrote on brewing, then, it was not the art of brewing but the state of public description which they were trying to improve.

Such literature could, of course, influence authors more concerned with practical improvement. The farmer turned author Gervase Markham (*c.* 1568–1637) took the *Maison Rustique* as a compositional model while reacting against many of its accounts of French practice, asserting the greater economies and efficiencies of English domestic culture.[28] Markham's *English Hus-wife* of 1615 gives the most extensive brewing narrative of its period, emphasizing the close association of small-scale private brewing with the womanly side of the conventional domestic division. Materials and apparatus are taken for granted, but there are explicit operational directions including signs to watch for. Hops, for instance, should be boiled in the wort 'till taking vp a dishful thereof you see the hops shrinke into the bottome of the dish'. Markham was writing with replication in mind.[29]

By definition, those who wrote to prescribe good practice did *not* represent their methods as being already general, although Markham gave the impression that his were established among right-minded brewers. Not so Thomas Tryon (1634–1703), Dissenter, ascetic, vegetarian and author on self-help and self-education. In 1690, Tryon published a small treatise entitled *A New Art of Brewing* in which he attacked hop-boiling – an almost universal practice – as unhealthy and inefficient, instead promoting a more gentle infusion. Commercial brewers accepted that this would save them money if it produced good beer, but unanimously declared that it could not; Tryon therefore sought alternative audiences in print, declaring that the trade's objections affected him 'no more than the buzzing of Flies about my Nose in Autumn'.[30] His ideas drew some attention but little support, and were the subject of ridicule in the trade for decades afterwards.[31] William Yworth, the only one of Tryon's contemporaries to publish at similar length, dismissed him as a 'Moddler' – a meddler, that is, in matters he had neither the practical know-how nor the philosophical insight to master.[32]

The very fact of this dispute, however, suggests the beginnings of a comparative experimental approach to the art, such as the virtuosi were seeking. In the first third of the eighteenth century, literature appeared which helped to cement this project. Unsurprisingly, the impetus came less from radicals such as Tryon than from sources closer to the philosophical mainstream: country gentlemen, apothecaries and physicians.

Sources for Systematizers

A major influence on the development of the systematic brewing treatise was *Directions for Brewing Malt Liquors*, a 23-page pamphlet first published in 1700, written from the perspective of a 'Countrey Gentleman' who personally superintends the brewhouse serving his estate. The author was apparently Jeffrey Boys (1646–1703), a Gray's Inn bencher who inherited, late in life, a rural manor at Betteshanger, Kent.[33] The process he describes is in many ways similar to Markham's of the preceding century. Both sources, for instance, describe making three mashes from the same malt, and managing the three worts separately through hopping and fermentation: the first to produce strong beer for storage, the second, 'good middle *Beer*' and the third and weakest, 'Small' beer for everyday drinking.[34]

The text differs from earlier writing, however, in two crucial respects: it covers the nature of the raw materials in detail, and it promotes experiment as a key part of brewing practice. The text considers water, malt and hops sequentially before turning to the stages of the brewing process, a structural arrangement which was to recur in the chapter divisions of the much more extensive works which followed. Boys assumes, as usual, a reader with significant prior experience in brewing, but frequently spells out the choices available and their likely consequences: pond water, for instance, makes a stronger drink than pump water, other things being equal. In his opening words, Boys affirms that he is writing particularly for 'They who are Curious in *malt Drinks*'. His sense of the word 'curious' is not quite that of the natural–historical *curioso*, who focused on the superlative and extraordinary in natural history.[35] Rather, he is signalling a preoccupation with enquiry, novelty, nicety and precision. Having a direct interest in the quality of his beer, he is not a dilettante observer; yet he is careful to position himself as neither a specialist nor a self-appointed expert. He addresses his reader as a peer, asserting the legitimacy of his advice through the shared opportunities and obligations of the landed gentry.

Like his contemporary Thomas Tryon, Boys published to promote methods which were not generally accepted. His reforms were milder than Tryon's: he proposed, for instance, an hour's vigorous hop-boil, against the common practice of low simmering for long periods.[36] Such prescriptions were the result of experiment. Boys noted that he had 'been very curious and sent some Miles' for samples of water from different localities, in a bid to repeat early successes, adding that perhaps 'the best *Water* in *England*' was that of the Devil's Arse cavern in Derbyshire.[37] Of course, there was no prospect of routinely conveying water some 250 miles to the Betteshanger brewhouse; indeed, we might well imagine the practically minded house servant grumbling at orders to carry water any distance at all to a site where plenty was already to be had. To Boys, however, it was quite legitimate to overrule common sense on the authority of systematic investigation. To thwart the mysterious contamination which rendered beer 'Sour', he dictated, brewers literally should not believe their own eyes: servants who 'by accident made use of any long-handled Jett, hand Jett, or Pail, with cold Water' must wash it in scalding water before continuing, even if it looked perfectly clean.[38]

For all its experimentalism, *Directions for Brewing* is not a philosophical work. Boys did not enquire into causes, or seek to form universal understandings: there is no attempt, for instance, to trace the varying properties of water to differences in its internal constitution. Nonetheless, his gentlemanly context and curiosity resonated with systematizing philosophers, even if they did not know him by name. Gentlemanly house-brewing underwent a significant resurgence as the War of the Spanish Succession (1701–14) curbed access and inclination to French wines and brandy.[39] In 1703, the year of Boys's death, the *Directions* reappeared with a new title page as *A Guide to Gentlemen and Farmers, for Brewing the Finest Malt-Liquors*; Hans Sloane, later to be President of the Royal Society, almost certainly acquired a copy of this reprint.[40] The *Guide* ran through five editions to 1727, the year in which Richard Bradley, Professor of Botany at the University of Cambridge, inserted it wholesale into his edition of Noël Chomel's *Dictionaire Oeconomique*; the text stood intact when Bradley's work was republished in 1758. Lightly edited, it served similarly for Nathan Bailey's *Dictionarium Domesticum* of 1736, among other works.[41]

The text simultaneously found its way into agricultural improvement literature, being added to the second (1703) and subsequent editions of *Englands Interest*, a miscellany attributed (probably incorrectly) to the mathematician Sir Jonas Moore.[42] This version came to the attention of John Mortimer (*c.* 1656–1736), a successful merchant turned dilettante agricultural improver who had been brought into the Royal Society under Sloane's patronage. Mortimer's *Whole Art of Husbandry* of 1707, which invokes Sloane's authority against those who 'have a slight Opinion' of agriculture as a topic for learned engagement, notes the production of '*English* liquors' as a neglected discipline in need of 'good management'. Mortimer's contribution under this heading, however, was simply to reproduce the *Englands Interest* text alongside another process narrative (which may, in fact, derive from it: the ratios of malt to finished beer in the two are notably similar).[43] Authors such as Mortimer remained unable, or disinclined, to investigate the production of beer directly.

Beer-*drinking*, on the other hand, attracted more authorial attention in the early eighteenth century. A few physicians and apothecaries attempted, as Tryon had done, to trace comparatively the dietetic effects of different varieties of drink, occasionally proceeding as far as 'curious' comparison of the materials used.[44] Beer was one of the foodstuffs included in the 1718 pharmacopoeia of the apothecary John Quincy (d. 1722), which attempted brief material explanations for its judgments: strong drink, for instance, carried 'viscid Parts' into the blood, provoking nausea and headache. Quincy similarly examined the consequences of high hop rates, highly dried malt (which resulted in brown beers, 'fitter to pass the several Strainers of the Body' than pale drink) and maturation.[45] Medical authors of the 1720s borrowed, endorsed or questioned Quincy's judgments, often in the context of drink in general, and particularly along the sidelines of public controversy on the medical properties of the waters of different locations.[46]

A further source of relevant literature was the spirits trade. The early stages of spirits manufacture were similar to those of beer-brewing: what went into the still, a liquid known as 'wash', was prepared by mashing grain before fermenting with yeast. As noted above, the distillery's overlap with pharmaceutical production made it relatively compatible with public philosophical enquiry, and some of those who published, including William Yworth, extended their considerations to beer. In his *Cerevisiarii Comes* of 1692, Yworth gave the first known atomistic account: strong 'double Ale or Beer' should made by applying the wort of a previous mashing (not fresh water, as was the norm) to fresh malt, because 'its particles are rendered Globical', aiding its keeping qualities.[47]

In the 1720s George Smith, a distiller of Kendal in Westmoreland, produced a general treatise on fermentation grounded in the material philosophy of Herman Boerhaave (1668–1738), the Leiden professor whose theories of chemical composition were becoming increasingly influential in the English-speaking world. Smith attempted a Boerhaavian explanation of the brewers' long-established observation that March and September were the best months to brew to avoid spoilage. Fermentation, he said, was a vigorous internal motion or 'vibration' in which the 'fermentable mixed bodies' are broken down, releasing elementary particles of spirit and air; March and September have the widest and most sudden changes in aerial temperature, allowing the most varied and thus complete fermentation.[48] Smith was probably also responsible for a publication of 1733 which integrated the practical, philosophical and dietetic strains of brewery writing. Presenting what was then a standard warning against brewing at the height of summer, the author essays a chemical justification: the atmosphere being 'too light and rarefy'd', the fermentation will progress too quickly; since this must 'cause the Spirits of the Liquor to exhale, the Ale must continue turbid, and soon turn flat and vapid'.[49]

Yet even this work by a practising distiller did not incorporate any direct evidence from commercial brewing: its only process account is openly derived from Gervase Markham's work of 1615. The author takes Markham, along with John Worlidge, as authoritative on practical matters, just as Boerhaave is his theoretical authority. Indeed, he positively forswears engagement with commercial brewers: the value of independent philosophical investigation is that it can guard the consumer against 'Fallacies us'd by the Makers and Retailers' liable to promote 'dangerous and fatal Consequences'.[50] No distiller would have written of his *own* trade in these terms, of course: the common brewers were fair game for this kind of attack, being outsiders to philosophy and to print.

Across the first half of the eighteenth century, accounts of beer-brewing in philosophical, medical and husbandry works were shaped by this handful of sources. Most authors borrowed from Boys's *Directions*, Quincy's dietetic investigation or the process narratives attributed to Markham; others used Yworth's treatise (usually cutting the alchemical theory) or Tryon's (although his reputation for waywardness pursued him through the decades).[51] Denis de Coetlogon's encyclopedia of 1745, for instance, combines Quincy, Moray's early work on malting and the *Directions* (via *Englands Interest*), bolstering its authority through a sage condemnation of the '*Chymical Brewer*' Tryon.[52] William Lewis carried Quincy's account verbatim and without comment into his influential pharmacopoeias, though he was himself the son of a Richmond brewer.[53]

What the Common Brewers Knew

How firmly, and how consciously, did the common brewers withhold their knowledge, and what was there to withhold? Most of the knowledge was confined by local patterns of oral communication, direct demonstration and guild affiliation.[54] Around 1740, a London apprenticeship might carry a premium of hundreds of pounds, even though the Brewers' Company did not control the right to brew: the fee bought a set of connections and a training not given to outsiders.[55] It is impossible to recapture the exact details of this training, which must have been based on observation and imitation in the various parts of the brewhouse.[56] We can, however, say something about how knowledge flowed, what kinds of people were in a position to control it, and how those people stood in relation to philosophers and public authors.

The large common brewers' practices had been diverging from small-scale private and victualler brewing since medieval times. Common brewers had been the first to introduce boiling with hops, an innovation imported by Flemish/German immigrants: this meant increased technical complexity but a product with better keeping qualities, suited to large-scale production and distribution.[57] Whereas gentlemanly house-brewing on Jeffrey Boys's pattern might be performed semiannually, around March and September/October, the common brewers' output was often near-constant, suspended only at the height of summer. By around 1700, a small group of 'great' common brewers was emerging as a class apart, enjoying unique advantages of scale and systematization.[58] In London, where the markets were most concentrated, the great brewers had a particularly distinctive production culture. The decade or so around 1720 saw the local brown beer, brewed for storage in bulk, evolving into 'London porter', a long-matured style whose reputation grew rapidly in domestic and export markets.[59]

Jeffrey Boys, the curious house-brewer, was blunt in his response to these trends: 'Common Brewers very seldom or never make good Drink'.[60] This, he said, was because they simply brewed too often, outpacing the brewhouse servants' ability to manage the fermentation correctly, and allowing agents of sourness to build up in the wooden vessels and taint successive brewings. Only in the small private brewhouse could the process be disciplined: Boys was obviously proud of the care with which he directed his servants to scour all the vessels with boiling water, and ensure they were dry, before the next brewing could begin.[61] The trade's failings, Boys hinted, were an inevitable consequence of its drive to magnify profit, which the private gentleman need not consider.

Improvement-minded philosophers, however, had good reason to be interested in the doings of the 'great' brewers, whose advantages were affirmed publicly in petitions to Parliament over the fairness of taxation around 1700. On one estimate, they could brew four barrels of strong beer from a quarter (around 290 litres) of malt, economies of scale giving them an advantage of half a barrel per quarter over smaller competitors.[62] In addition, London porter's fitness for long storage offered a valuable shield against fluctuations in the price of barley: brewers could build up stocks in cheap years to draw upon in years of scarce harvests.[63] It was popularly assumed that porter was impossible to produce on a small scale; certainly, its maturation required equipment and capital beyond the means of home brewers and the smaller firms.

Such technical knowledge, we should note, was not always central to the life of a successful proprietor. From around 1710, for instance, two branches of the Calvert family controlled major London brewhouses whilst living primarily as country gentry, ploughing their profits into Hertfordshire landholdings: this strategy was followed more famously by the first Samuel Whitbread, a gentleman-farmer's son who rose to the head of the London trade after 1750, eventually acquiring the Torrington estate in Bedfordshire.[64] Such men styled themselves not 'Brewer' but 'Gentleman' or 'Esquire' (unless, as occasionally happened, they were knighted), and might be better known for activity in Parliament or other public office than for their beer. Nevertheless, such men typically knew the brewing process at first hand. Most were sons or nephews bred to the trade, while some born outside it, including Whitbread, had served apprenticeships. There were exceptions. A partnership might take in silent investors, or a strategic daughter marriage might leave sole control in the hands of a newcomer, as happened when Sir John Hynde Cotton (1717–95), fourth baronet, found himself responsible for a leading London brewhouse in the 1750s.[65]

Suitably skilled or not, however, most owners delegated their technical responsibilities as capitalization and volumes increased. When one observer in the 1760s recorded a meeting with Henry Goodwyn, 'Brewer', and Mr Pyecraft, 'his Brewer', he was evidently distinguishing a proprietor from his salaried manager.[66] A 'brewer's brewer' was also sometimes termed a *clerk*, the same title used for the most senior employees in the counting house which was equally crucial to any large brewery's operation. In the largest concerns, the wages of such senior managers were considerably above the incomes of most private gentlemen. It was common for a clerk to invest a portion of this income into the firm, making him an unofficial or official partner.[67]

Indeed, the proprietor/clerk divide was notably permeable in both directions. Every ambitious young brewer in 1700 knew that the Anchor Brewhouse in Southwark belonged to its former head clerk, Edmund Halsey, who had married the proprietor's daughter. In turn, by 1730, it had passed to Halsey's clerk –his nephew Ralph Thrale – who did not inherit, but used his expertise as guarantee on a long-term loan to buy the brewery for £30,000.[68] Moving in the opposite direction, salaried clerks often came from families which had owned and lost breweries. Production failures, serious enough to destroy a business, were frequent and unpredictable, and by no means told fatally against the brewer's skill; bankruptcy was endemic in the trade, and brewers with long experience but no credit were often taken into the service of those in the converse position. Goodwyn's brewer Pyecraft, for instance, was probably of the family of John Pycroft, a brewer-proprietor in the Minories whose success (generating a fortune said to be nearly £30,000 in the 1730s) was not matched by his successors.[69] Oliver Slocock of Southwark, repeatedly bankrupt as a proprietor, found a more secure niche as 'Working-Brewer' to Alderman Parsons, who bequeathed £100 per year to ensure the training of Slocock's son.[70]

The brewhouse clerk himself oversaw a number of 'brewer's servants', often unapprenticed and unskilled. For the most part it was these labourers, who show up only fleetingly in the historical record, who literally brewed – although their supervisor might routinely get his hands dirty, and would often insist on controlling directly the most critical parts of the process. In a 1750s engraving depicting the interior of a 'large brewhouse' (Figure 1.1), the 'men' are engaged in stirring the contents of the mash-tun at lower right while the 'master brewer', wearing the apron which conventionally denoted his trade, oversees them from the wooden stage built out in front of the boiling coppers. The illustration is idealized and partly symbolic, but the central survey point, with views into all areas of the brewhouse, was an ideal which proprietors and clerks enacted as best they could.

The brewhouse clerk's role was primarily technical, like that of a colliery viewer or consultant engineer (though without the formalized professional status of the latter). While he might be understood as a general manager, his most prized skills were in manipulating a volatile bulk product, juggling the unpredictable constraints of supply and demand, and attending to the bureaucracy of unending surveillance by the gaugers of Excise. Nobody doubted that technical management could make or break a firm. One account ascribes Samuel Whitbread's meteoric success to his decision to brew large stocks of uncommonly strong beer during his first year of major operations (while he was, effectively, his own head brewer): a hotter than usual summer spoiled his competitors' stock, and Whitbread walked away with the trade.[71] Contrastingly, John Hynde Cot-

Figure 1.1: Brewhouse interior, from [J. Barrow], *A Supplement to the New and Universal Dictionary* (1754), plate 7. Reproduced courtesy of the Wellcome Library, London.

ton's reluctance to involve himself in brewhouse practicalities left him totally reliant on some bitterly fractious clerks. One such, James Daux, resigned in 1765 with a dire warning of the threats hanging over the brewhouse: systematic overcharging by unscrupulous Excisemen, the corruption of his fellow clerks, the ruinous souring of inadequately stored beer.[72]

The implications of these circumstances for the likelihood of public revelations were mixed. Managing brewers, like small-scale proprietors, possessed much knowledge drawn from systematic practical investigation, but were unlikely to disclose it. As a parting shot in his resignation letter to Cotton, Daux described a controlled experiment to prove his claim that open vessels caused sourness, the point being that he doubted Cotton's ability to comprehend it: Daux's wider stock of communicable knowledge was retained to secure his employability elsewhere. By contrast, some great brewhouse proprietors began to see the potential of an alternative stance: their success was largely a function not of secret knowledge, but of their scale of operations and probity of conduct, so some disclosure was beneficial (a logic which culminated in showpiece apparatus and royal visits in the later eighteenth century).[73] Leisured proprietors, particularly those who were university-educated, were also more likely to move in philosophically curious circles. Yet there is no clear trace of any common brewer (proprietor or clerk) contributing directly to the literature before the 1750s. There were, however, indirect routes for commercial brewery knowledge – or what was described as such – to pass into print.

William Ellis and the Trade Writer's Paradox

The London and Country Brewer, issued in four instalments from around 1735 to 1740, was far more extensive than any previous brewery publication, running to over 300 pages. The anonymous compiler, William Ellis (d. 1758),[74] was a Hertfordshire farmer noted for his prior publications on agricultural improvement, which had gained him a national network of correspondents including curious gentlemen and physicians. These books had appeared under Ellis's own name, but the new work identified him only as 'a Person formerly concern'd in a Common Brewhouse at *London*, but [who] for near twenty Years past has resided in the Country'.[75] The message was obvious: the author not only knew the common brewers' secrets at first hand, but was also in a safe position to reveal them. Ellis's authorship would not, in practice, have been hard to detect, since he was unsubtle about promoting his acknowledged agricultural writings in the text. The point of keeping his name off the title page was to ensure that he could not be proved as author, and also to meet his readers' expectations as to the precautions that would be taken by any author planning genuine disclosure of jealously protected knowledge.

It is not clear how far Ellis really knew the London trade. The bulk of his text deals with the agriculture of raw materials and with rural domestic brewing, drawing on gentlemanly sources such as the 'curious' Simon Harcourt of Pendley (1653–1724), mentioned in the opening quotation to this chapter, who was a near neighbour in the Hertfordshire countryside. Ellis's chapter on 'The *London* Method of Brewing', however, is presented as a first-hand recollection. In a later work, he mentioned that he had experienced London brewing whilst executing an uncle's will, an account which matches the probate record of Richard Ellis (d. 1713), common brewer of St James's, Westminster.[76] The chapter does not attempt to offer comprehensive instructions, but describes operations claimed to be particular to several commonly retailed styles. In the third instalment of 1738, Ellis added a short account of porter-brewing which was probably borrowed from a manuscript. These sources provide the earliest known published record of London practice.

Ellis delighted in the pose of undermining guild mysteries. He revealed, for instance, the shibboleth term *liquor* for brewing water, 'for it is Sixpence forfeit in the *London* Brewhouse if the word Water is named'. The methods described are distinctive: whereas house-brewers such as Boys fermented their three mashes separately into drinks of different strengths, some of the London beers are seen to be fermented 'Entire', from the combined worts of up to four mashes, with selective use of the weak fourth wort giving control over the product's strength. The text attaches much importance to the sequence of mashing heats, said to alternate between hot and cool.[77] How sensitive Ellis's revelations actually were is doubtful, however. His narrative, if it was accurate, described the production practice of twenty years previously. The intervening time had seen fundamental changes including a tremendous lengthening of maturation times, a point Ellis discusses only anecdotally.[78]

Ellis's general approach, indeed, faced problems of both theoretical and practical credibility. His husbandry texts illustrated the paradox of the trade writer all too well: racing to fulfil publishers' commissions, he not only resorted to space-filling trivia but neglected the running of his own farm. His reputation suffered badly as a stream of visitors, expecting model demonstrations of the new technologies and management schemes he so convincingly promoted, met with husbandry visibly poorer than their own.[79] The credit Ellis derived from publishing knowledge which had 'hitherto remained in private hands as a very great Secret', then, could easily be undercut.[80] Linnaeus's pupil Pehr Kalm, who sought out Ellis during his visit to England in 1748, pronounced him a disingenuous hack trading in dubious second-hand information.[81] Kalm exempted the brewing work from this judgment for want of knowledge, but we can see the pattern taking hold in the *London and Country Brewer*: the first instalment follows a systematic plan, but its three sequels dissolve into miscellaneous compilation with no scrutiny or comparative agenda.

Nonetheless, the collection was widely read, running through seven editions to 1759. Most subsequent treatises borrowed something from the structure of the first instalment, devoting individual chapters to particular materials (barley, malt, water) and operations (malt-grinding, boiling, fermenting). Moreover, it is possible that the mere semblance of revelations in Ellis's work was enough to weaken the perceived value of commercial brewers' silence, at a time when philosophers were making concerted moves to gain their cooperation.

'Commercial Chemistry' and Organized Trades Improvement

The systematizing and comparative agenda grew and changed as Hans Sloane, President of the Royal Society from 1727 to 1741, used his personal and institutional patronage to secure the careers of a number of specialists, mostly in agriculture and horticulture, who did not have conventional landed or professional backgrounds. The market gardener Philip Miller (1691–1771), for instance, was elected to the Fellowship in 1729: Miller's compendious *Gardeners Dictionary*, first published in 1731, answered botanists' demands for a 'general system of gardening' from an author 'who has added a considerable stock of reading to many years experience'.[82] The same period saw the growing development and distribution of systematic chemical treatises, which drew increasingly on artisan practice accounts.[83]

A key figure here was the physician Peter Shaw (1694–1763). A friend of the abovementioned John Quincy, Shaw established himself from the mid-1720s as a lecturer, translator and commentator on chemical theory.[84] Through editions of the works of Francis Bacon and Robert Boyle, Shaw placed himself in the tradition of philosophical engagement with trade practices: his particular agenda was 'to extend the Business of *Chemistry*, and render it applicable to the improvement of Philosophy and Arts'.[85] When Shaw, commentating upon Bacon, noted that the 'common Practices of Tradesmen might considerably enrich our present *operative Philosophy*', he had in mind an example from the brewery. Coopers (not, in this instance, barrel-makers, but brewery representatives who travelled around the public houses attending to the product) had begun to use '*artificial Acids*' to make cloudy beers clear. Inquiring into the causes underlying this technique might bring lessons for other fields of activity.[86]

Shaw's work confirmed a subtle revision of the dominant natural-philosophical identity. While paying every possible homage to Boyle, Shaw repositioned the search for wealth on a gentlemanly footing: self-interest could be submerged in an acceptable national interest. Better chemical organization, said Shaw, could create domestic surpluses through increased production, improve supply through better preservation techniques, and allow England (Shaw's usual term, though his reference was often to the extensive free trading zone of the young United Kingdom) to undersell rival nations. To this end, useful methods and processes 'kept as a Secret in few hands' must be brought systematically to a wider public audience.[87] Where necessary, premiums and other inducements should be used to nationalize knowledge: Shaw was among those who worked to raise, by public subscription, the £5,000 sought by Joanna Stephens for the recipe for her bladder-stone cure.[88]

In its broad outline, Shaw's position was not new: Thomas Sprat had sketched something similar in his early defence of the Royal Society, and there are comparable hints in John Beale's schemes.[89] Shaw's programme, however, was laid out in unprecedented concrete detail. Drink and its associated industries – now enjoying a prime and growing role in indirect taxation and various theatres of economic warfare – provided his most fertile source of examples. *Three Essays in Artificial Philosophy*, published in 1731, offered a prospectus of ripe areas for chemical investigation alongside two detailed case studies to illustrate the method: one on spirits manufacture, the other on concentrating alcoholic drinks for export. Shaw made similar points, with demonstrations accessible to a broad gentlemanly audience, in the public lecture course he delivered the same year. Of twenty lectures, four focused on malt, worts, vinegar, wine or spirits.[90]

Shaw's spirits essay best illustrates the novelty of his overtures to the trade community. Philosophers, he said, were concerned with the most 'perfect' fermentation process, namely that giving the maximum yield of spirit; men of trade were bound to seek the most 'advantageous', or profitable process, which was not necessarily the same thing. Commercial distillers, then, were excused from the philosophical search for purity, but their agenda was rational and open to reorganization based on experimental philosophy.[91] Gains in yield from a newly discovered process might, or might not, compensate for its cost in time, trouble and equipment: judgment of the issue would have to vary from case to case, because those in the 'small way of business' had different opportunities and obstacles from the 'large Dealers'.[92] The task of 'Improvement', then, was to apply comparative philosophical evidence selectively: Shaw showed how it could have a role in deciding how dilute to make the extracts, how warm to establish the fermentation and whether to add artificial fermenting agents.

Though this project addressed processes similar to the beer-brewers', and was later to inspire brewery work indirectly, Shaw apparently went no further than his predecessors in engaging directly with the beer trade. The most conspicuous early efforts to do so, based on his approach, came from the Dublin Society for Improving Husbandry, Manufactures and other Useful Arts and Sciences, active from 1731. The Dublin Society was founded by a distinctly Protestant community of wealthy landowners: their improvement agenda saw Ireland almost as a defective England, rich in resources but desperately short of sound management. The obvious way to correct this was to emulate the products and practices of English agriculture and commerce, guided by philosophical investigation and authors such as William Ellis. (An Irish edition of the *London and Country Brewer* appeared in 1735, not long after the London original, and bearing a recommendation from the Society on its title page.)[93]

In the Society's *Weekly Observations* of 1738, a correspondent, 'R—y W.', appealed to 'Gentlemen of character and fortune' to redress the Irish people's preference for imported wine over domestic beer. Improvers should 'make proper observations and inferences from the different appearances', codify the inferences, and generalize operations based on the results.[94] The author provided detailed procedural descriptions of various brewery operations, and referred to a qualitative comparison of Kentish and Irish hops, based on destructive distillation. The Society also put into practice Shaw's model of offering cash premiums for the production of strategically important commodities, and for technical improvements. Premiums for 'Irish hops' were recorded frequently from 1744, with further premiums announced occasionally for sales of finished beer, the establishment of new breweries or the production of malt 'in the *English* Manner, and equal in Goodness to *English* Malt'.[95]

This premium model was re-imported to England through the foundation, in 1754, of the Society for the Encouragement of Arts, Manufactures and Commerce ('Society of Arts').[96] Its earliest premiums were awarded for extracting or producing valuable commodities, such as cobalt and madder, wherever members saw potential to undercut overseas competition.[97] Though Peter Shaw was not active in managing the Society, his manifesto strongly inspired its work. Growing interest in the improvement agenda inspired Shaw to have his *Chemical Lectures* reprinted in 1755; the *Three Essays* reappeared in 1761, now titled *Essays for the Improvement of Arts, Manufactures, and Commerce*.[98] Into the second half of the century, the Society's initiatives to collect and distribute production narratives allowed some conscious division of labour with the Royal Society, which was increasingly excluding practical process accounts from its publications.[99]

By the mid-1750s, then, several factors had increased the prospects for circulating ideas and practices between experimental philosophy and the commercial brewery. Peter Shaw and his sympathizers had moved subtly away from the trades historians' less-than-fruitful search for comprehensive process disclosures: 'commercial chemistry' focused on the search for universal insights, which might bring general advantages to the trade, while stressing that particular applications remained a matter of practical art (and could thus be kept private). Correspondingly, paths were opening to make testimony from the brewhouse directly credible to some sections of the philosophical elite: the commercial motive need not be associated with self-interest and secrecy, but could be subordinated (at least in rhetoric) to a managed culture of public disclosure serving patriotic interests.

It remained to be seen whether commercial brewers would prove either willing or able to fashion explanations fit for outsiders. William Ellis, in the 1730s, was able to publish on the commercial brewery precisely because he was working at a significant remove from it, a point which weakened the value of his testimony. By the early 1750s, however, at least one commercial brewer had formed a plan of publication. William Reddington (d. 1755), who brewed at Windsor in Berkshire, collected more than a hundred 'instructions and precautions' – 'Why Beer requires more Hops in the Summer than in Winter', 'How to proceed when there is a Scarcity of Yeast', and so forth – with a view to having them printed up for sale, 'Induced by a Desire of promoting the Good of others'. Reddington apparently intended to release the material in his own lifetime, though the collection appeared only posthumously, in 1760.[100]

By this point, however, a treatise had already been issued by a brewer who was very much alive. Michael Combrune positioned his *Essay on Brewing* as a direct response to Peter Shaw's entreaties, and could soon count Shaw as his intellectual patron. Combrune was fundamentally concerned to produce an account which spanned particulars and universals. Like George Smith, he grounded his general statements in the chemistry of Herman Boerhaave; Combrune went much further, however, in developing a grand theoretical scheme to underpin the management of all parts of the brewing operation. In the next chapter, I will examine Combrune's publications alongside various responses from philosophers, useful-knowledge enthusiasts and public and private brewers. Combrune's theoretical scheme was largely disregarded, and yet his chief practical innovation – comparative thermometry – was quickly assimilated into commercial brewing, and drew applause from his harshest critics. The fact of his publication, far more than its content, was crucial to this acceptance.

2 THE THEORIST AND THE THERMOMETER

> COMBRUNE (*Mr. Michael*), the first who has, with any degree of accuracy, applied the thermometer to use in brewing, V, 3. His table of the medium heat of the air in and about London, in each season of the year, *ibid*. Great pity that he did not endeavour to render his book intitled 'The Theory and Practice of Brewing' more intelligible than it is, 111.
>
> J. Mills , from the index to *A New System of Practical Husbandry*[1]

Michael Combrune's *Essay on Brewing* of 1758 looks like a convenient milestone. It is the first brewery publication structured as a chemical treatise, and the first to discuss thermometry. It is also the first whose author was an active commercial brewer: indeed, it is dedicated to 'the Master, Wardens, and Members of the Worshipful Company of Brewers in *London*', to which Combrune belonged. It is tempting, therefore, to see the *Essay* as signifying a shift in brewers' attitudes away from the closed conventions of craft mystery, towards the systematic public development of useful knowledge. This temptation is one we should resist. The Company's dominance was waning by the 1750s, with corporate leadership of the trade passing to well-capitalized gentlemen-proprietors from diverse backgrounds;[2] and, although Combrune had his imitators, few brewers published anything before 1800.

Nonetheless, Combrune's publication is an important marker of two significant changes in philosophical culture: a revised trades improvement agenda and the increasing availability of systematic chemical texts. Specifically, Combrune presented his writings as a contribution to Peter Shaw's programme for 'commercial chemistry' (chapter 1), which aimed to abolish perceived gaps between systematic philosophical and artisanal knowledge. Combrune based his work both theoretically and structurally on Herman Boerhaave's *Elementa Chemiæ*, the most influential chemical text of the period. The *Essay* and its sequel, *Theory and Practice of Brewing* (1762), were chiefly an attempt to articulate Boerhaavian thermometric universalism in the context of malting and brewing.

In itself, this is not particularly surprising: numerous historians have underlined how blurred, in practice, were the lines between artisanal knowledge and philosophy in Combrune's day.[3] The investigators addressed in most of this literature, however, enjoyed forms of credibility which Combrune did not. Philip Miller, for instance, was no longer a market gardener when he had the running of Chelsea Physic Garden; John Theophilus Desaguliers was nominally a priest when he undertook engineering commissions for the Duke of Chandos. Such experts were systematizers of trade, but not tradesmen; it was their business to address the universal, sometimes by reference to particulars. Combrune, as a brewery proprietor, was by definition concerned with particulars; his attempt to generate universals lacked a clear precedent, and took most of his audience by surprise.

Combrune's credibility was open to question on five loosely distinguishable levels. First, could beer-brewing knowledge be reduced to chemical universals at all? (It might be that the processes were too diverse and inconstant for systematic analysis.) If so, second, could these theoretical results translate into any useful revisions in practice? (Tradition, and the evidence of unsystematic experience, might remain a surer or more efficient guide.) If so, third, had Combrune actually achieved such useful results? (His theory or observations might be flawed.) If so, fourth, had he revealed his success frankly in his publication? (As an active brewer for profit and Company member, he might be selective about circulating information liable to bring competitive advantage.) If so, finally, could others gain the knowledge needed to emulate him by reading his print narrative? (Crucial techniques or interpretations might need to be passed on by direct demonstration.)

Combrune was, in fact, challenged on every one of these levels, by various observers interested in brewing, chemical philosophy or trades improvement, and in the general reviewing press. Compared to later brewing theorists, he did little to defend himself against such challenges; in fact, he probably had little opportunity to do so. Ultimately, few of those concerned with policing the boundaries of natural knowledge admitted Combrune as a philosopher, or his work as a valid contribution to chemical philosophy.

Nonetheless, Combrune's legacy is worth pursuing for two reasons. First, he provided a conceptual and structural model for later attempts to reduce the order of the brewhouse to print. The widespread distribution of his writings led chemically minded brewers, and a few brewery-minded chemists, to conclude that theory (if not *his* theory) represented a productive avenue for exploration. Second, Combrune popularized the brewhouse thermometer. Though his Boerhaavian theoretical underpinning soon collapsed, the practical innovation endured, and led to closer brewery engagement with philosophical instrument-makers and the associated techniques of measurement and recording. Combrune's enshrining as a pioneer in theory and instrumentation, which began within a decade of his death, was genuinely a factor in the emergence of a category of brewery science.

This chapter begins, therefore, by examining the thermometer's role in the eighteenth-century trades improvement agenda through which Combrune claimed legitimacy. Suggestions that the thermometer was self-evidently useful came only with its general acceptance. Its initial appeal was to a limited, yet significant constituency of trade reformers, who valued it less for the accuracy of its readings than for its promise of a criterion for universal comparison, analysis and synthetic emulation. This was Combrune's key concern: I go on to chart how he was enlisted in the commercial chemistry programme, his largely independent evolution of an arithmetical–thermometric system of brewing, and his novel invocation of a tabletop experiment to support claims about full-scale brewhouse work. I then address the several challenges to the epistemic status of his claims and, the survival of parts of his programme notwithstanding, the lack of any clear answer to these challenges.

I conclude with an examination of Combrune's limited but influential role in harnessing publication to private instruction. To his detractors, Combrune's position was hypocritical or incoherent: while pretending to publish natural knowledge in the common interest, he maintained the necessary proprietary privacy concerning brewhouse methods on which he was, self-evidently, commercially dependent. Combrune saw the matter differently: while it was obvious that direct personal attendance was needed for full communication, print publication was a useful adjunct, and could, indeed, be combined with private instruction on commercial terms.

The Thermometer as Chemical Index

As noted in the previous chapter, early modern writers on brewing were highly selective about how much of the process to attempt to reduce to rules on paper. They often confined themselves to weights and volumes (of malt, of water, of finished beer), timings and brief descriptions of any unexpected apparatus or procedures. They also, increasingly, recorded directions on the heat of the materials. ('Heat' was the common term to describe the intensity of hotness in matter – what today would be called temperature – throughout the period under consideration. 'Temperature' usually referred to the concept of a particular blending or regulation of 'heats'.) Heat values were important at several points in the process. Most malted barley was dried by kilning: the degree of heat determined its colour and influenced the taste of the beer. In the brewhouse, the heat of the water used to mash the malt had to be carefully regulated. The progress of fermentation, too, depended on atmospheric heat, running faster in warmer weather; either too fast or too slow a fermentation was liable to spoil the beer.

The case of mashing heat demonstrates brewers' increasing concern with these questions. Sixteenth-century topographic accounts describe the water being brought to the boil and poured slowly onto the malt in the mash-tun.[4] In the seventeenth century, it became common knowledge that using water near the boiling brought a serious risk of 'setting': the mash would clump together like porridge, preventing the valuable fermentables from being drawn off. The water therefore needed to be cooler than boiling – but if it was too cool, the process would not work at all. Early attempts to find a happy medium employed the so-called 'hour-glass' method, bringing to the boil before cooling for a fixed time.[5] By 1700, however, this had apparently given way to purely sensory judgment. Most written sources advised the brewer to let the water cool from near-boiling until the steam subsided so far that he could look down onto the surface and see his own face reflected.[6] Common language could capture (albeit imprecisely) the more elaborate procedures of the commercial brewery, as in the account of a five-step mashing sequence which follows 'Milk-warm' water with 'Scalding hot' before 'Blood-warm, then hot, and then lastly, cool'.[7]

Otto Sibum, in studies focusing on the brewery thermometry of the 1830s and later, has plausibly described a process whereby such qualitative traditions, based on bodily sense, give way to quantitative understandings based on instrument readings.[8] From this, we should not infer any general polarization between the chemical laboratory and the brewhouse, particularly in the eighteenth century. Experimental chemistry, as Lissa Roberts has noted, was itself a highly sensuous business: only with the promotion of intense reliance on the evidence of the balance, under the banner of Lavoisier, did the chemist's bodily involvement begin to lose its privilege.[9] Commercial brewers, for their part, were habituated to precise and accurate quantification, at least as to volume, through the constant scrutiny of the Excise gaugers.[10] It was thus possible to represent mashing liquor heats in terms of mixtures between specified volumes of cold and boiling water: the posthumous work of the Windsor brewer, William Reddington, proposes distinct values. The method was not, however, seen as reliable enough to overrule the individual brewer's expert judgment.[11]

Commercial brewers of this period certainly knew what a thermometer was. Bracketed alongside the barometer and hygrometer (often literally, in a wall-mounted case), the device had become a widespread article of gentlemanly consumption by the 1730s. Thermometers were typically enjoyed as ornamental conversation pieces, though the curious farmer or physician might buy 'weather instruments' to probe trends underlying the progress of crops or fevers.[12] The graduated scale of the thermometer had provided a favourite analogical plaything for satirists since at least 1710, when Richard Addison devised the 'Ecclesiastical Thermometer' by which the 'different degrees of heat in religion' could be measured at various coffee houses.[13] Thermometric values featured increasingly routinely in newspaper accounts of meteorological curiosities, as, for instance, in the bitter winter of 1739–40.[14] Less well known, however, was the principle of applying the thermometer to particular bodies. William Reddington used an atmospheric thermometer to direct the timing of brewing operations but, it seems, did not consider adapting the device for topical use.[15]

Topical thermometry became widespread, however, as the chemistry of Herman Boerhaave spread across Europe. Boerhaave conceived the medium for all chemical action to be a subtle elemental fire, distinct from common flame and not reliably revealed to human sense. Since one of its intrinsic tendencies was to expand matter, however, the progress or quantity of the fire could be revealed using a well-calibrated thermometer. Boerhaave's chemistry was highly operative, focused on the management of processes: elemental fire was depicted as the chief instrument of chemical change, and the thermometer, as Jan Golinski puts it, a 'second-order instrument' to make the fire amenable to human will.[16] Since the fire inhered in all kinds of matter, Boerhaave promoted vigorous topical application. Thermometers were to be thrust into philosophers' air pumps, apothecaries' vessels and medical patients' bodies to gauge and trace the processes in action. This mode of application, highly amenable to public display, was spread by Peter Shaw and other lecturers during the 1730s. The system of chemistry in which Boerhaave embedded this view was acclaimed across Europe for its unparalleled breadth, and for its integration of theoretical doctrine with practical process accounts. English-speaking audiences learned of it through Peter Shaw and Ephraim Chambers's *A New Method of Chemistry*, a translation of a pirated edition of Boerhaave's lectures at Leiden; to forestall further piracy, Boerhaave released an approved edition, *Elementa Chemiæ*, which was translated by Timothy Dallowe in 1735 and by Shaw, again, in 1741.[17]

Adherents of thermometry, whether atmospheric or topical, usually promoted it as a superior substitute for human sense in existing operations. Brewing was an obvious prospect: process narratives had long referred to 'degrees of heat', defined with reference to expert tactile judgment or visible signs. The *London and Country Brewer*, for instance, advised heating the mash liquor to 'three Degrees this Side boiling', a state marked by the appearance of white steam clouds: the ensuing degrees were signified by the appearance of small bubbles rising to the surface, the 'breaking' point heralding the onset of boiling and the state of continuous boil.[18] It was from such qualitative degrees that the numerical scalings of thermometers developed: the scale adopted by the instrument-maker Daniel Gabriel Fahrenheit, which became prominent through Boerhaave's patronage, arose by subdivision and extension of seven qualitative degrees, originally intended for meteorological work.[19] Conceptually, it was easy to define how the thermometer could pass into trade use.

There were more formidable barriers in practice. Judged as a substitute for readily available human skill, thermometers were fragile, expensive and easily dismissed as unnecessary. Indeed, as Golinski has emphasized, their readings frequently conflicted with the sensory judgment of experienced artists.[20] Peter Shaw, in 1731, deemed them invaluable in 'philosophical and chemical' investigation, but accepted qualitative degrees as sufficient 'for common, or œconomical [domestic] uses'.[21] In the environment of the large commercial brewery, there were additional difficulties: techniques for applying thermometers to the ambient atmosphere, the human body or apothecaries' preparations did not transfer easily to the vast bulk of the materials involved. Even reading the scale was a challenge when the thermometer was plunged into a steaming mash-tun, to emerge caked in sticky malt. At least the mash was stirred throughout the process, so that its heat was presumably spread evenly; the same assumption could not be made for fermenting tuns, which could be probed thermometrically only near the surface. Most importantly, there was no evident sense in which the thermometer offered brewers a crucial improvement on their current individual practice. Many had brewed consistently enough to make large personal fortunes before thermometry was ever suggested – and, for that matter, much brewing continued to go 'out of order' decades after it became general.[22]

What propelled the case for thermometry in the practical arts overall, however, was not so much accuracy, precision or robustness, but the prospect of a universal value system. Its chief enthusiasts were the trades improvers: those who prescribed emulation to cure economic deficit, such as the Dublin Society (chapter 1) and those seeking centralized philosophical authority over production, such as Hans Sloane and his satellites in the Royal Society. Philip Miller, who visited Leiden in 1727 and established a friendship with Boerhaave, came to regard thermometry as crucial to well-ordered botanical translocation. Miller's *Gardeners Dictionary* mentions the thermometer repeatedly, with a section on 'Fire' derived largely from Shaw and Chambers's 1727 edition of Boerhaave's chemistry.[23] The Dublin Society recommended the device in its guidance on hop-planting, first published in 1733, while botanically informed growers such as Timothy Sheldrake built on Miller's work to cultivate tropical plants in England, using greenhouses and artificial heating regulated thermometrically.[24] The new centrality of the thermometer was acknowledged in the frontis illustration which appeared in the *Gardeners Dictionary* from the sixth edition of 1752 onwards (Figure 2.1). While Nature stands aloof from Britannia, withholding the meagre contents of a rather unwieldy cornucopia, masculine Industry delivers a bounty of exotic fruits, newly producible on British soil, directly into Britannia's lap. Science stands at Britannia's right shoulder, watching over Industry: her thermometer regulates the application of his watering can and spade. Prominent in the foreground is a book of directions.[25]

Figure 2.1: Frontis illustration to Philip Miller's *Gardeners Dictionary*, 6th edn (1752). Reproduced courtesy of the Wellcome Library, London.

This message propagated swiftly to the brewers. In a 1735 paper to the Royal Society, the physician Cromwell Mortimer (*c.* 1693–1752), a pupil of Boerhaave, and another protégé of Hans Sloane, extended the rhetoric of thermometric management to the useful arts more generally. Mortimer's concern was the difficulty of replicating the successful results of experiment, even when the materials and operations were equivalent. This he ascribed to differences in the 'Degree of Heat made use of'. Replication, therefore, could be guaranteed by recording, transmitting and applying thermometric values: what worked for the 'curious Gardener' could be applied in any industry based on chemical transformation. As a principal example, Mortimer singled out the brewery and its allied trades of malting, distilling and vinegar-making:

> [B]y Thermometers placed in different Parts of the Heap of wetted Malt, the proper Heat for its sprouting might be determined, and then regulated: The same for the Heat of the Kiln when the Malt is spread on it. By Thermometers the Brewer may ascertain the Heat of the Water when he pours it upon the Malt, the Heat of the Wort when he sets it to work, and the Heat while working.[26]

Perhaps not coincidentally, it was also around 1735 that a London instrument-maker, John Clarke, made the first known attempt to sell heat-gauging apparatus to brewers. Clarke's instrument was not, strictly, a thermometer: it was a hydrometer, or density measure, adapted to register the rise in heat of water through its expansion as it neared boiling. The approach relied on the density of all water being roughly uniform at a given temperature; it could not be extended to fluids of obviously variable density, such as beer or worts. Understandably, Clarke found it difficult to execute the device reliably. Though he claimed the endorsement of 'several eminent Brewers', and the device was taken up for assessing wort strength (a principle I will return to in chapter 4), there is no evidence of its passing into use as a heat gauge.[27]

The extent of brewery experimentation with true thermometers around the same time remains obscure. Other instrument-makers must surely have courted the brewers, but thermometry is not mentioned in the *London and Country Brewer* or other print accounts, and Peter Mathias's extensive survey revealed no thermometers in contemporary valuations or inventories.[28] Michael Combrune, in presenting himself as the pioneer, stated that many practitioners had toyed with thermometers as part of a general push for quality around 1740, but that all had abandoned them – himself excepted – for want of reliable results.[29] I have discovered only one case of a brewhouse thermometer in use before Combrune's first publication: Henry Goodwyn of Deptford, Kent was by 1758 recording values from an unusual thermometer, marked with 15 degrees between the freezing and boiling points of water.[30] It is possible, though by no means certain, that others likewise employed thermometers privately, evolving their own techniques of use.

Nonetheless, the thermometer's arrival was clearly not a self-evident business of locally replacing sensory judgment with a superior quantitative instrument. Had it been so obvious, thermometric recording would have been general in large-scale brewing for profit before 1750. The work which made brewery thermometry a matter of general consciousness, and thereby of general use, was Combrune's; and Combrune's innovation was firmly in the tradition of philosophical improvement, and explicitly guided by Boerhaave and Shaw.

A Brewer Writes

Michael Combrune had no known brewing ancestry. His parents were wealthy Huguenot refugees settled in Westminster, and most of his family followed merchant or overseas trading careers. Most probably he, or those around him, had simply identified brewing as a trade in which new entrants with significant capital could prosper – the same judgment made by Samuel Whitbread's mother, who opened the path to her son's meteoric rise by apprenticing him into the London porter trade in the mid-1730s. Combrune, however, served no apprenticeship, gaining his freedom of the Brewers' Company by redemption in 1743. We find him three years later owning a brewhouse on Pond Street, Hampstead, then a village separated by three or four miles of open country from urban London.[31] This made him, officially, a 'country brewer', subject to the distinct excise regime outside the traditional limits of the metropolis. He had some experience in London of porter-brewing, however, and continued to supply city customers from Hampstead.[32]

Combrune's situation at the intersection of 'London' and 'country' production gave him a keen interest in what underpinned the characteristic products of different localities. As he later put it:

> each Town carried on the Process with such Malts and Heat in the Water, as they guessed best would answer their several Purposes; Tradition and Experience were their only Guide; and Places became noted for peculiar Beers and Ales, as they were different from others in Taste, and sometimes in Colour ... Prejudice has never a firmer Hold than when it rests on Mystery. The Opinion was readily embraced, and speedily propagated, that each Sort of Drink was to be brewed but in particular Places.[33]

To attempt to prove otherwise was nothing new. William Ellis, as we have seen, boasted of disclosing London methods to a national audience. Conversely, in London, where the ale of Derby, Burton and other towns in the English north and midlands enjoyed a reputation, there were rumours that much of the 'Burton ale' was brewed locally, and early miscellanies carried a recipe which purported to emulate northern drinks.[34] Combrune's goal, however, was not merely to imitate, but to profile particular beers analytically, reducing their production to uniform rules.

On his own account, Combrune had two particular mentors, 'the late learned Dr. Barker and my friend Dr. Maty', who directed him to Boerhaave's chemical works as a model to systematize the art of brewing.[35] John Barker (1708–49), physician to the Westminster Hospital around 1746, had studied under Boerhaave in the 1730s.[36] Matthew Maty (1718–76) – like Combrune, the son of Huguenot refugees – held doctorates in both philosophy and medicine from Leiden, and had written a laudatory life of Boerhaave.[37] The London Huguenot community may also have given Combrune access to other leading natural philosophers. He was, like many Huguenot professionals and tradesmen, a Freemason at a time when the fellowship included many influential members of the Royal Society, serving as a steward to the feast of the Grand Lodge in 1740 in the presence of John Theophilus Desaguliers, Martin Folkes and, as Deputy Grand Master, the Leiden-trained medic William Graeme.[38] He does not appear, however, to have absorbed a wide range of philosophical literature: his published work remains firmly grounded in the chemistry of Boerhaave's *Elementa*.

The appeal of Boerhaavian topical thermometry to the brewer is easy to understand. Reflecting his medical orientation, Boerhaave framed the thermometer as a diagnostic device. 'Of what infinite use' (runs Dallowe's translation, which Combrune consulted) 'are *Fahrenheit*'s mercurial Thermometers? How certainly do they point out to us the danger that arises from the Heat in acute Diseases?'[39] This chimed with brewery concerns: during fermentation, in particular, fluctuating or excessively high 'heats' could ruin the operation, and beer, like human subjects, could suffer from 'diseases' of obscure, but apparently atmospheric origin.

Combrune, by his own account, began his thermometric investigations in 1741, and must therefore have kept them private for more than fifteen years.[40] Two developments probably determined his decision to publish. The first was the republication of Shaw's *Chemical Lectures* in 1755: inspired by the notion of 'commercial chemistry', Combrune adopted the *Lectures* as a compositional model, and occasionally as a source for borrowing.[41] Secondly, public discussion of brewery practice had become inevitable during a period of scarcity and extreme pricing of grain, blamed variously on poor harvests and extortionate dealing, around 1757. Brewers campaigned to suppress the production of spirits from malt, and to raise the price of porter in London above threepence per quart (a cherished customary value, protected by law since 1722), and were challenged by pamphlets and newspaper articles castigating them as engrossers and adulterators.[42] In the brewers' defence, several publications pointed to the changed circumstances of the trade, thereby bringing to light elements of an internal history which had hitherto remained closed to outsiders.[43]

In 1757, a letter along these lines, probably composed by Combrune, found its way into print – with or without Combrune's consent – under the title *Some Account of the Rise, Progress and Present State of the Brewery*.[44] Alongside the usual justifications, the *Account* offers a novel plan for the reform and improvement of the trade. It positions brewing as a branch of chemistry, 'more neglected than its Mother Science', lacking philosophical principles or laws. Were these to be established, the patchwork of local specialisms could be abolished, and any desired product produced in any location given the appropriate ingredients and conditions of heat:

> Success must become universal, whenever the Rules of the Art shall be discovered by Study, found agreeable to Reason and Experience, and uniformly followed, by Means of more proper and more general Use of the Thermometer. Then will Mankind more generally enjoy the Benefits which Providence has almost equally dispersed, and a constant happy Practice be both the Effect and the Proof of a solid and experimental Theory.[45]

Combrune had already, by this point, developed what he considered to be such a theory. This he committed to print in the *Essay on Brewing* of 1758, and expanded considerably for the *Theory and Practice of Brewing* of 1762. Perhaps Matthew Maty had a hand in the *Essay*'s emergence: the publisher, Robert Dodsley, was a long-time stockist of Maty's works and a friend of his primary patron, the Earl of Chesterfield.[46] By the time of publication, however, Combrune had acquired a more valuable public supporter in the shape of Peter Shaw himself, now physician to George II. Combrune had sent the manuscript of the *Essay* to Shaw, who advised him to publish; in a letter reproduced in preface to the published version, Shaw notes that he 'should be glad to see some other Trades as justly reduced to Rule'.[47]

The focus of Combrune's 'Rule' is amply indicated in the extraordinary frontis illustration to the *Essay on Brewing* (Figure 2.2), perhaps inspired by that of the *Gardeners Dictionary*. Above hop poles and sheaves representing barley for malting, and with breaking light in the background, a soaring eagle bears, rather incongruously, a thermometer. The instrument has a Fahrenheit scale: following Boerhaave, Combrune refers exclusively to Fahrenheit's thermometer, 'as it is the most perfect, and the most generally received'.[48]

Combrune promoted the thermometer as useful in determining hop rates, the quantity of yeast to employ, the due boiling times for worts and, in particular, the heat of the mash. In the most elaborate version of his system, every quantity, except volume, became expressible in degrees of Fahrenheit's thermometer. Combrune proceeded from a conviction that the material composition of the malt, mash, wort or fermented beer, at any stage, could be wholly determined given sufficient thermometric precision. 'If curiosity should lead us so far', he wrote,

> we might ... determine, by it, the particular strength of each Wort, or of every Mash; for if Water boils at 212 degrees, Oils at 600, and Worts be a composition of Water, Oil and Salt [sweet fermentable matter], the more the heat of a boiling Wort exceeds that of boiling Water, the more Oils and Salt must it contain, or the stronger is the Wort.[49]

AN

ESSAY

ON

BREWING.

WITH

A View of eſtabliſhing the Principles of the ART.

LONDON:

Printed for R. and J. DODSLEY, in *Pall-Mall.*
MDCCLVIII.

Figure 2.2: Title page to Michael Combrune's *Essay on Brewing* (1758). Reproduced courtesy of the Wellcome Library, London.

Combrune did not pursue this determination of strength, however: his focus in the *Essay* was on finding thermometric signatures for barley malt. In Boerhaave's chemistry, vegetable matter is composed chiefly of acids and oils, whose proportions and intermixture determine its character. Grapes, for instance, are very acidic when they are unripe; as the growing season progresses, the heat of the sun draws out oily principles, and the acids are 'smoothed over' by the oils to produce sweet, fermentable 'acid salts'.[50] Malting, said Combrune, is an artificial analogue of this change. Ungerminated barley, 'viscous and replete with acids', is not suitable for brewing sound liquor: its acids promote uncontrollable fermentation, leading to putrescence.[51] Once germinated and dried in the malt kiln, however, 'the grain hath fewer acids in proportion to its Oils; and, at the different stages of dryness, obtains different properties', which are variously suited to various brewing procedures.[52] Combrune, therefore, related the quality of dried malt to the heat of the malt kiln, as established by the Fahrenheit thermometer.

To secure this point, Combrune introduced an experiment. He recounted filling 'an earthen pan, of about two foot diameter, and three inches deep', with pale malt, and stirring it over an open fire until the contents were reduced to cinders. He did not collect the products given off: instead, he observed the colour changes in the malt, through orange and brown to black, and tabulated them against readings from the thermometer. Using the table, he asserted, the brewer could assess thermometrically the character of the malt he purchased.[53] In Combrune's system, malt dryness determines the times at which beer made from the various malts will fall into – and out of – drinkable condition. The palest malt, at 120°, is dominated by acid principles and ferments rapidly, so that the beer is ready to drink in two weeks; but, if stored much beyond that point, will become putrid and sour. Drying to 152° (a high brown), by contrast, draws out oils to retard the fermentation, so that the beer may keep sound for many years, though it may take fifteen months' maturation before it first becomes drinkable. Combrune summarized this information in another table, relating thermometric drying heat directly to maturation time.[54]

This is the first known narrative of a laboratory-style experiment in brewery literature. In separating 'Practice' from 'Theory' in the *Elementa*, Boerhaave had scrupulously distinguished 'processes', productive of useful materials, from 'experiments', productive of natural knowledge.[55] Combrune had grasped the import of this distinction: to the philosophical chemist, it was legitimate to depart from common practice in search of insights which might be adapted to inform practical innovation indirectly. Peter Shaw had affirmed this point in his reconstruction of the Boylean tradition: Robert Boyle's greatest accomplishment, to Shaw, lay in the seemingly relentless experimentation with which he had 'scrutiniz'd and tortured' nature.[56] Combrune, accordingly, supported his thermometric calibration with a single detailed account of a tabletop experiment (recalling those in Shaw's *Chemical Lectures*), most unlike the real operation of malt-drying. If he performed any experiments to validate these results in the practical maltkiln or brewhouse, he left no trace of them in the *Essay*.

Combrune greatly expanded the investigation in his second treatise, completed under Peter Shaw's guidance and published in 1762. In a clear nod to the structure of Boerhaave's *Elementa*, *The Theory and Practice of Brewing* was composed of two sections: 'Theory' was the *Essay* unamended, while 'Practice' was entirely new. Combrune's 'Practice' did not, however, offer process accounts like those of the *Elementa*: it differed from the *Essay* chiefly in giving concrete indications of possible thermometric values throughout the brewing process. These Combrune derived from numerous computations representing relationships between malt character, mashing heats, hop rates and fermentation times. Slightly modifying the chemistry of the *Essay*, Combrune related maturation time to a composite figure, the 'governing medium heat': this he defined as the arithmetical mean of the malt-drying heat and the heat of the mash liquor, slightly modified to account for the 'power of the hops'.[57] This simple arithmetical relationship determined the optimal heats to apply throughout the process.

Combrune presented calculations for four distinct modes of brewing, representing the most popular beer styles as rational consequences of his chemistry. Pale keeping beer, for instance, '[f]rom its name' (and also to maximize strength) required the palest possible malt, associated with a malting heat of 119°. Maturation time must be the longest possible which still allows spontaneous fining: this, by Combrune's empirically derived tables, required a governing medium heat of 138°. Beer in this style must be well-hopped: Combrune rated the 'value of the virtue' of ten pounds of fine hops, presumably also from empirical evidence, at 3°, thereby reducing the effective governing medium to 135°. The best mashing heat was therefore that which averaged with 119° to give 135°, namely 151°.[58]

The expanded theory demonstrates Combrune's attempts to reconcile Boerhaave's chemistry with brewery convention. Brewers had traditionally believed that some of the heat of the malt kiln was stored latently in the malt, and made its presence felt in the mash-tun. As a good Boerhaavian, Combrune dismissed this explanation: 'subtile' elemental fire could not be contained by solid bodies.[59] Yet Combrune's arithmetical scheme carefully preserves its operational consequences: if the malt is dried a degree higher, the mash liquor must be taken a degree cooler to get the same end result (as embodied by the governing medium heat, the mean of the two). Only the basis of explanation is shifted, from the heat itself to its effect on the oil-acid chemistry of the malt.

At the same time, Combrune was attentive to newer brewery methods. London porter brewers, most importantly, used both hot-dried brown malts and higher mash heats than for pale beer, which meant that the governing medium also had to be high. Neatly, a high governing medium was associated with indicated long storage and artificial fining, and could be augmented by heavy hopping – all characteristic features of the London porter that had evolved only a few years before Combrune began his researches. For all that it was shaped by the rhetorical conventions of contemporary philosophy, Combrune's scheme was carefully attuned to brewery concerns in its accommodation of urban innovations which seemed to go against traditional wisdom.[60]

'Far Fetched Theories': Interpreting the Brewer's Philosophy

The first public response to Combrune's *Essay* was less than positive. 'His principal design (besides making and selling a book)', noted an unidentified reviewer in the *Gentleman's Magazine*, 'seems to recommend the thermometer, which might as well have been done in 10 words as in 214 pages'.[61] This point was somewhat self-negating: an octavo treatise demanded attention from the reviewing press, a brief oral recommendation did not. As Peter Mathias has suggested, awareness of Combrune's project grew chiefly through periodicals such as the *Gentleman's*, which included an extensive synopsis of the *Essay*, and subsequent commentary in a variety of publications.[62] Combrune's interpreters – brewers, reviewers, improvers, chemists and curious gentlemen – differed widely in their estimates not only of Combrune's degree of success, but of his actual intentions. In combining formal and theoretical appeals to Boerhaave and Shaw with brewhouse directions, the *Essay* and *Theory and Practice* were ambiguous. They could be read as practical brewing manuals, attempts to reform the trade on philosophical principles, or contributions to chemical philosophy in their own right. Insofar as they spanned these categories, the results could be painted as productive or incoherent.

The whole project seemed rather pointless to John Noorthouck, the miscellaneous writer who generally covered publications on brewing, anonymously, for the *Monthly Review* (alongside several other industries; his background was as a bookseller and stationer). Noorthouck recognized Combrune's philosophical aspirations but deemed them irrelevant to the brewer's business: 'an inquisitive brewer or baker [*sic*] may collect a smattering of natural philosophy from it – It will never make a brewer of a philosopher'.[63] The *Gentleman's* was yet more negative, since its starting assumption was that the *Essay* should provide concrete process directions. Finding them absent, the reviewer concluded that Combrune had practised conventional trade secrecy. That the Brewers' Company had sanctioned publication counted against Combrune: if, indeed, his method held any practical value, he had perhaps applied prudent self-censorship to key values, such as the optimum malting and mashing heats. If so, he made himself ridiculous by seeking the profits of publication and the glory of a discovery whose essential purpose was its universal communicability. A later piece expanded on this theme:

> The success of brewing depends wholly upon the heat of ... the first mash ... If a thermometer was used the first time, and the brewing succeeded, the great point would be ascertained with the utmost precision, and the first trial that succeeded would become a standard for ever. This is the arcanum which the brewers, by profession, keep to themselves, and which a late treatise on brewing, that explains every other particular of the process, leaves wholly undetermined.[64]

The belief that public knowledge of thermometric values might de-skill the operations and disempower the operators was apparently shared by some in the trade, given the cultures of secrecy that grew up around records in later years. Yet the assumptions of the *Gentleman's* reviewer were misplaced: Combrune was not reticent to publish numbers. There are, indeed, no mashing heats in the *Essay*, which focuses on malting, but Combrune responded to the reviewers' complaints with a profusion of values in the *Theory and Practice*. He stressed, however, that they must be understood only as representative examples. A given value would be a 'standard for ever' only if the whole process were kept perfectly uniform, and in practice this could not happen. Mashing heats 'must be properly varied according to the driness [*sic*] and nature of the Malt; to its being applied either in the first or last mashes; and in proportion also to the time the beer is intended to be kept'.[65] Maltsters' and brewers' skill, on their own understanding, lay in adapting their practice to the inevitably variable: the character of the malt and hops, atmospheric conditions, Excise duties and procedures and the tastes of the drinking public.[66]

Otherwise, the *Gentleman's* reviewer was correct in presenting Combrune as bound to discretion by a number of personal and corporate interests. In private correspondence, Combrune noted that he had first offered all his work 'to be inclosed within the walls of Brewer's-hall', for the attention of Company freedmen only; he was seemingly surprised when the Company affirmed him 'at full liberty to dispence of it to the public'.[67] Whether or not the Master and Wardens of the day saw merit in Combrune's approach, they evidently trusted that he had not revealed anything prejudicial. He is, at least, notably silent on the practicalities of the thermometer. He chronicles its history, but says little of its present state; endorses the Fahrenheit pattern, but recommends no specific maker; and is altogether silent on the operations of taking readings (where to position the bulb, the time to obtain a stable value, care of the device when not in use), how to make sure the nominated heats are reached, and how to maintain them consistently across the extended body of a malt kiln or fermenting tun. Combrune's 'timorousness' continued with the *Theory and Practice*, which was originally printed with tables showing the length (volume of beer made per quantity of malt) to be drawn at various malt and hop prices, 'with the profits accruing'. Ongoing controversies over alleged engrossing behaviour by the brewers gave Combrune second thoughts: at what must have been considerable expense, he ordered the plates to be reworked and the relevant sheets reprinted, with a few very general indications in place of the offending data.[68]

Omissions, however, were not only a matter of trade security. In removing the particulars of time, place and physical extent, and addressing all his materials as though they were thermometrically homogeneous, Combrune emulated the universalizing rhetoric of his philosophical authorities. He omitted numerical heat values from the *Essay* for the same reason that Boerhaave had omitted them from the *Elementa*: the aim was to universalize, not the heats themselves, but the rules of management based on thermometric recording.[69] Those reviewers most disappointed in Combrune's works were looking for an intelligent practical man's advice on how to brew; Combrune was rather striving to establish himself as a philosopher in his own right. The 1762 *Theory and Practice*, whose production in quarto format is almost unique in brewery literature, represents a clear bid for credibility. Whereas the 1758 *Essay* had retailed at 3*s*.6*d*., the *Theory and Practice* was advertised at 10*s*.6*d*.; the printer, Johann Christian Haberkorn, was a noted specialist in lavish productions.[70] Here was not a cheap, portable manual of practical advice for literate artisans, but a treatise fit to grace the shelves of a gentleman's library.[71]

Combrune may, indeed, have weighed his prospects of attaining a Fellowship of the Royal Society, to which he presented copies of both publications. Beside Peter Shaw, his connections among the Fellows included William Allix, a near-neighbour in Highgate and also a second-generation Huguenot, who conveyed the donation of the *Theory and Practice*; and his old friend Matthew Maty, who held an increasing role in the running of the Society as part of the 'Hardwicke circle' which gathered around Philip Yorke, Earl of Hardwicke, from 1764.[72] In the 1770s, Maty proposed John Glen King, Combrune's son-in-law, for membership. Yet there is no evidence that Combrune was proposed at all, or otherwise involved in the Society's activities.[73]

The election of such a man as Combrune to the Royal Society, at this stage, was uncontentiously possible, yet practically improbable. Some with trade backgrounds – notably, the instruments retailer Benjamin Martin – had stumbled badly in their quest for Fellowship, presenting it as just another commercial proposition. It is unlikely that Combrune would have made any such error: his correspondence demonstrates a mastery of conventional rhetorical modesty.[74] On the other hand, the issue was perhaps less pressing to Combrune: his livelihood, unlike Martin's, depended on the judgment of customers to whom the initials 'F. R. S.' were not a consideration. The Society, conversely, had no evident strategic use for him. Though commercial men of rather lower social standing had been brought into the Fellowship under Hans Sloane and Martin Folkes, these were mostly instrument-makers with precision manufacturing skills, such as John Dollond, or, more rarely, well-read operatives with extensive systematic knowledge, such as Philip Miller. Combrune's only public offering was his modified Boerhaavian chemistry, and this, despite Peter Shaw's endorsement, was not particularly compatible with the dominant culture of natural philosophy.

At the time of Combrune's research, the main common ground between philosophy, artisanry and organized commercial improvement lay in chemical extraction. The search for the fundamental principles of matter and the agenda for synthesis and substitution both relied on circulating accounts of processes to distil, separate and otherwise transform materials.[75] Boerhaave had collated scores of process accounts in the second volume of the *Elementa*. Yet Combrune's reduction of the materials of brewing to their parts was purely conceptual: he did not seek to characterize or assay the extracts of malt or hops, preferring to rely entirely on thermometric indications. His experiment in the earthen pan was apparently derived, loosely, from the technique of analysis by dry distillation: he did not engage the non-destructive analytical approaches, especially those involving water and spirits as solvents, which were increasingly prevalent in the mid-eighteenth century.

To the agriculturalist and FRS John Mills, Combrune's limitations in exploring his materials were compounded by an 'affected shew of science'. His commercial status as a brewer was not in itself a problem: quite the reverse. The great pity, said Mills, was that the only tradesman to write on the subject happened to be afflicted by an 'imagination which is perpetually in pursuit of far fetched theories'. The central flaw lay in Combrune's programme to determine heats and their interrelation by calculation, where direct experimentation was not only more straightforward, but the only method likely to succeed. The malt-heating experiment in the earthen pan, Combrune's sole public experiment, was simply too artificial: Mills was astonished that a man who presumably dealt directly with maltsters, as part of his daily business, had not taken his thermometer to an actual malting kiln.[76]

What crucially affirmed Combrune as philosophically marginal, however, was change in chemical understanding more broadly. Dallowe's translation of Boerhaave and Shaw's *Chemical Lectures* – though still widely read – reflected the state of chemistry three decades prior to Combrune's publications. Around 1760, Joseph Black was invoking the method of mixtures to demonstrate that heat content and bodily expansion were not generally proportionate, and could not be conceptually conflated. This secured the downfall of Boerhaave's elemental fire: chemical processes could no longer be codified in terms of the instrumental action of heat alone, and the rise and fall of a thermometer was no longer directly representative of chemical state.[77] The specifics of Combrune's system were steadily unpicked: malts, one later critic pointed out, mellow during storage, and so their character cannot be represented by thermometry at the drying stage.[78] Meanwhile, the intensive thermometric quantity, increasingly dubbed 'temperature', was gradually disentangled from extensive 'heat', a far less easily quantifiable proposition in a working brewhouse. Perhaps thermometry would provide an 'excellent index' of the progress of fermentation, if only the internal temperature bore any determinate relationship to the quantity of heat evolved: in practice, most of the heat was simply lost to the surroundings.[79]

Even when deemed valid in theory, Combrune's scheme might be faulted as unworkable. John Noorthouck in the *Monthly Review* considered the *Theory and Practice* a great improvement on the *Essay*, but again queried the point of the exercise. Combrune's elaborate thermometric arithmetic might ornament the knowledge of the (gentlemanly) 'public', but was unlikely to be taken up by practising brewers: 'calculations and rules *so philosophically, so critically, nice*' would baffle the less adept, while the 'expert artist' would find them far more time-consuming than his tacitly established, 'mechanically shorter' and perfectly effective modes of management. The life and attainments of the artisan, Noorthouck argued, were simply too different from those of the philosopher.[80]

Combrune's manifesto appealed more to those chiefly concerned with national improvement, in a variety of nations. His guiding motivation to abolish the 'prejudice' of geographical superiority resonated with David Loch, a Leith merchant promoting the domestic substitution of imports to Scotland, including those from England: the rise of porter, said Loch, had brought an influx of London product which was totally avoidable, as the drink was already being emulated well enough in Edinburgh and Leith. The 'sensible' Combrune was co-opted as a philosophical buttress by Loch, who wrote always 'merely as a man of business, not as a scholar'.[81] Indeed, we might expect to see Combrune recognized among the public culture of improvement, perhaps in the Society of Arts, which increasingly issued publications dealing with agricultural and industrial topics less palatable to the Royal Society after 1760.[82] Here, again, it appears that his supporters (both Shaw and Maty were Fellows) did not secure his election: Combrune did, however, develop some form of advisory role to Sir Charles Whitworth, an early vice-president and leading figure in the Society, for whom he composed a historical analysis of grain prices.[83] Combrune retired from the brewery around 1767, and spent his last years compiling economic and legislative statistics in support of retail price control and Excise reform.[84]

Though Combrune was largely rejected as a contributor to chemical philosophy, however, his methods survived, and his reputation, if anything, grew after his death. Just as the process accounts in Boerhaave's *Elementa* had gained a life independent of the theoretical precepts (with one German translation of 1782 omitting the 'Theory' section altogether),[85] so the operational prescriptions in Combrune's treatise could be unpeeled from the chemistry that supposedly underpinned them. In particular, the innovation of thermometry drew applause from even the harshest critics.[86] Further, not all readers agonized over the implications of Combrune's commercial interest. It could be taken as a simple indication of horse's-mouth reliability, and often was, particularly among audiences less concerned with philosophical principles. Combrune's personal reputation, and that of the thermometer, became entrenched in tandem, as the next section reveals.

Philosophical Brewing without the Philosophy

We might assume that, once the word got about that there was merit in applying the thermometer, Combrune's work had no necessary role. The commercial brewer – skilled in the use of sometimes delicate equipment, and inured to endless gauging and tabulation, courtesy of the unloved Excise – was well placed to take an instrument like the thermometer and develop techniques on a localized, internal basis. He might heat his mash liquor, for instance, using customary methods, register a value with the thermometer, then set the heat thermometrically on the next brewing. If the result was not as desired, he would modify the value, or other parts of the process, based on his established understanding of the broader situation.[87] Such practices aided the reliability of the process, along the lines Cromwell Mortimer had sketched in the 1730s. This intensely localized reliability did not, however, produce the definitive reduction to general rules which Mortimer and Shaw had hoped for, and which Combrune had tried to enact.

Nonetheless, Combrune's treatises provided a channel to the generality of reading audiences, and he quickly began to be regarded as an important technical authority. The content of his work mattered less than its aspirations, its structure and the crucial fact that Combrune's name and vocation were spelt out on the title page. Editors and compilers who had merely raided anonymous productions such as the *London and Country Brewer* acknowledged Combrune in full. In the years following his death, Combrune's name reached many readers by way of an entry in the second (1778–83) and third (1797) editions of the *Encyclopaedia Britannica* which drew heavily on the *Essay on Brewing*.[88] It was the physically impressive *Theory and Practice*, however, which gained a reputation as a standard work. Samuel Bentham, pursuing Prince Potyomkin's wish to introduce English-style brewing to Russia, ordered a copy at Krichev; so, decades later, did Thomas Jefferson at Monticello.[89] German-language publications drew on Combrune from the 1770s, with a full German translation in 1796. This prompted a French translation in 1802 which proclaimed the work '*le guide de tous les brasseurs*' in England, and suggested that English methods might profitably be applied to the beer-brewing regions assigned to France under the Treaty of Lunéville.[90]

Combrune's willingness to nominate mashing and fermenting heats in degrees Fahrenheit held a particular appeal for those readers who believed there must be universal optimum values. A correspondent to the *Monthly Magazine* of 1797, writing on behalf of 'many families in the country', felt sure that the ideal figures were 'well known to all public brewers', but found them frustratingly absent from the available works in print. 'Combrun [*sic*] is become so scarce', he concluded plaintively, 'I have not been able to procure it, or possibly might from that have obtained the information'.[91] In 1804, such demands were met by a 'new edition' of the *Theory and Practice*. This was sharp practice on the part of the publishers: Combrune had died in 1773, and the text seems to have been set, without updates or commentary, from a superficially revised version he had compiled sometime after 1762.[92]

New readers were clearly either untroubled or unaware that commercial brewing had undergone countless technical changes since the *Theory and Practice* was compiled, or that Combrune took no account of post-Boerhaavian chemistry. Indeed, a handful of brewers accepted and extended his modified Boerhaavianism in the early and even the mid-nineteenth century; one, George Blake, reproduced the account of fermentation from Shaw's 1727 edition of Boerhaave verbatim in 1817.[93] More typically, Combrune's readers discarded his conceptual scheme; what they were interested in was the practical thermometry. One feature which must have been attractive to salaried brewers was Combrune's denial that optimum heats could be standardized: this safeguarded the need for the individual brewer's expertise wherever the thermometer was employed. Those readers who attacked Combrune's chemical theory, meanwhile, did not extend their criticisms to his directions for brewing, which, as philosophical gentlemen, they could not judge. Indeed, John Mills's objection to Combrune's theory was not that it would lead brewers astray, but that it offered no new direction at all, being a jumble of ideas adjusted *ad hoc* to validate existing practice.

The published presence of Combrune's work was, nonetheless, crucial to the establishment of brewery thermometry, both commercially and in private households. In 1768 an anonymous manual, *Every Man His Own Brewer*, was published, the structure of which suggests that it was drafted prior to, but revised in light of, the thermometry of Combrune's *Theory and Practice*. This was the first in a profusion of brewing and malting manuals, by diverse brewers, hacks and chemists, which urged the use of thermometers at all scales of production.[94] The 1768 author suggested that the large London porter breweries had already adopted the device, a supposition borne out by Benjamin Martin's promotion of a specialized 'Brewer's Thermometer' from 1766 or earlier.[95] Thermometry was probably general in the commercial trade before the end of the century.[96] Only after 1800, however, did the thermometric authors routinely present recipes and directions featuring distinct Fahrenheit values, as Combrune had done in 1762.[97]

In focusing on the accounts of thermometric operations in Combrune, brewers increasingly abandoned the search for generalized systems of management which had secured his initial, philosophical patronage. The trend in later publications was towards much deeper procedural description of particular operations. The following, from an anonymous 1790s manual, has no counterpart in Combrune:

> I take my first observation by means of a thermometer, whose stem is so long that the bulb reaches to the middle of the mash; there I leave it standing till the time of running off the extract, or, as brewers term it, slacking ... I note down the heat at this time, and as this first mash stands three hours in winter and two in summer, I take another observation in the intermediate space, and a third at the time of slacking ... When my mash is half spent I apply my thermometer to the worts at the cock of the underback, and find there are only three degrees [Fahrenheit] of difference between this observation and the one I made when just slacking, my thermometer being then in the mash. This is a true sign of good malt and of a fair mash; bad malt causes the heat of the mash to vary considerably.[98]

These later narratives located the thermometer more clearly in relation to the brewer, his vessels and the act of recording. Authors emphasized that their materials were heterogeneous and unruly, pointing out that cold liquor added to warm tends to sink straight to the bottom, and so must be roused. Some embedded their directions into prescriptions for the brewhouse labour regime: the brewer reading the thermometer index must expect an instant response from his 'man' working the tap.[99] Increasingly, they also discussed the form of the thermometer itself. Specialized adaptations had emerged in the immediate wake of Combrune's publications, as the instrument retailers locked onto their market: George Adams, probably responding to his rival Benjamin Martin, differentiated 'Fahrenheit's thermometer', 'Ditto larger' and 'Ditto for the brewery' in the mid-1760s.[100] Conventional nineteenth-century equipment included 'tun-thermometers' three feet long or more, so that, with the bulb sunk into a fermenting wort, the scaled portion would rise above the billowing yeast head.[101] Some writers recommended two, or even three thermometers, physically adapted to the different environments of the coolers, fermenting tuns and boiler: these should be checked carefully to ensure agreement, and never interchanged. John Levesque, a Combrune-inspired brewer of the 1830s, advertised his own thermometers at £2, or £2.10*s*. 'in the Form of a Bamboo Walking-Cane'.[102]

The ultimate material expression of brewhouse requirements was the 'blind' or 'private' thermometer, available from Atkins of Fenchurch Street by 1802. This device featured a detachable scale, the body of the thermometer bearing only a travelling index marker. This marker would be set to a particular value, such as a mashing heat, with the scale in place; with the scale removed, the thermometer could be applied to set this heat without the privileged value being discernible by anyone – including the brewery servant actually performing the operation, a significant consideration in a competitive market with a high turnover of staff.[103] This safeguard was apparently commonplace well into the nineteenth century.[104]

The authors themselves, contrastingly, stepped increasingly into public view. Combrune had demonstrated that open identifiability was a plausible strategy. William Ellis's responsibility for the *London and Country Brewer*, carefully preserved as a penetrable secret in his lifetime, was publicly confirmed by his son in 1761; an anonymous treatise of 1760 reappeared in 1767 bearing the name of George Watkins, 'Who has practised Brewing in all its Branches, upwards of Thirty Years'.[105] As we will see in chapters 3 and 4, such authors promoted themselves keenly as practical brewers, often working directly from records of brewings they had performed themselves. The paradox of the trade writer, however, meant that in doing so, they often had to be careful to avoid implying that they made a living from brewing at the time of publication.

Combrune, unlike most of his immediate successors, acknowledged himself as an active brewer-proprietor. He published information which, were it genuinely valuable, he could have exploited by keeping it private: the credibility of this position relied strongly on his appeal to the core values of philosophy and a public culture of useful knowledge. Other published works by self-identified proprietors followed Combrune's in focusing on general principles. Authors who narrated specific processes, by contrast, were more likely to claim employee backgrounds, allowing them to emphasize long records of *experience* in successful breweries which did not amount to *interest*. Some found a neat resolution by forging new, specialist identities as brewers' engineers or consultant brewers, particularly after 1800.[106] Only then did the legacy of guild-protectionist concerns, to which Combrune had responded so carefully, evaporate. Combrune's own publications contributed to their decline, as the wide circulation of general statements fed demand for information about particulars.

Private and Public Authorship

What would it mean to disseminate, as Combrune claimed to be able to do, the 'Theory and Practice of Brewing' in print? Some readers were clearly hoping for directions to allow an established brewer to replicate the trade practices (particularly, those for London porter) which Combrune personally knew. Others, more attentive to Combrune's self-presentation, expected a philosophical enquiry into fermentative and related processes, leading to determinate principles and fundamental laws, not confined to brewhouse processes, but fit to explain and perhaps modify them. Both groups, as we have seen, were largely disappointed. In the end, Combrune's work was chiefly valued as an exhortation to systematic improvement and a compositional model.

Interestingly, no respondents judged Combrune's writings by whether they could, in isolation, teach a non-brewer to brew. Above the level of small-scale domestic brewing, it was universally taken for granted that no book could achieve this. Beyond the common assumption that a self-interested author would withhold some knowledge, there was awareness that much of the knowledge required was communicable only by practical guidance. Combrune's thermometric scheme, and its successors, did not challenge this position. Routinely, those seeking to emulate successful products such as London porter turned directly to brewers. Typical was Thomas Jefferson's experiment in small-scale brewing at Monticello, which turned out to require an imported English brewer as well as textual accounts. 'I much doubt', he wrote to a correspondent, 'the operations of malting and brewing could be successfully performed from a reciept [*sic*]. If it could', he added paradoxically, 'Combrune's book on the subject would teach the best processes: and perhaps might guide to ultimate success with the sacrifice of 2. or 3. trials'.[107]

Publication could, of course, operate alongside personal instruction, and could foster demand for it. Combrune, by his own account, did not initially intend this strategy: as the first easily identified author-brewer, however, he began to receive requests for direct assistance. Our limited knowledge of this side of Combrune's activities comes largely from his dealings with James Best, an established brewer of Chatham, Kent, a maritime town beyond the fringes of the London market. Best, like Combrune, was strongly focused on emulation: the motive in his case was presumably to pick up naval or export contracts for London-style porter. In mid-1762, Combrune announced to Best his decision to take on some consultancy work. Concerned for his philosophical reputation, he was careful to couch this enterprise in the language of service to the brewery community, offering better instruction than 'the bare printing of a book on an art can possibly convey'. Combrune strove to maintain the transaction on a gentlemanly footing, emphasizing that the work would be carried on in his leisure hours only.[108] We find him, in 1766, advising Best on porter-brewing in person at Chatham. Directions for the positioning of the thermometer, absent from Combrune's publications, appear in Best's notes of this meeting.[109]

The overall extent of Combrune's practice is unknown, but he continued to consult in retirement. In 1772, bringing the emulation agenda full circle, Combrune's proposals for 'communicating a regular and certain Method of Brewing good Porter and other Malt Liquors' reached the Dublin Society. The Society enthusiastically opened correspondence via the military engineer Charles Vallancey, another Huguenot improver.[110] Echoing the initial manifesto of *Some Account*, Combrune assured his Dublin audience that there were no inevitable barriers to emulation of the London drink. The materials could be copied; differences in the heat of the air could be allowed for thermometrically; and the action of fire, of course, was universal. Combrune supplied samples of the fiscal guidance he had erased from his public work, computing the profits on London brewing as an aid to assessing the viability of Dublin operations, and a manuscript of guidance on porter-brewing.[111] These acts might have troubled the London brewers – had they known of them – far more than any publication: it was through such programmatic emulation that several Irish breweries rose, in the nineteenth century, to compete with London in the production of the staple London product, Guinness of Dublin ultimately attaining international pre-eminence in porter.[112]

Despite these activities, it is clear that Combrune saw publication as the key to philosophical recognition, and the primary duty of the theorist. His demonstration of a market for brewery instruction, however, led other chemists and brewers to attempt both public and private endeavours, usually in parallel. Publication could serve, as it did for Combrune, to announce and promote consultancy, lectures or printed directions circulating under promise of secrecy. In the next chapter I look in detail at the activities of Humphrey Jackson, a manufacturing chemist and projector who saw the brewery as a principal market for his services. Where Combrune failed to gain admission to the philosophical elite, Jackson was largely successful in establishing himself as a theoretically aware public chemist. His attempts to assert authority on brewery questions from outside the trade, however, left him vulnerable to attack from the rising generation of theoretical brewers who followed in Combrune's wake.

3 BREWERY INSTRUCTORS IN PUBLIC AND PRIVATE

> The brewer is satisfied to find, that by taking his liquors at a certain heat, with a certain kind of malt, and conducting his fermentation in a certain manner, to a certain period, he can produce a certain kind of beer, other particulars being regulated accordingly. The chymist goes a much nicer way to work. With his *chymical key* he unlocks every door of the principles of brewing. He can walk at large in a kernel of malt, like Shakespear's Queen Mab, sail over the surface of a boiling wort, like the Nautilus, on a hop-leaf, and securely visit every corner of a fermenting must, in an air bubble.
>
> [J. Richardson][1]

The would-be communicator of brewery knowledge could operate by publication, by private instruction or by selective combination of the two. Each approach carried both opportunities and dangers for the promoter's credibility. Publication, as illustrated by the case of Michael Combrune in chapter 2, could raise suspicions that what was on offer was dilettante conjecture, not valuable enough to withhold. Too much privacy, on the other hand, left clients with little means to judge the instructor's reputation – if, indeed, he even came to their attention in the first place. Though some instructors wrote personally to every significant brewer they could identify, it was far easier to promote their names by appearing in booksellers' lists and the monthly reviewing press. Most, therefore, took the middle course: partial disclosure could achieve the advantages of publicity while stimulating a market for private guidance.

Partial disclosure was also, however, the approach most likely to draw criticism from established theorists. From the time of Robert Boyle, most philosophical gentlemen had professed a commitment to the general circulation of all useful knowledge, defining themselves – at least rhetorically – in opposition to mercenary 'secretists'.[2] This picture was not challenged by the eighteenth-century agenda of Peter Shaw and the Society of Arts (chapter 1): though concerned to accommodate the aims of trade, Shaw's approach was to offer cash premiums in exchange for full disclosure.[3] To insist on withholding products and processes from scrutiny raised the automatic suspicion that they relied on mistaken principles or (more likely) conscious deceit. Early modern medicine offered a vivid vocabulary for characterizing the secretive professional deceiver – the 'Quack, Empirick, Mountebank, Nostrum-Monger' – which soon found its way into chemical and other philosophical texts.[4]

This chapter describes how the various authors who emerged in Combrune's wake tried to balance commercial and philosophical credibility. Central to my analysis is the distinction between those instructors who were trained brewers in their own right, and outsiders whose claim to expertise rested on a knowledge of wider chemical principles. We might expect the brewers to be largely 'commercial' in their approach, and the chemists largely 'philosophical', but the evidence shows that all practitioners found it necessary to work with both sets of commitments. Nonetheless, the two groups were sharply and consciously distinct in their opportunities and priorities, and there was little love lost between them.

I begin with the case of Humphrey Jackson, a manufacturing chemist who had already learnt the value of strategic partial disclosure outside the brewery. Jackson proclaimed himself an expert on the causes and effects of brewhouse processes: this was an entirely unprecedented claim for an acknowledged non-brewer to make, and inevitably met deep scepticism. Jackson's defence invoked the universal nature of chemical principles, and the experimenter's independence from customary practice, to assert that non-brewers were if anything better placed than brewers to improve the trade. Some brewers were clearly satisfied with these arguments, but the approach was risky: I go on to contrast Jackson's significant success as a public chemist, involved in import substitution and other useful-knowledge projects, with the sudden collapse of his private reputation.

An influential challenge to Jackson's credibility came from John Richardson, one of his former clients. Richardson closely emulated Jackson's promotional strategy while dismantling his authority: only a brewer, said Richardson, could advise a brewer. Borrowing from the parodic conventions of contemporary pamphlet warfare, Richardson undermined his rival by making public what Jackson had held private. Richardson's intervention spawned a small genre of self-consciously 'practical' manuals by experienced brewers, all of whom used publication to promote private communications. These brewer-authors were often under no less pressure than the chemists to demonstrate their integrity: explicit reference to the public/private division were often helpful in convincing their trade brethren. Since the approach was usually to publish on generalities while consulting privately on specifics, however, such texts often sought to lay down universal principles, and drew increasingly on established chemical literature. In codifying a 'science of brewing' through public and private instruction, Richardson and his imitators sought to define a brewhouse ethic that was as philosophical as it was practical.

The Possibilities of Partial Disclosure

Humphrey Jackson is recorded as an apothecary's apprentice in Stockton, County Durham, in 1735.[5] By 1743 he had opened a chemist's shop in East Smithfield, London, close to the variety of manufactures fringing the Thames, including the brewhouses of St Katherines and Southwark. He first came to brewers' attention when he developed a substitute for isinglass, a substance used chiefly in managing beer. Dissolved in acid and added to the cask, isinglass had the remarkable effect of rendering cloudy beer 'fine' (clear) within hours, speeding up a process which might take months to occur naturally. Beers brewed with high-dried malt were particularly prone to cloudiness, sometimes refusing to become fine at all without artificial aid.[6] The importance of isinglass to the London porter trade made it a key target for import substitution, since it was generally imported from Russia in its finished state: British merchants knew that isinglass was a by-product of sturgeon fishing, but had been unable to discover the process. The growing captive market of industrial porter-brewing meant that the price had skyrocketed by 1760, when Jackson patented his substitute.[7] He claimed to have analysed samples of brewers' finings and concluded that the substance was prepared from the sounds (swim bladders) of the fish, and could be replicated using Newfoundland cod and other fish available in British waters.[8] The London porter magnates, it seems, underwrote this work collectively through the Brewers' Company, which gifted Jackson 'a considerable sum of money' to improve the technique.[9]

Jackson's commercial agenda, however, was not that of useful-knowledge pioneers such as Peter Shaw, who had preached systematic disclosure. Like most operative chemists, Jackson applied publicity and privacy selectively, as tools of his trade. From 1744, he had published a series of chemical pamphlets tied to topics of current interest which were guaranteed attention from the reviewing press. The first was inspired by George Berkeley's promotion of the medicinal properties of tar water: where the Bishop promoted a simple preparation, suitable for all, Jackson offered 'a more efficacious Remedy' – at a slight premium. It was unavoidable, he stressed, that his insight should be commercialized in private hands: publishing it would be pointless, since, unlike Berkeley's method, it depended on complex techniques which 'none but Chemists can perform'.[10] Jackson made a similarly strategic contribution to the controversy over the popular 'fever powder' patented by Robert James and retailed by the publisher John Newbery, in the 1750s. A rival cure's distributor had launched an infringement action, recruiting the public lecturer, Erasmus King, to demonstrate analytically that the two products were equivalent. Jackson was called in to support James and Newbery with a pamphlet describing his own analysis. Shrewdly, Jackson presented enough information to affirm that James's powder was *not* his rival's, without revealing enough about what it *was* to permit replication.[11] Jackson approached his new-found brewery connections, then, with a sharp awareness of the power of publication to keep knowledge private, and to promote private knowledge.

His first step, in 1762, was to propose a specialist course of chemical lectures. Malting and brewing had earlier featured in the lecture syllabus of Peter Shaw (chapter 2), and also of Erasmus King.[12] Jackson, however, was probably more inspired by the instrument-maker Benjamin Martin, whose daughter he married the following year. Public lectures, with demonstrations built around instruments offered for sale, had easily proved their promotional worth to Martin and his rivals; still wider audiences could be reached by printing the texts. Jackson's prospectus focused on explaining 'the real Causes of Cloudyness in Malt Liquors', and was presumably designed to raise awareness of the isinglass substitute beyond his established London base. He also proposed, however, to cover malting, management of finished beer, the analysis of malt, hops and yeast, 'Hints for an easy practicable Method of Brewing Porter in Summer' and 'a Description of a most useful new improved Thermometer adapted to the Brewery' – probably Martin's device.

Jackson offered his lectures on a subscription basis similar to those of Shaw and Martin, but with one crucial modification: these were not to be *public* lectures. Stating 'that every Subscriber will duly consider the Importance of the Discovery, and the Extent of his Business', he advertised as follows:

> As soon as an adequate Sum is subscribed, which shall be determined by the Subscribers, the Proposer will execute his Proposals, by way of Lecture, supported with Experiments, at any Time and Place agreed on; after which he will print them at large; and by leaving certain blank Spaces in the Book, to be filled up afterwards in Writing, or by References printed separately, they will be rendered useful to the Subscribers only.[13]

This was a variation on an old approach sometimes used to protect guild knowledge. In the seventeenth century, for instance, the Company of Distillers in London had circulated printed recipes to its freemen, with crucial information disguised by symbols which became intelligible once handwritten keys were added. The system had one obvious drawback: within a few years, the distillers' text had been pirated and republished with the keys included, 'for the Publike good'.[14] Jackson, however, was looking to force the pace of technical change in the brewhouse, and perhaps reckoned on being able to show up any leaks of his work as obsolete.

Partial printing, like the blind thermometer discussed in chapter 2, combined standardized information recording with total control over access to the information. This was not its crucial advantage, though: the same could have been achieved more cheaply by having a manuscript copy made for each subscriber. Manuscripts would not have had the rhetorical effect of a gap-filled publication, circulating 'at large', in making the missing details conspicuous by their absence: the approach emphasized Jackson's guidance as systematic enough to be reduced to print, yet commercially sensitive enough to be worth concealing.

Before he could commence his lectures, however, Jackson found an arrangement which apparently suited him better. Perhaps via the isinglass initiative, he had encountered Henry Thrale (1728–81), son and successor to the former clerk who had bought out the immensely profitable Anchor Brewery in Southwark. The younger Thrale had been raised on an extravagant allowance, mixing with aristocrats, but this social mobility had drained the firm's capital: when he inherited it in 1758, production was falling in absolute terms, whereas competitors such as Samuel Whitbread were pushing far ahead in the expansion of their output and plant. Output figures for 1760 rank Thrale's as the eighth brewery in London, its annual production less than half the leaders'.[15] Thrale was intensely conscious of this decline. Though far removed from Whitbread's almost puritan work ethic, he was no absentee proprietor; living part of the time in a house on the brewery site he became, as his widow Hester was later to put it, 'mad with the noble ambition' of outdoing his rivals.[16] Deciding that the surest path to success lay in chemical reform, he took on Jackson as a private consultant.

The association between Thrale and Jackson formed one strand in a web of commercial, medical and literary connections. Thrale was a witness when Jackson married Martin's daughter, whose name is variously given as Mary or Maria, in 1763; the other witness was John Newbery, promoter of Robert James's abovementioned fever powder. The same year, Jackson apparently arranged Thrale's own marriage to Hester Salusbury, whose mother was later attended by James. Samuel Johnson, an old schoolfriend of James and a frequent author for Newbery, was introduced to the Thrales in 1765, becoming a member of their Streatham Park household and a regular travelling companion. Johnson was also an early and active member of the Society of Arts, to which Newbery, James, Martin, Jackson and Thrale were all subscribers by 1762.[17]

There was no obvious precedent to inform the terms of Jackson's service to Thrale. Brewers often consulted professional engineers on large technical projects, but none had yet engaged a chemist on a similar basis. It seems that Jackson provided Thrale with privileged access to his new research in exchange for capital and materials. This, however, was no more than a sideline for Jackson, who remained in business as an operative chemist and continued to supply the open market with his isinglass substitute, working his patent as a monopoly. This unusual position created opportunities of a kind which appealed strongly to Jackson's mentality. If partial publication in the literal, fill-in-the-blanks sense was now off the agenda, he saw more subtle ways to synthesize the contrasting appeals of public and private knowledge.

In 1765, Newbery published for Jackson a long pamphlet, or short treatise, titled *An Essay on British Isinglass*. At one level, this was a straightforward promotional device for Jackson's substitute, presenting a detailed account of the best method for applying it to beer. More than half the text, however, addressed brewing more generally, along the lines of the 1762 prospectus. Jackson hinted, for example, that he had discovered the causes and remedies of more stubborn cloudiness, beyond the reach of isinglass. He gave no concrete detail, however, merely mentioning that findings 'too nice and speculative' for publication *might* be communicated. Yet he did not define terms for private communication or repeat the proposal to lecture.[18]

The *Essay*, then, was pitched to boost Jackson's opportunities to trade in both materials and knowledge, without jeopardizing his relationship with Thrale. He described, for instance, a pyramidal observation jar of his own invention, intended to bring a measure of laboratory discipline into the brewhouse cellar: Thrale could hardly object to the sale of apparatus unaccompanied by conclusions drawn from its use.[19] Jackson was less open about materials that might just as easily have been obtained from his rivals: the suggestion that his finings' effect could be improved by a 'certain saline Matter easily procurable' could only be read as an invitation to enquire privately. Withholding from print what could easily be printed in this fashion was unsafe. An otherwise sympathetic notice in the *Monthly Review* – probably by Thomas Dawson (1725–82), who reviewed primarily on medical topics – observed quizzically that 'it would have been more to his reputation' had Jackson not here 'appeared a little in the character of a *nostrum-monger*'.[20] Such imputations could have serious consequences, as we will see later.

The summer of 1770 saw a change in strategy. Writing to the Chatham brewer James Best, Jackson announced that 'Mr Thrale has at length agreed that I may read Lectures upon Brewing to Country Brewers on certain Conditions'.[21] In practice, if the surviving evidence is representative, the conditions were chiefly imposed by Jackson himself, and concerned audience rather than content. Jackson effectively offered each country brewer a local monopoly on instruction, restricting its geographic scope by population size. He dealt preferentially with those prepared to pay the most, it appears, and offered distinct terms for information on various parts of the process.[22] Perhaps these constraints truly helped to reassure Thrale, in that they kept knowledge out of the hands of his direct rivals in London. Their chief purpose, nonetheless, was surely to maximize revenue for Jackson; and they served as a convenient promotional device, highlighting the commercial sensitivity of the arrangement and definitively resolving the trade writer's paradox. To have any chance at all of succeeding on these lines, however, Jackson first had to defuse another major hazard to his credibility.

'How should a Chemist understand Brewing?': The Uses of Chemical Authority

Conventionally, at all levels of production, guidance was founded on long service within the brewhouse. Anonymous authors such as William Ellis, as noted in chapter 1, could embroider the extent of their personal experience; Jackson, whose scheme depended on drawing potential buyers to the warehouse at East Smithfield, could be nobody but himself. So obvious was the awkward question that Jackson was forced to pre-empt it: '*how should a Chemist understand Brewing who never saw the Side of a Copper?*'[23] His answer was grounded in the universality of chemical knowledge. Brewing must be understood as 'a chemical Process throughout ... for it is Chemistry alone that teaches us to analize Vegetables, and resolve their Components into simple Parts, to separate those fittest for our Purpose, and reject the rest'.[24]

The experimental methods and theory needed for improvement, said Jackson, were precisely those left undiscovered by a commercial brewer's practice: advice from a non-brewer was not only legitimate, but necessary. There are strong echoes of Peter Shaw in this vision of the philosopher scrutinizing trade knowledge, reducing it to essences, synthesizing optimal approaches and returning them to the trade.

Jackson was, of course, following the standard strategy of drawing attention to the philosopher/artisan division in order to settle himself on the dominant side of it. His status, in fact, was insecure. Thanks largely to the emergence of systematic chemical treatises, a well-connected chemist could increasingly represent himself as part of the philosophical elite by mid-century;[25] yet the acknowledged chemical authorities – Georg Ernst Stahl, Boerhaave, Caspar Neumann, Shaw, William Lewis – were not *trading* chemists, as Jackson was. Professionally speaking, they were mostly physicians with influential courtly or university appointments: their authority came from lectures and treatises written for others, including apothecaries and manufacturers, to follow. Jackson was unavoidably a trading chemist; he had published no systematic work, and his pamphlets were arguably mere advertisements for his wares.

Jackson's response – recalling his father-in-law, Benjamin Martin – was to press on regardless. If he could not dictate terms to the philosophical establishment, he could surely direct brewers, laying down an optimal process for 'a considerably greater yield of essential Matter with Certainty'.[26] Departing from Shaw's distinction between philosophically 'perfect' and commercially 'advantageous' practice (chapter 1), Jackson insisted that there must be 'one scientific Method, only, of conducting the Business to Advantage, whether we regard Dispatch, Perfection, or Profit'.[27] The scheme meant reducing the product to chemical or physical essences: the 'essential Matter' was simply the spirit in the beer, which could be quantified by distillation. The preliminary 'saccharine Matter' obtained from mashing prompted a series of investigations: does it pre-exist within malt, or is it somehow generated by the malting process? If it is generated, can the process be modified to raise the yield? Jackson's point was that such questions, so materially important to the maltster and brewer's livelihood, could only be answered by chemists.[28]

In readily admitting that his reforms would invalidate customary patterns of brewing, Jackson took a very different line from Michael Combrune. As a brewer himself, Combrune was concerned to show that established trade practices followed as legitimate consequences of his theory. Jackson used the *Essay* to take a few public potshots at his only established rival as 'Professor in Brewery', who had earned his hostility by pronouncing the isinglass substitute ineffective. Like John Mills (p. 50), Jackson accused Combrune of putting up an affected show of theory, having only half digested his authorities. Seizing on Combrune's doubtful use of the term *sapo* to refer to malt sugar, Jackson derided him as a '*Saponaceous Brewer*' who had fallen sprawling in the gap between theory and practice.[29] So far as Jackson was concerned, trading chemists knew chemistry, but trading brewers did not: effective chemical reform from inside the brewery was impossible.

When he re-mounted his lecture scheme around 1770, Jackson made analysis and the search for essences yet more prominent. Much of what he described was geared firmly to the laboratory, addressing costly and fragile instruments designed for small-scale experimentation. To measure wort strength, for instance, he recommended the hydrostatic beam balance, a device far too sensitive to operate reliably in the brewhouse. So far as his syllabus reveals, Jackson followed earlier conductors of polite public lecture series in presenting limited, didactic demonstrations in a managed environment. Often, as in the 'attempt to extract worts from malt without mashing', the syllabus confines itself to general principles or possibilities, leaving practical application to his clients. Nonetheless, Jackson still promoted a general reform of the trade around a chemically determined optimum. A first step would be to bring the geographical distinctions of customary practice under a universal rule. Different brewing waters, for instance, might be analysed, described (presumably through quantitative recording of mineral salt content) and synthesized, so as to make, as appropriate, 'hard waters soft, and soft hard'.[30] Jackson likewise proposed to analyse 'the most celebrated beers' of the English regions, to determine 'the proportion of their spirituous with their grosser parts compared, with a view to trace out what and wherein consists their different character and merit, with respect to their strength, flavor, and colour'.[31]

Jackson proposed concentrating the saccharine matter from malt extracts, and extracting the active principles from hops: these could conveniently be stored against future scarcity, and could be added in small quantities to allow precise adjustment of flavour. Similarly, he proposed 'preserving nurseries or stores of yeast' over the summer, when brewing could not be undertaken, and 'curing' yeast for export.[32] He also outlined structural and mechanical innovations including a method for 'compressed' fermentation in closed vessels fitted with a pressure valve of his own specification.[33]

Jackson was not the only non-brewer to assert his authority through chemical credentials in this period. John Long (d. 1807) was a protégé of John Beresford (1738–1805), MP for Waterford and for many years First Commissioner of the Revenue in Ireland, a position which brought him regularly into dispute with the Dublin brewers.[34] Beresford, like the agricultural improvers of the Dublin Society (chapters 1 and 2), argued that Ireland's beer production was in dire need of reform through comparative study of English practice.[35] Accordingly, Long made a tour of England's major breweries and distilleries in 1789.[36] Rather than borrowing established English technology, however, Long devised a new system of brewing, which he patented in 1790 and promoted privately to the leaders of the trade.[37]

Long's manifesto was almost equivalent to Jackson's: defining brewing as 'a series of chemical operations', he contrasted improvements grounded in philosophical study with the hidebound 'custom' of received brewery practice. Most of his innovations involved adding submerged pipe coils, similar to distillers' worms, to the brewery vessels: circulating hot or cold water through these, he proclaimed, would improve the controllability of heating and cooling, which had become an increasing concern since the arrival of the thermometer. More radically, Long deemed the customary hop-boil inefficient and proposed to abolish it altogether, instead boiling the wort briefly without hops and mixing it with a hop infusion prepared separately.

The message of these systematizers was a decidedly tough one: what the brewers knew counted for little, and they must reorganize under chemical authority or face commercial extinction at the hands of rivals willing to do so. Long went beyond Jackson in suggesting, in his manuscript presented to potential clients, that most brewers (themselves presumably excepted) were not even fit to brew. It would be in the drinker's best interest, he felt, if the government introduced a system of examination which would deprive incompetent practitioners of the right to trade.[38]

It is impossible to judge how far such messages resonated at first, precisely because of the carefully private channels through which they were communicated. The best surviving account appears in the papers of Combrune's ex-pupil, the inquisitive James Best, who sought testimonials from Jackson's past customers at the end of 1770. By this point, Jackson's scheme appears to have been well developed, with a number of clients across the south and east of England. A brewer in Cambridge and another on the Isle of Wight professed themselves highly pleased with their tuition, though the latter respondent noted that he could not be more specific 'without saying more than would be agreeable to Mr Jackson'.[39] Meanwhile, of course, the chemist retained his relationship with Thrale.

Within a few years, however, Jackson's entire project had collapsed in embarrassing circumstances. As the remaining sections of this chapter show, many brewers were keen to engage with philosophical approaches, but would not tolerate, and could effectively resist, outsiders' attempts to overrule their authority.

The Public Rise and Private Fall of Humphrey Jackson

Humphrey Jackson's career across the 1770s sheds interesting light on relations between the cultures of polite knowledge-making and trade. An important source of support for chemistry related to the manufacturing industry in this period was the Society of Arts, which had kept up its founding focus on domestic productivity and import substitution.[40] Jackson's isinglass substitute was just the kind of innovation which the Society aimed to support, and the *Essay on British Isinglass* is strongly flavoured with the appropriate useful-knowledge rhetoric.[41] Jackson seems to have taken no institutional role in the Society, however, and by the time of the *Essay*'s publication he was no longer a member: a 1764 proposal (not, eventually, enacted) to make members ineligible for premiums may have prompted his withdrawal.[42] Jackson received several premiums from the Society, including £50 in 1766 for what seems to have been an alternative approach to the isinglass problem, importing (presumably pickled) sturgeon from the North American colonies.[43]

Jackson was likewise attracted by the Society's repeated premiums for a 'composition for securing ships' bottoms from worms'.[44] His proposal was to develop a chemical solution in which timber could be steeped or boiled to harden it: this would not only solve the worm problem, but would allow fir from the American colonies to be substituted for German hardwoods. His chief audience for this work was always the naval commissioners, who gave him several lucrative contracts,[45] but he also saw opportunities among his brewery clients. Brewers were among the heaviest and most conspicuous users of coopered vessels, and Jackson used both the 1762 lecture advertisement and the *Essay* to drop hints of the possible savings to be made.[46]

In the event, Henry Thrale bankrolled the development of the timber process, and seems to have enjoyed its sole use for a time. Hester Piozzi (as Thrale's widow became known when she married for a second time) recalled a huge custom-made copper vessel, 'the very metal of which cost 2000*l*.', being used to prepare the timber for twenty vats of 1,000 hogsheads (i.e. 1,500 barrels), plus ten more of 1,000 barrels each.[47] Such gigantic containers had first been introduced by rival London brewers in the 1730s, in place of the traditional arrays of butts (three-barrel casks) on stillions, but it appears that Thrale did not move wholeheartedly to vatted storage until prompted by Jackson's proposals.[48] An account of 1774 presents Thrale as the leading user of the technology, with 23,000 barrels' worth of vats.[49] At the end of 1768, Jackson obtained a patent on his process and began to promote it more widely. When he remounted his lecture scheme for provincial brewers, he linked the two projects, soliciting brewers to carry on the timber process under license: their large vessels, he argued, would be ideal for steeping timbers during the summer season, when brewing was discontinued.[50]

Jackson's simultaneous self-promotion to naval and brewery audiences led, perhaps by chance, to his greatest status-building triumph. This came about through what Chris Lawrence has called the 'Cook–Pringle–Admiralty nexus', a powerful knot of interests which directed the medicalization of the Royal Navy during the 1770s. The physician John Pringle had proposed that scurvy, the most debilitating of the diseases afflicting sailors who were at sea for long periods, might be treated using the antiseptic properties of fixed air. Pondering how to provide a ready shipboard source of this gas, David MacBride, an Irish naval surgeon and useful-knowledge proponent, noted that it was produced in large quantities during fermentation, and suggested brewing on ship.[51] Everyone recognized, however, that it would be hopeless to attempt the conventional process at sea. The most plausible approach was to mash on a large scale on shore, and concentrate the wort by evaporation: the concentrate, a thick syrup or glassy solid, could then be rediluted and fermented shortly before drinking.[52] Conceptually, this was a well-established approach, but it had drawn little interest from the trade. The Admiralty was more enthusiastic, and directed the Commissioners of Victualling, responsible for onboard provisioning, to look into the matter.

A consignment of malt wort was sent, alongside lemon juice, with James Cook (the future Captain Cook) on the exploratory *Endeavour* voyage in 1768. Joseph Banks, sailing with the *Endeavour*, reported favourably on the trial. In 1771, the Admiralty approved shipboard brewing experiments on the scale of a few dozen barrels, but the Commissioners of Victualling found they had no suitable vessels to perform the evaporation. Significantly, rather than approaching any of the several brewers who regularly brewed under contract to the Navy, they turned immediately to Jackson: we may suspect that the idea first entered their heads via the chemist, whose shop and workshops were near the Victualling Office.[53] Samples of his product were taken on board the *Resolution* and *Adventure* for Cook's second circumnavigation in 1772.[54]

Later that year, Jackson was elected a Fellow of the Royal Society. The election certificate cites Jackson's 'discovery of the method of making Isinglass from British Materials, and likewise for his invention of preserving Naval Timber from speedy decay'.[55] Jackson's involvement in the wort trials also weighed in the balance, to judge from the list of sponsoring signatories. It includes the naturalists Joseph Banks and Daniel Solander, who had both sailed with Cook on the *Endeavour*, and several of their circle: John Ellis, a particular proponent of transplantation projects; John Hunter, who had made his name as a military surgeon; and the civil engineer, Robert Mylne.[56] Jackson, then, was co-opted for his presumed usefulness to State-sponsored philosophy on the 'Cook–Pringle–Admiralty' model – John Pringle became President of the Royal Society in the time between Jackson's nomination and election – and to the useful-knowledge movement more generally.

Jackson's FRS should have been the crowning glory of his public reputation: the Stockton apothecary had succeeded where his father-in-law, Benjamin Martin, had stumbled, and had outdone even the strenuously philosophical Combrune. Having successfully couched his research in terms of the public interest, he could now hope to enjoy the commercial benefits. What is striking about Jackson's case is how little his public status availed him when, concurrently, his private schemes began to collapse.

Thrale's brewery lost an entire year's production in 1772. The contents of one of the great maturation vats were lost outright when it burst apart, and the beer in the intact vessels was found to be 'spoiled'. The brewhouse clerks were united in blaming Jackson. From the surviving details, found mostly in the later recollections of Hester Piozzi, it is not clear whether the clerks held his timber-hardening process, his innovations in beer management, or both, responsible, or with how much justice. What is more certain is that Thrale, under Jackson's influence, had over-committed himself financially to the hardening project. As the combined shortage of beer and capital reserves brought Thrale's to the brink of bankruptcy, Hester Thrale and the head clerk, John Perkins, developed an uneasy alliance to recapitalize the firm and banish the chemist.[57] Samuel Johnson, who became increasingly involved in brewhouse affairs in Henry Thrale's declining years, took Hester's side, and Jackson was permanently dismissed from their circle around 1773.[58]

Having separated from Thrale, Jackson was, of course, free to continue instructing others, but it seems that the incident put a permanent dent in his credibility. He had made at least two powerful enemies. John Perkins followed the pattern of clerkly upward mobility, drawing on wealthy Quaker connections to buy a partnership in the firm when it was sold after Henry Thrale's death in 1781: as Barclay Perkins, it became the greatest brewery in London.[59] Samuel Johnson, meanwhile, had any number of philosophical and useful-knowledge connections. Did he, for instance, communicate his displeasure to his friend Joseph Banks, then about to begin his rapid rise to dominance of the Royal Society?[60] We can only speculate: somehow, however, Jackson's promising career was sharply curtailed around the end of 1775. Robert Thornton, who had been Jackson's apprentice, announced himself 'Successor' at the chemical warehouse in that year, and took over the wort concentrate project.[61] John Appleby's detailed survey of Jackson's activities revealed one further patent, in 1778, but no further premiums, contracts or public chemical activities of any kind.[62] Though Jackson's methods drew the attention of Robert Brakspear, a brewer of Henley, Oxfordshire, as late as 1784, there is no clear evidence the two were in direct contact; Brakspear judged Jackson's system 'fallacious & absurd in the extreme'.[63]

Jackson's downfall was more than a brewery affair, but it was crucially influenced by his incautious approach to the brewers. Late in 1775, a remarkable pamphlet appeared titled *Observations on the Art of Brewing, in a Series of Strictures on a Secret System inculcated in a Private Course of Lectures*. The author, 'A Practical Brewer', presented a vigorous attack on 'H. J––, *Esq; F. R. S.* ... Grand Master of the *Secret Society of Brewers*, founded by himself, and by his *art* supported'.[64] The Practical Brewer was himself, he explained, one of the '*Sir Epicure Mammons*' (referring to the principal dupe in Jonson's *The Alchemist*) who had suffered by Jackson's expensive advice.[65] His ultimate aim, however, was not only vengeance, but something more dangerous to Jackson: imitation. The final pages lay out plans for an alternative course of instruction, soliciting contacts via the pamphlet's three London booksellers.

The perpetrator of this ingenious subversion was, beyond reasonable doubt, a brewer named John Richardson (1743–1815).[66] By 1777, when Richardson began publishing under his own name, he was mimicking Jackson's strategy directly: while selective publication fostered a market, more specific guidance, detailed in a privately circulated 'Advertisement addressed to the Common Brewers', could be communicated under conditions of secrecy.[67] There was one stark difference, however, between the rival instructors. Richardson defined himself, very firmly, as a brewer. His claim to credibility rested not on chemical abstraction, but on a wide first-hand knowledge of the commercial practice in London and other towns.[68] In this way, he was able to construct for himself a safer identity as a brewery instructor.

Publicity through Parody

The 'Practical Brewer' employed the characteristic techniques of contemporary satirical pamphleteering, subjecting Jackson's hitherto private directions to a sardonic line-by-line dissection in mock-heroic tones. His favourite device was to pretend to support Jackson's philosophical authority unstintingly against brewery tradition, cheerfully drawing the most absurd conclusions: a brewer, he boasted, 'may scald his fingers in his worts, and I will not believe they were hot, unless he can *demonstrate* that there was *heat* in them'.[69] Mockery of natural-philosophical adepts and projectors was, of course, a well-established literary theme, and the text nods explicitly to the Academy of Lagado in *Gulliver's Travels*. There are also allusions to Shakespeare and Addison, and quotations from Boileau and assorted classical authors in the original, all serving to accentuate the bathos whenever the author is plunged reluctantly back into brewhouse reality. Crucially, the Practical Brewer's stance is not that of a capable but ignorant workman, who might be merely uncomprehending of Jackson's high-flown language: rather, he is a gentleman sufficiently erudite to know when erudition is not enough.

The pamphlet also turns Jackson's public stature as a chemist against him. '[E]mployed his whole life-time in distillation, calcinations, pulverization, culling of simples, transmuting of metals, and commixing of acids and alkalis', Jackson is literally an outsider to malting and brewing, having 'scarce seen the inside of three malt-houses in his life'.[70] Improvements and productive generalizations, said the Practical Brewer, must be founded on a sound first-hand knowledge of the *status quo*. Jackson, as noted above, had foreseen such objections, and argued that chemistry's (and thus his own) special authority came from its status as a universal art: through '[e]xperiments in the small Way', the brewer might 'safely proceed to the large, for whatever Phœnomena occur in the Glass, will turn out exactly the same in the Butt'.[71] Richardson did everything in his considerable rhetorical power to deny this transfer. As any time-served brewer knew, the disparity between large and small scales was no simple matter of managing ratios. Synthetically emulating local waters might be feasible 'in a snug room, with a few half-pint phials'; in the quantities needed for porter-brewing, it could be ruinous.[72] *Experiments*, in the philosophical sense, were no substitute for *experience*.

In concluding with a short syllabus of his own, Richardson systematically opposed the brewer's perspective to the chemist's. Unlike Jackson, he said, he would emphasize those areas of particular commercial significance, such as brewing for export. He disdained analytical methods based on chemical extraction: rather than reducing the hop to '*parts*', he would advise only on how to select among the various sorts available. Such decisions must depend upon the desires of the local customers: there was no point in trying to establish a 'general uniform taste' as optimal. Where Jackson offered to furnish written testimonials from satisfied clients, the Practical Brewer offered the most practical testimonial: samples of his own beer.[73]

Why did Richardson begin his consulting career with anonymous sarcasms? As a public rejoinder to a wholly private project, the *Observations* provoked some bafflement in the reviewing press;[74] as an advertisement for Richardson's services, it was evidently less successful than the conventional approach he adopted soon afterwards. Perhaps its strategic value lay in forcing Jackson's communications into public view, where the chemist could not easily manage them. One of Jackson's attempts at damage limitation certainly backfired to striking effect. In noticing the *Observations*, the *London Review* disclosed a letter it had received, signed by 'a practical Brewer' and enclosing a copy of Jackson's syllabus:

> I am a Brewer and amongst many more of my Bretheren find my Self much hurt by these Lectures for in a little Time there will be no Such Thing as getting an Apprentice with a Fee if such pretenders to our Art be permitted to instruct any Person whatever that applies for a Course of Lectures from the within named Person: he has instructed many Country Brewers, who run away with all the Trade, where they reside to the Ruin of others Several of them have increasd [*sic*] their Trade 2 and 3 Times more than what they did, as Everard at Lynn, Temple at Portsm[outh], and Cook at the Isle of Wight and many more, but as He will not instruct two in a Place I have no Chance amongst them therefore hope you will say all you can in your next Review, to put a Stop to these Lectures.

This unlikely plea was, of course, concocted (surely by Jackson himself) in a bid to make the reviewer do the opposite – that is, to praise Jackson for demolishing a cozily protectionist established order – and to alert uninstructed brewers among the readership. Unfortunately for Jackson, the *London* reviewer enjoyed the Practical Brewer's learned mockery, and was too well versed in the mechanics of bluff and puff to take the bait. Instead, he reprinted the letter for the reader's entertainment, twitting the chemist for his tactical error.[75]

The strategy of killing a private scheme by publicity was emulated a few years later by another consultant, George Blake. On his own account, Blake had spent ten years from 1765 brewing in Ringwood, a Hampshire town noted for its ale; by the late 1770s, he was travelling around Britain and Ireland as a consultant.[76] Blake's self-promotion was, at first, entirely private. His 1780 approach to the Commissioners of Victualling is probably representative: for a salary of £400 per year, Blake offered to tour their several brewhouses in sequence, spending a month in each directing the brewers, prescribing rules to be followed in his absence, and instilling a culture of journal-keeping. He would keep nothing secret. Blake freely accepted that he would thus make his own services unnecessary, perhaps within a year, but indicated his confidence that he would then find a market for his services elsewhere. The Commissioners, whose brewhouses were then suffering badly from spoilage, accepted the proposal and were apparently well satisfied, granting Blake a glowing testimonial certificate in 1783.[77]

Blake was not moved to publish until 1791, when the abovementioned John Long persuaded the Victualling to implement his patent system. Intrigued by rumours of the new method, Blake had tracked Long down in London, and Long, who probably knew of Blake's sound reputation, proposed a kind of partnership: Long would introduce Blake to clients as 'his brewer', and Blake would oversee the introduction of the patent apparatus. On finally receiving the specifications of the system, however, Blake was appalled. Long's proposed heat values, he considered, were absurd, while his heating and cooling coils would waste vast quantities of water, prove impossible to clean and probably contravene the Excise-enforced laws on brewery apparatus. Concerned that his own reputation would suffer by association, Blake rushed to the Victualling Office to disclaim any endorsement of Long (there encountering the man himself, who responded, according to Blake, in a manner 'inconsistent with the character of a genteel man'). Within seven days, Blake had worked his initial notes up into a detailed pamphlet and sent it to the printer's.[78]

Blake's *Strictures on a New Mode of Brewing* was almost certainly modelled directly on Richardson's *Observations*.[79] The text is built around selected quotations from Long's manuscript proposals, parodying his high-flown verbiage and tendency to stray from the point. Blake contrasted his own twenty-eight years of experience with Long's upstart status: Long, said Blake, underestimated both the complexities of practical brewing and the sophistication of commercial brewers' established knowledge. For instance, Long (like Jackson) proposed to maximize the conversion of wort sugars to spirit, not realizing that some ales were highly valued for their less complete fermentation.[80] Elsewhere, Long correctly stated that malt was better crushed than ground, but missed the point that competent brewers already knew this: if they still ground their malt, it was because their sites could not accommodate crushing rollers.[81]

It is instructive to compare these brewers' position with that of the shipyard craftsmen discussed in Simon Schaffer's study of the competing knowledge cultures of Georgian naval architecture. Shipwrights such as William Shrubsole stressed that their regular direct experience provided a crucial element which the 'professors of abstractions' could not match; their point was certainly not that theory was useless, but that it had largely been developed by craftsmen, who best understood how far it could really be applied.[82] Shrubsole was responding to a central authority which ultimately resorted to imposing its norms through daily measurements of shipwrights' work. The would-be chemical authorities who sought to master the brewers, by contrast, had no official sanction (or, in Long's case, only the protection of a temporary contract which the Victualling might come to regret). Arguably, they were as much tradesmen as the brewers themselves; and the nature of their trade – as the brewers were keen to emphasize – was distinctly more dubious.

In 1775, Richardson had placed Jackson in a class of his own for audacious quackery; Blake, sixteen years later, hinted at a whole tribe of 'sanguinary and enterprising professors of chemistry' attempting to extract comfortable livings from brewers. Over the following decades, several more authors with brewery training were to press the same judgment. Though never a self-defined community – indeed, they were often rivals – these consultants studied the contents of each other's courses and publications, and in so doing fostered the development of the brewing manual as a generic form.

The Practical Brewers' Posture

The first brewing publication to assert its practical credentials, in fact, slightly predates Jackson's efforts. Initially anonymous, *The Compleat Brewer* of 1760 was the work of George Watkins, on his own account a commercial ale-brewer with thirty years' experience. Although the book drew structural elements from William Ellis's *London and Country Brewer*, Watkins clearly differentiated himself from Ellis, who, though practically minded, was hopelessly digressive and prone to 'idle opinions'. The more systematic Michael Combrune, on the other hand, 'deserves all the praise his patron has bestowed on him; but he is philosophical more than practical'. There was a need, therefore, for a text which was both practical and systematic.[83]

Watkins presented his directions as based on real-life techniques which, though established and confirmed by experience, were not yet generally known: most importantly, he proposed, porter-brewing could be extended beyond London. This approach steered between the perilous zones of common knowledge and abstract theoretical 'improvement'. Throughout, using frequent concrete examples, Watkins emphasized that successful brewing was a matter of careful management of men and plant. The fires for the boiling copper, for instance, should be arranged so that smoke rises directly up the chimney: thus, 'the master may look in from time to time without disgust', reducing the risk of his workmen misbehaving.[84]

Later authors similarly grounded their authority in professed brewhouse experience. Their title pages styled them not merely 'brewer' but 'practical' or 'common brewer', and pointed to twenty, twenty-five, thirty, thirty-five or 'upwards of forty' years' brewhouse experience.[85] Though their texts were sometimes derivative, they usually contained some first-hand description of commercial brewing life: the relative convenience of different layouts, the legality and illegality of various procedures, the direction of servants. In the 1820s, for example, John Tuck, a brewery proprietor and consultant whom I will discuss further in chapter 5, gave precise instructions for the control of the stopcock on the brewing copper as follows:

> When the brewer gets his liquor to a proper heat, and has ordered his fire to be stopped, he calls out, 'Turn on;' the man answers, 'Here's on'. So when the brewer finds, by his float, as much liquor as is required has run off into the mash tun, he calls 'Off;' the answer must be instant, 'Here's off'.[86]

Such practical brewer-authors in general echoed Richardson's appeal to *experience* as superior to *experiment*. When Blake, for instance, addressed John Long's proposal to start the first mash thirty degrees lower than was conventional, then raise it almost to boiling, he simply asserted that any competent brewer would 'shudder' at the thought.[87] Among brewers, the method was so ridiculous that even to consider testing it might threaten credibility. Implicitly, of course, a consultant such as Blake *had* experimented, long and hard: his experience was private, however, and presented fully formed to his brewing clients, who would usually demand to judge its value for themselves. To begin to set out evidence on paper implied exactly that conflation of the brewhouse and the laboratory which the practical school found unsafe.

The practical brewers also attacked chemists for offering unwieldy statements of the obvious. Jackson, for instance, had noted that the brewer should 'extract the necessary contents of malt by infusion, and hops by decoction': Richardson, tickled by the high-flown technical expression of a point too basic to mention, archly congratulated his enemy on putting straight any disciples who were planning instead to boil their malt and mash the hops.[88] In beginning with first principles, Jackson was merely following the convention of public lectures, but this became a *faux pas* before an audience of experienced brewers. The objection might appear trivial, but putting a price on common knowledge might be a prelude to more serious quackery. If the chemists' addiction to specialist terminology was a vice, the practical brewers cast their plainness as a virtue, writing self-effacing prefaces which begged their readers to excuse their lack of verbal elegance. By this device, of course, they hoped to give the impression of lives spent more by the copper-side than at the writing desk.[89]

For all the attacks on the excesses of theory, however, the overriding distinction was between brewers and outsiders: Michael Combrune, though far from 'practical', generally escaped censure. The chief objection to the chemical brewery instructors was that their professed identity was incoherent. As portrayed by Richardson, Jackson – engaged in the 'duple profession of beer-lecturer and isinglass-monger' – was duplicit in every available sense. He distracted the brewer with schemes for '*everlasting wood*'; he proclaimed an absolutely reliable technique to prevent disorders, while also promising to cure them; he pronounced himself a friend to all the trade, but withheld his secrets from his clients' neighbours.[90] At least Jackson was an acknowledged public tradesman (albeit in the wrong trade): Long, as Blake observed, had seemingly arrived from nowhere. Though he clung to chemical authority, it was not obvious that he was a chemist at all: Blake painted him as an all-purpose 'philosopher', sending obvious danger signals.[91]

Finally, the practical brewers also exploited their insider knowledge of how brewers behaved as a community. This was particularly useful in setting out what distinguished them from past advisors. Blake argued that the chemical lecturers' imposture contained the seeds of its own destruction: their authority depended on the belief that brewers could not explain how to brew, which was not only contrary to experience but an insult to the target market. The chemists, therefore, were increasingly giving up on approaching the brewers directly: instead, as in Long's case, they sought out agencies such as the Victualling Office, which controlled brewhouses but was run by non-brewers.[92] (This, in effect, was a version of the timeless charge: shady contractors prefer to milk the government.) Furthermore, Long's claim to have toured the principal breweries of England and been told the methods of each – in the course of developing a rival system – beggared belief. The simple explanation for Long's more bizarre assertions was that mischievous sources had fed him false information.[93]

Justifying Practical Publication

The practical brewers themselves faced a different problem. If what they knew truly gave them a commercial advantage, how 'practical' was it to publicize it? Among the reviewing press, the default assumption was that real trade secrets stayed secret, and all revelations were suspect. The *Monthly Review* summarily dismissed George Watkins's book because of an internal contradiction: Watkins claimed to have practised brewing 'in all its branches' for thirty years, but later admitted that he did not know porter-brewing at first hand, relying instead on 'a gentleman once concerned in this trade' who, 'having left it off with a fair fortune, is above deceiving me'. To the reviewer, this was no guarantee of anything.[94]

Later practical authors were more prudent, emphasizing that they could *not* reveal the whole of the method in print. John Richardson's first publication under his own name, *Theoretic Hints on an Improved Practice of Brewing*, states frankly that

> [t]he design of the Author ... is rather to intimate the probability of *a complete system of brewing*, than to promulgate one. Were he prompted by inclination to the undertaking, his interest would forbid the execution of it. The ordinary emoluments arising from the sale of a book ... would be a compensation very inadequate to the fatigue of many years wearisome attention to the multifarious practice of a brewhouse.[95]

Richardson's strategy, undoubtedly inspired by his great enemy Jackson, was to advise publicly on generalities – the means to form rules – and privately on the rules themselves, by manuscript directions and by personal attendance. When acting anonymously, he had used his booksellers as an intermediary; upon revealing himself in 1777, he directed all inquiries to Joe's Coffee House in Mitre Court, on the south side of Fleet Street. Joe's combined the traditional coffee house function of poste-restante collection point with an easily identified meeting place, at which Richardson made himself personally available during his limited stays in London: the times of his availability were occasionally advertised in the daily press.[96] Publication and the postal system carried his message to a national audience; *Theoretic Hints* advised brewers that they could receive Richardson's private advertisement postage-free by dropping a line to Joe's or, in Ireland, to Richardson's appointed booksellers in Dublin or Cork.

Richardson's private manuscript, published many years after his death, consisted largely of operational directions. These include apparently universal directions, such as thermometric values, but leave some authority in the hands of the client, instructing readers to vary the process according to local circumstances. Richardson's terms for the manuscript, by later report, were a fee of 150 guineas – well over a thousand times the price of the 2*s*. *Theoretic Hints* – and twenty years' secrecy.[97] His policy, he stressed, was very different from that of Jackson, who had created local monopolies: Richardson was willing to instruct anyone prepared to pay.[98] This stance, significantly, drew no censure from the reviewing press.[99] If Richardson was mercenary, he was at least not duplicitous: and trade attitudes were more palatable from a working brewer than from a Fellow of the Royal Society.

Other practical authors took a different tack, arguing that full disclosure in print was literally not possible: unlike chemical 'nostrums', which could only be concealed or revealed, the legitimate techniques of practical brewing were complex, and the success of print transmission was bound to vary.[100] George Blake stressed that he opposed all 'reservation' of useful information, but felt sure that there would remain a market for the kind of practical demonstrations and on-site assistance that could only be delivered privately, as his own successful business showed.[101] John Tuck worked on the same basis: his published directions were, he claimed, transcribed directly from instructions he had given to a brewhouse proprietor. Tuck made a show of reproducing another author's private directions, in order '[t]hat no improvement may be withheld'.[102]

All the brewer-authors after Watkins, in fact, had something openly to sell beyond their books. Blake, who had first published to block the challenge of John Long, recognized the opportunity for promoting his own consultancy in subsequent work.[103] Tuck advertised as an instructor, valuer and plant consultant; his manual also recommended particular firms involved in ancillary trades such as coopering and copper-setting, presumably for a consideration.[104] Others promoted specialized equipment: John Levesque, in the 1830s, published in support of a thermometer of his own design.[105] If such promotions invited comparisons with the private proposals of outsiders such as John Long, the practicalists could ward them off by embedding their promotional copy in comprehensive, public treatises which made clear their brewing credentials.

The rhetorical position of the practical brewers was a strong one, and led some instructors without brewery backgrounds to revise their approach. John Long put the lessons of his encounter with George Blake to work in a pamphlet on malting, addressed to the Dublin Society. Long shrewdly highlighted the fact that his audience must include both maltsters and philosophical gentlemen, forestalling objections of offensive didacticism on a process 'thought to be so universally understood'. Canvassing the approval 'of the chymist and of the unprejudiced', he made a show of deference to established practice, before stating that custom must not be allowed to override 'reason and experience' in finding the most profitable procedure. Richard Shannon, another non-brewer, struck an identical pose five years later: he 'would be very sorry it should be supposed he meant to hurt the feelings of professional men, by attempting to teach them their business, his wish being merely to facilitate it, by removing difficulties that lay in their way'.[106]

By these reassurances, such authors sought to remount Jackson's central authority claim: reform of brewing practice must depend on ideas imported from outside it. The practical brewers, however, were meanwhile finding ways to turn the same sentiment to their own advantage. If Combrune had been 'philosophical more than practical', his successors were determined to be both.

Towards the 'Science of Brewing'

The brewer-authors defined themselves in opposition to those non-brewing philosophers who claimed authority over the brewhouse, but this did not mean they rejected philosophical theory. Their position was usually that a skilled brewer knew best when formal rules could be framed, and that often, the complexities of the case meant that it was safest to rely on the traditional evidence of the brewer's experience.[107] Where appropriate, however, a literate brewer could philosophize as well as anyone. Richardson, in his 'Practical Brewer' guise, suggested that a systematic theory of brewing might be founded, by a suitably qualified insider, but would be far more complex than Humphrey Jackson had naïvely imagined. Indeed, he faulted some of Jackson's zealous philosophizing on philosophical grounds: careful investigation would have told Jackson that his process to conserve the saccharine content of malt extract had a ruinous side-effect, removing 'a preservative principle he has unfortunately overlooked'.[108]

A shift in tone came with Richardson's *Theoretic Hints* of 1777. Without abandoning his 'practical', trade-minded framing, the no-longer-anonymous Richardson wrote in an increasingly philosophical register. His goal, he said, was to steer between the extremes of those who 'jumble together a number of recipes, in the stile and manner of a *cookery-book*' (Ellis, and perhaps Reddington), and one 'who had the misfortune to be too abstrusely philosophical for common comprehension, and refined upon trifles till his theory frittered away the intention of his practice' (clearly Combrune).[109] This was probably a modification of George Watkins's practical manifesto, discussed in the preceding section; it is also a presumably deliberate echo of the 'middle course' taken by Francis Bacon's bee, a long-standing totem of those who seek to control the interaction between philosophy and practice.[110]

Subsequent brewer-authors were similarly careful about their positioning. George Blake claimed an asymmetry: 'a brewer by close study and application, may be a chemist ... a chemist, however eminent, cannot be a brewer, until he will condescend to quit his elaboratory, and attend the malt and brewhouses for some years'. Until the two worlds were properly combined, it was reasonable for the brewers to continue along traditional paths; managing the combination required respect for chemical knowledge but was a task best left to the brewers.[111] Blake, accordingly, described his own research on gravimetric methods, and cited Stephen Hales on the role of fixed air in natural processes; his mysterious adversary John Long, he hinted, was actually as incompetent in chemical as in brewery matters.[112]

A philosophical tone might also help to make partial publication legitimate. Richardson was understandably concerned to make sure his 'Hints' were understood as expressions of underlying principles, not the teasing descriptions of chemical wares which had fuelled Jackson's reputation as a nostrum-monger. This manoeuvre proved hard to manage: in formalizing general definitions, Richardson resorted to just the kind of verbiage he had complained of in Jackson, defining fermentation as '*a spontaneous internal motion of constituent parts, which occasions a spontaneous separation and removal from their former order of combination*'.[113] When he sent out an advertisement by post to a wide variety of country brewers, some were distinctly unimpressed. Richardson later recounted, '*verbatim & literatim*', one retort from a Lincolnshire brewer:

> Sr, I know Not who or what you are or what Profitchion you are off I have Rec[eive]d, a long Rodda mantade about Nothing ... you Point me to Nothing how I am to avoid these actidents that may happen So upon the whole I looke on it as a pese of quaekere [quackery] ... Pray Sr, what would you have thaught of me if I had Sent you Such an apistle about My 45 years Expereance and Say as much upon it as you have dun on your *there* [theory].[114]

Richardson dismissed the objection as merely illiterate: gentlemanly standards of written communication were required for admission to the community of competent practical brewers, as he defined it. The dismissal was superficial, of course. Transcribed into conventional English, the Lincolnshire man's complaint is strikingly similar to Richardson's own heckles against Jackson. Richardson's 'Profitchion' was, indeed, not evident: there was no precedent, in 1777, to show that a man who advertised as he did could belong to any professional community capable of guaranteeing his good faith. This was not a problem that could be solved by publication alone. George Blake relied on testimonials from past consultancies to pre-empt it; in Richardson's case, it was ultimately mitigated by his evident ability to practise what he preached. From around 1783, Richardson managed his instructional career in tandem with partnership in a brewhouse in the East Yorkshire port of Hull, which, under his management, became the most successful in the area.[115]

By the 1790s, the position of the philosophically attuned brewer, equally comfortable with books and brewhouses, was increasingly unproblematic. Aping Richardson and Blake, would-be instructors contrasted their prowess in 'the important science of Brewing' with the meanderings of 'men totally ignorant of the basis on which the system is founded', or offered testimonials from 'Gentlemen of the first scientific abilities'.[116] Such authors often found their own twists on the principle of partial publication. Around 1792, the London papers carried advertisements from one 'who has ever made it his whole and peculiar study to investigate the principles' of fermentation, offering to instruct by attendance or correspondence, and promoting thermometers and tables for their use, available via a bookseller in Romsey, Hampshire. The instructor was presumably William Ploughman, an established brewer in Romsey, who shortly afterwards published a short manual ostensibly aimed at home brewers. Ploughman's preface emphasized that there was enough difference between domestic and commercial brewing that, should the pamphlet 'fall into the hands of the common brewer', it would not be a suitable guide: commercial brewers, therefore, should apply to Ploughman by letter for private guidance.[117]

In 1788, Richardson collected his works under the title *Philosophical Principles of the Science of Brewing*.[118] As in other fields at this time, the word 'science' was used to assert that brewing had become a system of practice founded on a codified body of knowledge, as opposed to an 'art' consisting only of acquired skill. Subsequent practicalists adopted the same position. John Tuck, in the 1820s, acknowledged that mischief could arise from 'a reliance on science instead of practical acquirement', but stressed that if a brewer united the two, bringing routine observation and recording to bear on his practice, he could not help but profit.[119] The 'scientific' brewer-authors of this time, then, were more concerned with systematic working than specifically with importing approaches from chemistry, hydrostatics or the other disciplines which were beginning to be talked of as 'sciences': these fields, nonetheless, provided the brewer-authors with much of their material.

In his anonymous assault on Jackson, Richardson had sought to emphasize the gap between the malthouse or brewhouse and the laboratory. Now, with Jackson removed from the picture, he endeavoured to close it on his own terms. He found opportunities, for instance, in the work of Edward Rigby, a Norwich physician who considered sugar as a compound of an acid principle with phlogiston. Noting that fuel burning in a malt kiln should provide a constant stream of phlogisticated air, Rigby suggested that it might be possible to take the spent grains from the brewing process, dry them again and synthesize further sugars. Rigby's frank acknowledgment that he could not practically try the experiment was just what Richardson needed: the physician's deference to 'some one who could do it upon a proper scale' safeguarded the brewer's authority. At the expense of ten quarters of malt, Richardson reported that a hitherto unexpected yield of fermentables was, indeed, achieved. Richardson took issue with Rigby's chemistry, stressing that the effect might be purely mechanical, but nonetheless thanked him for a promising suggestion.[120]

While careful to avoid Humphrey Jackson's error of dismissing established trade knowledge out of hand, the 'scientific' brewers proposed reforms and reinterpretations along lines which Jackson would have found very familiar. The chemist, for instance, had claimed to be able to analyse the beers of different localities so that they could be synthesized elsewhere; Richardson proposed much the same project in the *Theoretic Hints* of 1777.[121] This marked more than a rhetorical shift. In his anonymous pamphlet of 1775, Richardson had mocked the attempt to separate analytically the 'spirituous' and the 'grosser ... constituent parts' as a pointless abstraction; now, he was at pains to distinguish 'oleaginous' and 'subtile' elements from the 'grosser mucilage' in a fermenting wort.[122]

In chapter 4 I discuss how this trend in Richardson's thinking resulted, a few years later, in a comprehensive scheme to determine the value of beer, and thus the validity of brewing practices, according to universal criteria. Key to this scheme was the introduction of new philosophical instrumentation: in Richardson's unashamedly commercial enterprise, instruments retail joined the established resources of publication and private consultancy in an ambitious bid to achieve, as far as possible, a monopoly on the emerging 'science of brewing'.

4 THE VALUE OF BEER

> A man who has practised brewing all his life, and who may generally make good beer, will not readily submit to the correction of one who tells him he has always worked at random, and who proposes to subject his future operations to the test of instruments he does not understand ... But should the instruments he recommends be established by an experience of their utility, formidable as they now appear, familiarity will simplify the application of them to a new generation of brewers, who may then be tempted to exult over their predecessors.
>
> [J. Noorthouck][1]

The previous chapter introduced John Richardson as the first author to draw together the powerful rhetorical positions of the practical commercial brewer and the philosophically literate theorist. In 1784, Richardson extended this approach in a new direction by issuing a 250-page treatise, *Statical Estimates of the Materials of Brewing*. The title recalls Stephen Hales's *Statical Essays* of 1727–33 on pneumatic chemistry: one of Richardson's interests, as we will see below, was the contribution of gases to the character of drink.[2] The book's real focus, however, was indicated by its subtitle: *A Treatise on the Application and Use of the Saccharometer*. This was a floating device for measuring wort and beer strength numerically: it provided, said Richardson, a universal indicator of the value of brewery products and practices.

Richardson's project invites comparison with Michael Combrune's grand claims for the thermometer a quarter of a century earlier (chapter 2). Indeed, many sources since the nineteenth century have yoked together Combrune and Richardson as the two most important harbingers of scientific progress, each having introduced a crucial instrument to the brewers of his generation.[3] In fact, the two men's intentions were very different. Combrune brought into the brewhouse an established instrument, scaled in the already conventional Fahrenheit degrees. Richardson claimed the saccharometer as his own invention, manufactured only to his specifications; the quantity it recorded, the pound per barrel of extract, was newly defined and, said Richardson, the outcome of long investigations to create a measuring regime tailored to the needs of his fellow brewers. Where Combrune published to seek philosophical validation, Richardson sought a commercially useful monopoly over both production and interpretation.[4]

This chapter examines how Richardson introduced and secured his instrument, quantity and overall scheme of management. The broad project of reducing beer, conceptually speaking, to parts which could be compared according to a universal standard had been a goal for Combrune, Humphrey Jackson and several chemical analysts, but it was Richardson's practically attuned interpretation which became entrenched in the minds of commercial brewers, with lasting consequences. The saccharometric project also deserves study as a defining influence on the commercial strategies of those who proclaimed a 'science of brewing'. To the emerging complex of public theory and private consultancy, Richardson added a new mode of instruments retail. The saccharometer and the *Statical Estimates* were mutually interdependent: the brewer who bought one would (or so Richardson intended) find need of the other. This strategy provided a pattern for the promoters not only of later instruments, but, as we will see later in the book, of both patent machinery and chemical consultancy.

Richardson needed to take three particular measures to secure his goal. First, he had to demonstrate that saccharometry was useful to his fellow brewers. This was a case of refining the philosophical-practical stance of his earlier writings: care was needed in urging change without seeming to disparage the reader's established skills. Second, Richardson had to show that the innovation was novel, to secure his conceptual monopoly. This was not straightforward, because the general principle had circulated privately between some brewers and instrument-makers since the 1730s: the rhetorical devices of the *Statical Estimates* were needed to affirm novelty in a way that the manufacture of the saccharometer itself could not. Third, Richardson had to clear away possible objections – from chemical philosophers and brewers alike – that the valuation scheme defined by the saccharometer was not compatible with established understandings. This, again, depended crucially on the circulation of Richardson's book.

Saccharometry, then, was by no means a simple linear application of contemporary physical theory to brewing. I proceed to illustrate the individuality of Richardson's project by contrasting the efforts of a rival theorist, James Baverstock of Alton. To Baverstock, there were no 'philosophical principles of the science of brewing' in the sense Richardson intended. Philosophical principles were (by definition) universal; their brewhouse applications could be deduced by any brewer able to understand them; hence, everything valid in Richardson's method had already been achieved, without ceremony, by himself and others independently. All the brewer needed to accurately measure beer strength, said Baverstock, was an ordinary flotation hydrometer and a little education: Richardson's proprietary instrument, quantity and methods involved an unnecessary mystification.

Richardson, I conclude, judged the audience of brewers better than his adversary. His saccharometric definition of value became widely entrenched before 1800, and he prospered as a result. He did not, however, gain the monopoly on saccharometers he had been hoping for, as resourceful competitors were quick to adapt to his definitions. Combrune's advocacy of instrumentation had outlived his theories; parts of Richardson's theory, as laid out in the *Statical Estimates*, far outlived the instrument it had been framed to support.

Introducing the Saccharometer

Historically, brewers had defined strength in terms of the volume of malt used to make a given volume of finished beer (chapter 1). This allowed clear quantitative recipes, but they were understood to need constant empirical adjustment. The yield of sweet fermentables from malt varied tremendously, depending on the quality of the harvest and the method of management: even apparently identical samples from the same maltster might turn out to give noticeably different yields. It was part of the brewer's skill to adjust the malt bill or the output volume accordingly. Taste was his chief guide, as it was for his customers. Too strong a beer, and the brewer would realize he was spending unduly on malt; too weak, and drinkers would complain. Between these limits, the strength could fluctuate in a way that – in profitable times, at least – kept both parties happy. A key goal for the philosophical brewers was to replace this qualitative approach with a quantitative procedure that could be reduced to paper directions and replicated. Michael Combrune had sought to do this by thermometry alone, but had not been able to realize a method.[5]

An alternative possibility involved the *gravity* – density, in modern parlance – of unfermented wort. The sticky, fermentable malt sugars which dissolved into the liquor during mashing made it much denser than ordinary water: if the gravity rose in direct proportion to the fermentable content, it could serve as an index of potential strength. The approach was undoubtedly suggested by developments in the distillery, where a different form of gravimetric valuation was well established. Pure alcohol having a much lower density than pure water, the alcoholic strength of a commercial spirit sample – assuming it could safely be approximated as a mixture of the two – could be found from its gravity. In the laboratory, this could be determined precisely by weighing a sample of known volume on a large beam balance. This method, however, was too slow and delicate for routine assessment by commercial distillers, or by the travelling Excise officials who assessed their wares. A convenient approximating device was the hydrometer, a floating device with a graduated stem and a weighted ball which, by Archimedes's principle, rose or sank depending on the sample's gravity. Recording the weighting needed to float the hydrometer, and the point at which the liquid cut the stem, gave a tolerably precise indication, and Excisemen had long carried hydrometers for distillery work.

John Richardson, it seems, first learned of the hydrometer's possibilities through the efforts of instrument-makers competing for this spirit trade. The national Excise project depended crucially on having hydrometers constructed to a reliably standard pattern: an official monopoly belonged to the successors of John Clarke, who had introduced the first hydrometer under the patronage of John Theophilus Desaguliers in 1725. Rivals of the Clarke firm made repeated attempts to usurp this lucrative monopoly. The most vociferous, as so often the case, was Benjamin Martin, whose efforts included a 1762 pamphlet dedicated to the Excise, claiming the superiority of his own hydrometers.[6] The approach proved unsuccessful, however, and Martin began to seek alternative markets. A minor adaptation allowed the existing spirits weights to be attached to the top (outside the fluid), rather than the bottom of the stem, to allow readings for heavier-than-water liquids. The modified device might, Martin suggested, be applied to salt-water, blood, pharmaceuticals and urine, and also '*Worts, Beer, Ale, Punch, &c*'.[7] As with the thermometer (pp. 53–4), Martin was quick to see the potential of approaching commercial brewers: they had much the same quantitative literacy as distillers, and there were far more of them.

Richardson was a natural early target. Mitre Court, his base of operations in London from 1777 to 1783, lay almost directly opposite Martin's shop on Fleet Street, and Richardson was committed to thermometry even before his work took its philosophical turn: the 'Practical Brewer' manifesto of 1775 (chapter 3) accepted it as a no-nonsense aid to reliable production.[8] At some point in this period, Richardson read Martin's advertisement and equipped himself with a hydrometer.[9] He did not, however, discuss wort hydrometry in the philosophically toned *Theoretic Hints* of 1777; indeed, it is not clear that he included gravimetric assessment among his privately communicated methods until the publication of the *Statical Estimates* in 1784.

By this time, Richardson had moved to Hull, an east-coast port town and a promising base for exports. He took on the North Brewery in partnership with John Dobson, formerly a merchant in Liverpool and the brother of Matthew Dobson, natural philosopher and physician to the Liverpool Infirmary. Richardson involved himself closely in the emerging philosophical culture of his adopted town: he later belonged, for instance, to the 'Society for the purpose of Literary Information', a short-lived (1792–9) forerunner of the Hull Literary and Philosophical Society, meeting at the home of one of Richardson's booksellers. It was Richardson who instigated the Society's attempts at regular publication of papers read, on the model of the Manchester Lit and Phil.[10]

The saccharometer, however, was clearly an outcome of Richardson's time in London. The *Statical Estimates* provides an illustration (Figure 4.1) and detailed description of the device, stating that it was manufactured to Richardson's specifications by the sole supplier, John Troughton. In 1782 (coincidentally, the year of Benjamin Martin's death), Troughton had bought out the business of Benjamin Cole, who rivalled Martin in supplying a wide variety of philosophical instruments; by 1800, rather usefully for Richardson, Troughton's firm had developed a strong reputation for the accuracy of its equipment, and was widely seen as pre-eminent among the Fleet Street makers.[11]

Yet the majority of Richardson's market consisted of provincial brewers whose brewhouses, like his own, lay far from Fleet Street. A note in the *Statical Estimates* clarified that 'Country brewers may be supplied by means of their booksellers, who have correspondents in London'.[12] This was a well-established approach, the distribution channels created by London booksellers having proved useful for a variety of high-value items such as philosophical instruments.[13]

The *Statical Estimates* was therefore very much a promotional device for the saccharometer. Whereas the book was priced at 5*s*., the instrument cost three guineas, more than twice the £1.5*s*. charged for Troughton's ordinary copper hydrometer.[14] Instrument and treatise were in fact designed to require each other. Although the saccharometer itself was sent out with a small pamphlet of directions, this made frequent reference to the *Statical Estimates*, requiring a complex process of calculation which would have made little sense without the detailed descriptions given in the treatise.[15]

Richardson began the *Statical Estimates* by describing the calibration process which had led him to scale the saccharometer in distinct, brewery-specific units. He began, he said, with an ordinary half-barrel for beer, 'as exact in its guage [*sic*] as we could procure', which he filled with water and weighed, presumably using a beam balance. He then refilled the half-barrel with wort from his own brewhouse, and found it to be 19½ pounds heavier. On a scale taking water as zero, the value of the wort's gravity was therefore 39 pounds per barrel. Richardson next placed his saccharometer in distilled water and established a zero, then transferred it to a sample of the wort and added weights to the top of the stem until the zero was regained. This added weighting, he proclaimed, directly represented the 39 pounds per barrel gravity of the wort. Repeating the process with other samples from different brewings convinced him that the stem weighting rose linearly with the wort's gravity.[16]

The point of staging this scene, of course, was to convince Richardson's audience that saccharometric measurement had literally begun in the brewhouse, through operations that made sense to the trade. Yet he was equally concerned to give the technique some grounding in a space of philosophical exactitude.

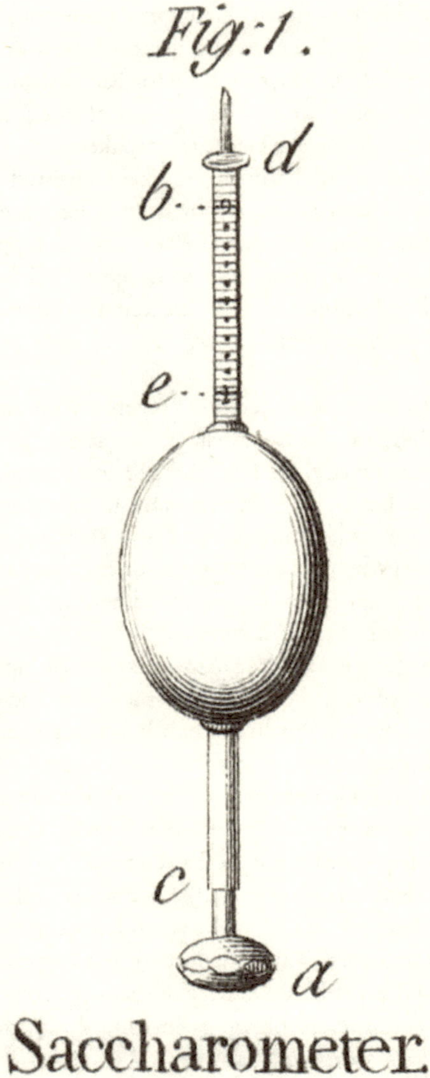

Figure 4.1: The Richardson–Troughton saccharometer, as illustrated in J. Richardson, *Statical Estimates of the Materials for Brewing* (London, 1784), frontispiece. The instrument's 'regulator', discussed below, consists of the sliding telescoped tube between points *c* and *a*. The frontis illustration does not depict the weights, which were attached at *d*. Reproduced by permission of the National Library of Scotland.

To 'unite the brewhouse with the laboratory', however, was an ambition he had decried as absurd in the work of Humphrey Jackson, so Richardson instead invoked a division of labour. He parcelled up his observations and instrument and sent them to a 'friend' in London, who performed more precise experiments in a laboratory setting, using trial worts prepared on a miniature scale from paper directions supplied by Richardson. Between them, the collaborators arrived at a revised pattern for the saccharometer.[17]

The saccharometric project therefore built on the philosophical-practical stance Richardson had defined with the *Theoretic Hints*. Richardson went further than before, however, in insisting that all valid practice must be 'scientific', a term which recurs through the *Statical Estimates*. The customary mild censure of the inefficiencies of home-brewed ales is replaced by loud condemnations of 'obstinate stupidity' and 'the empiricism of ridiculous old women' – a remarkable departure from the 1775 attack on Jackson, which had praised the 'fine, untheoretical ale' of a canny female brewer.[18] Richardson – unlike most brewery authors, as we will see later – was prepared to insist on brewhouse operations and mathematical adjustments as complex as those found in contemporary laboratory analysis.

Complexity, indeed, was now part of Richardson's proprietary strategy. The methods he described involved a series of numerical corrections for temperature and other factors, for which the brewer needed to consult empirically derived tables of figures supplied in the pamphlet of directions. Each pamphlet was signed by Richardson with a number in manuscript corresponding with the instrument's number, suggesting that the tables might vary with the calibration of individual instruments. This was probably not the case, as the tables were printed and do not appear to have varied.[19] By drawing attention to the possibility, however, Richardson and Troughton took a basic precaution against imitation once the device became established. To reach that point, of course, Richardson had to show that the complexities of the innovation were worthwhile for working brewers, and here, again, the *Statical Estimates* was crucial.

Selling Saccharometry

From the first page of the *Statical Estimates*' preface, Richardson presented the saccharometer as a device to increase the brewer's profits, reducing costs without jeopardizing any established notions of quality. Saccharometric measurement would let the brewer determine, quantitatively, the lower edge of the acceptable strength range and steer his production very precisely towards it, achieving savings of 10 per cent or more.[20] It was also a powerful diagnostic: a lower than expected measurement could signify poor barley, an unskilled maltster, an inappropriate malting process or inappropriate extraction by the brewer during the mash. Brewers in general, said Richardson, did not extract all the fermentables that were

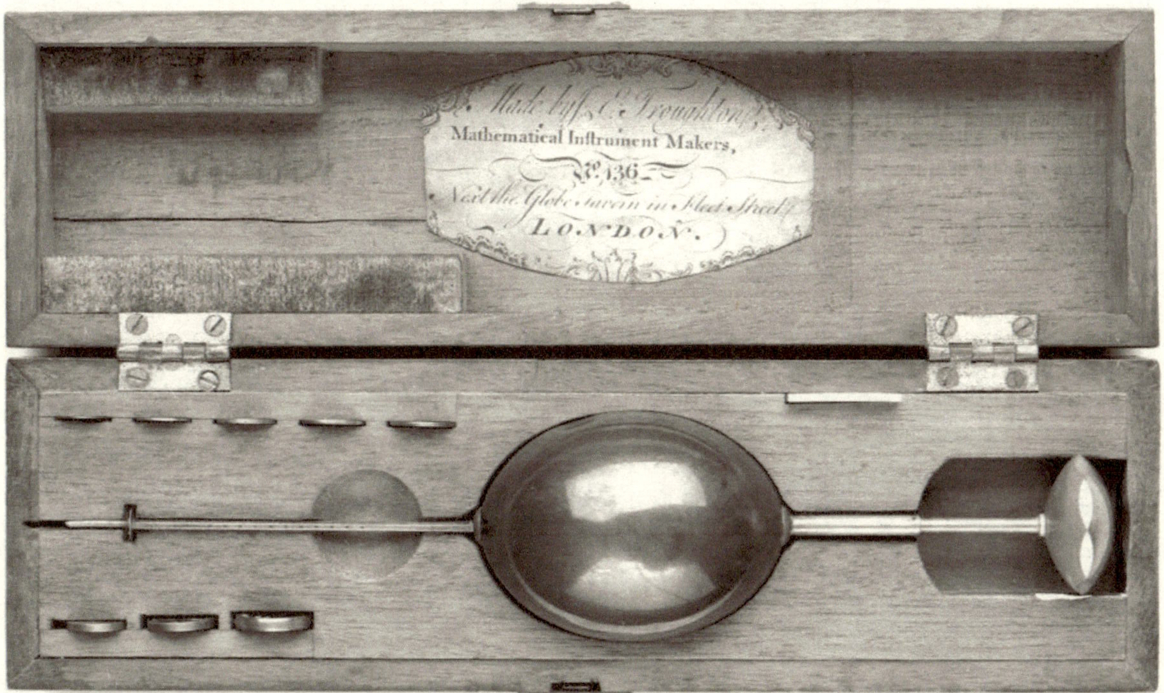

Figure 4.2: Richardson–Troughton saccharometer in its original case, with eight weights. From the collection of the Science Museum, object number 1954-381. The stem is engraved 'Richardson *inu*' on one face and 'Troughton *fecit* 261' (the serial number) on the other. Reproduced by permission of the Science and Society Picture Library.

available: in private consultation, he could advise on how to maximize the value. Richardson's terms for this service, he was careful to point out, did not involve a fixed fee payable in advance, but a proportion of the savings enjoyed by the brewer, over a specified time period, and subject to the usual conditions of secrecy.[21]

Here was a plausible solution to the paradox of the trade writer. Saccharometry was a rising tide that would lift all boats, at least for brewers. The losses implied by their gains would fall elsewhere: on the drinkers (who, however, would receive a more consistent product), the maltsters (but only in that they could no longer pursue fraudulent or inefficient practices) or the proverbial pigs who ate the used grains after mashing. It was all to the good that Richardson was an established brewer himself, and partner in an increasingly successful operation. Releasing his knowledge, he could claim, was demonstrably no threat to his livelihood: rather, it provided an honest sideline in instrument sales and consultancy.

Richardson was particularly ingenious in setting himself up as an opponent of conservative practice without alienating his market in the way Humphrey Jackson had done. He was careful to state that the image of the ignorant rustic, still gauging heat by rule of thumb, did not truly describe a country brewer, but a 'country fellow, dignified by the appellation of *brewer*' who was presumptively a servant. The servant's ignorant contempt for new devices, Richardson hinted, was not 'unmixed with *cunning*': a universal measure such as the thermometer might become a 'rule in the master's hand', providing an independent index of his subordinate's competence.[22] Richardson cast his reader firmly as the master in question. This prefatory device was to be emulated repeatedly by the brewer-authors of the nineteenth century.[23]

Similarly, the saccharometer promised mastery over the maltsters. Malt had traditionally been graded and priced qualitatively by its physical appearance, its response to biting and its ability to float in water.[24] Richardson's saccharometer revealed that the invisible differences were highly significant: some apparently 'indifferent' barley was more extractible than 'well-made' samples produced in different years or locations. By the saccharometer, said Richardson, a brewer could establish 'the intrinsic worth of every kind of malt, to the very great precision of the one-thousandth part of the fermentable matter extracted from every quarter employed'. By analogy with the distillers' reference standard of proof spirit, he suggested fixing as '*par*' a wort gravity of, say, 70 or 80 pounds per barrel achieved from a quarter of malt: other samples could be described as so far above or below par. Richardson even proposed an 'apparatus for ascertaining the value of malt to the purchaser or maker, independent of the consumer'. This prospect, though Richardson was never able to bring it to market, attracted enthusiastic enquiries from several brewers.[25]

Most importantly, Richardson promised to put the brewer on a more even footing with the Exciseman. Since the seventeenth century, the two principal taxes had been a duty on the malt used – assessed by volume, irrespective of quality – and a beer duty, assessed by retail value alone. Beer valued at 6*s*. per barrel or less before tax was deemed 'small beer' and lightly taxed, while all dearer beer was classed as 'strong', and paid a much higher rate (in 1784, 8*s*. per barrel), whatever its strength. The legislature had never adjusted the 6*s*. watershed for inflation, however, and good 'Table Beer', used for everyday drinking and by no means intoxicating, typically sold for 10*s*. the barrel. The letter of the law was therefore unworkable. In practice, since around 1730, most Excise officers had applied a common-law adjustment known as 'the Tenth', notionally interpreting 10*s*. beer as a composition of one-tenth strong beer and nine-tenths small. Such legal fudging led to frequent disputes, but the Commissioners of Excise held firm against brewers' pleas for a formalized solution.[26]

Richardson's take on this situation was that the legislature was mired in just the same 'darkness', mixed with self-interest, which afflicted the rustic servant.[27] The uniform beer tax was a blunt instrument with a hopelessly obvious flaw: a brewer might make extra-strong beer, pay duty at the fixed 'strong' rate and then dilute it down to a respectable retail strength with lightly taxed small beer or water. Officially, such tax avoidance was forestalled by a seventeenth-century ban on all mixing of worts and beers by brewers and publicans, a measure which assumed not only incorruptible but incessantly vigilant Excise surveyors.[28] Attempts to minimize abuses through an intermediate category of 'table beer' (from 6*s*. to 11*s*.) merely encouraged further mixing, and were eventually scrapped.[29] The final court of authority in detecting frauds, more often than not, was a 'vague and barbarous appeal' to the Exciseman's sense of taste.[30]

The solution, Richardson said, was simple: abolish the uniform bands and tax the unfermented wort in direct proportion to its gravity, as measured by the saccharometer. With authority vested in the instrument, rather than the Exciseman,

> altercations would immediately vanish, the revenue would be increased, the brewer would be at liberty to make, alter, or compound his liquor into as many and as various sorts, as he has palates to please, without subjecting himself to the interference of the officer, or the lash of the law.[31]

A 'standard gravity' would serve as a reference point: if beer containing, say, 25 pounds per barrel of extract were sold at 25*s*. per barrel before duty, beer of 40 pounds per barrel should be at 40*s*., and so on proportionately.[32] Richardson proposed the standard not as an optimum, but as a conventional device to help brewers produce the various customary drinks more reliably, recording and replicating the strengths most appropriate to their markets.[33]

Richardson, then, proclaimed for the saccharometer what Combrune had sought in the thermometer: the means to reduce local practices to points on a universal scale. While reassuring brewers that their established practices were valid, he tempted them with mastery of the practices of other localities. The 'brewer of Edinburgh', for instance, might seek to undercut the London porter-brewers with a similar product. If he relied on recipes based on malt volumes, the enterprise would fail because of variations in the malt supply; knowledge of the pounds per barrel drawn for porter would provide a much surer way to success.[34] In his privately circulated directions, Richardson offered gravity values for Burton ale and porter, alongside mashing heats and other replicable details. Occasionally, as in a very particular outline of the physical appearance of the casks used ('a bar is put across each head, and the brewer's initials or name, with B or BURTON at length, are branded in front in letters of about an inch and a quarter high') Richardson's guidance appears to move beyond emulation towards passing-off.[35]

Ultimately, of course, Richardson's chief intended beneficiary was himself. Making the legislature accept saccharometric valuation according to the theory laid out in the *Statical Estimates* would mean establishing the Troughton saccharometer as the official Excise instrument, with the same monopoly status as Clarke's hydrometer for spirits. Besides creating a dependable stream of instrument sales, this would firmly cement Richardson's position as a consulting authority. The manoeuvre, however, demanded a good deal from his audience. Installing the saccharometric quantity as the sole determinant of economic value in beer meant marginalizing the scope for independent judgment by Excisemen, maltsters and brewers alike. Richardson's ambition threatened to conflict with established understandings which he could not afford to dismiss. It also depended on his ability to clear from the scene several prior conceptual innovations, some of which were suspiciously similar to his own.

Making the Saccharometer Novel

There is something odd about the *Statical Estimates'* opening narrative of calibration experiments in the brewhouse and laboratory. Richardson writes as though the instrument used were the saccharometer itself, yet at the same time he clearly states that the underlying principles of saccharometry were not yet established: it was through the experiments that Richardson devised the device manufactured by Troughton. What instrument, then, was he using to start with? The obvious candidate is Benjamin Martin's hydrometer, which had provided his first introduction to hydrostatics. Martin's account of recalibrating the spirits instrument for heavier-than-water regions seems to be the model for Richardson's narrative.[36]

Richardson insisted, however, that his work owed nothing to Martin, because Martin had never truly grasped the particularities of the brewer's needs. His mistake, said Richardson, was to assume the hydrometer should be applied only to fermented beer, just as it was applied to spirits at the end of the production process. This approach, which William Reddington had also investigated, using a home-made device,[37] was fatally flawed when it came to comparing beer samples made by different methods. Beer, unlike distilled drinks, contained both spirit (less dense than water) and residual, unfermented malt extract (denser): the proportions varied with the production process, so that the combined effect on the gravity did not correlate to any existing notion of 'strength'. On Richardson's account, Martin was so mystified that he gave up and recanted his claims for the device: it was Richardson himself, with his brewhouse background, who perceived that the assessment should be applied *before* fermentation, to give an unambiguous reading based on malt extract alone.[38]

Richardson's account is highly improbable.[39] Even if we buy the story of Martin – that most adaptable of retailers – abandoning such a promising commercial prospect, the principle of wort hydrometry had certainly been proposed repeatedly since the 1730s: Peter Shaw, for instance, had outlined the concept in his address to distillers.[40] It had even been developed in practice, and for beer-brewers, by John Clarke, the proprietor of the official spirits hydrometer. In the 1740s, Clarke had marketed a parallel device for the denser-than-water region, 'of great Use In Brewing, And Discovering the *Goodness* or *Quality* of Malt' (Figure 4.3). At the request of various brewers, Clarke said, he had devised volume measures and a miniature mash-tun, so that different malts could be mashed, and their worts evaluated, under standard conditions. His published prospectus included determinations for a variety of samples, expressed in the arbitrary units of his device (Figure 4.4).[41]

Martin's efforts were more conspicuous, however, and led rival competitors for the distillery Excise market to extend their instruments to the denser-than-water region. In 1780, for instance, John Dicas of Liverpool gained a patent for 'hydrometers for sliding-rules, to ascertain the strength of spirituous liquors, malt worts, and wash for fermentation'.[42] Richardson is conspicuous by his absence from this competition. Though his consultancy was well established before 1780, he neither published on hydrometric management nor, so far as we know, mentioned it as a topic for private consultancy until the *Statical Estimates* of 1784. Indeed, when Humphrey Jackson had offered guidance on using the hydrometer, Richardson (as the 'Practical Brewer') had merely retorted that such devices were obvious enough to need no instruction.[43]

The onus was therefore on Richardson to establish that the saccharometer was truly a novel invention of his own devising. Affirming this novelty would help both to establish his authority and to minimize associations with the disputes over the existing distillery hydrometers' reliability which had been stoked by Martin and his rivals.[44] For this reason, he not only cut Martin out of his narrative of the instrument's development, but made a show of withholding from print some objections his London correspondent had raised to Dicas's hydrometer.

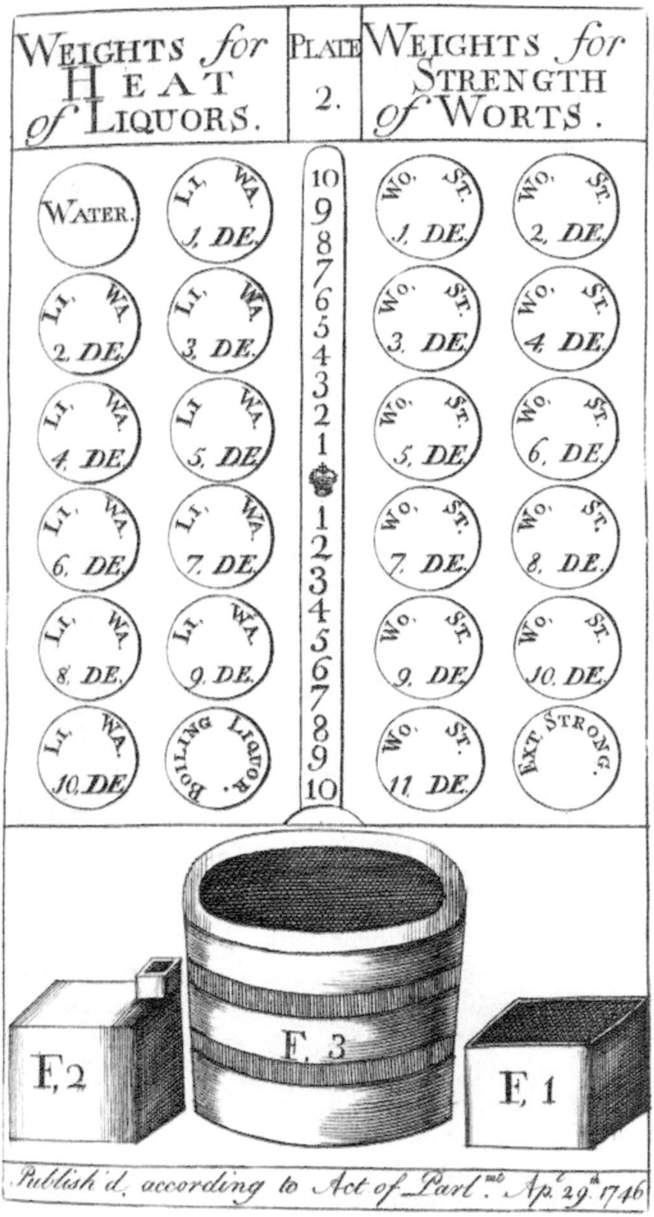

Figure 4.3: Clarke's wort hydrometer stem and weights (top), trial mash-tun (bottom centre) and standard volume boxes for malt. R. Clarke, *Notice is Hereby Given, to all Dealers ... that the Hydrometer, or Brandy-Prover ... is now Brought to its Utmost Perfection* (London, [1746]), end matter. Reproduced by permission of the National Library of Scotland.

Weight of ye Malt.				PLATE 3.	Strength of the Several Worts.			
Exp	Sorts of Malt.	℔ Pint, Oz. Dr.	℔ Bushel ℔, Oz.	℔ Sack ℔, Oz.	1st Wort.	2d Wort.	3d Wort.	4th Wort.
1st	Pale	9 : 12	39 : –	156	10, DE.	Bare 4. DE.	2, DE.	1, DE.
2d	Pale	10 : 1	40 : 4	161	7. DE.	Between 2,&3.	Strong 1, DE.	Between 1, & Li.
3d	Pale	9 : 6	37 : 8	150	8, DE.	Bare 3, DE.	Between 1 & 2	Between 1. & Li
4th	Pale	9 : 11	38. : 12	155	7. DE.	Bare 3. DE.	Very Strong 1. DE.	Between 1 & Li.
5th	Pale	10 : 5	41 : 4	165	6. DE.	Very Bare 3. DE.	Very Strong 1. DE.	Between 1. & Li.
6th	Brown	8 : 8	34 : –	136	Bare 8 DE	Strong 3, DE.	Bare 2, DE.	Bare 1. DE.
7th	Brown	8 : 9	34 : 4	137	Very Bare 6. DE.	Between 2. & 3.	Strong 1. DE.	Very Small
8th	Brown	8 15	35 12	143	6, DE.	Bare 3, DE.	Very Bare 2, DE.	Between 1, & Li.
9th	Brown	9 : –	36 : –	144	7. DE.	Bare 3, DE.	Bare 2. DE.	Between 1. & Li.
10th	Brown	9 : 1	36 : 4	145	5, DE	Between 2 & 5	Between 1, & 2.	Very Small

Publish'd according to Act of Parliament April 29th 1746.

Figure 4.4: Wort strength determinations for ten samples of malt. Clarke, *Hydrometer*, end matter. Reproduced by permission of the National Library of Scotland.

'The mode of applying [my] instrument', Richardson wrote, 'is so entirely [my] own, that it cannot possibly receive a recommendation from the depreciation of any other'.[45] It may also be significant that Richardson chose Troughton, one of the few high-profile makers not publicly involved in the scramble for Excise and distillery business, as sole producer and retailer.

The new name *saccharometer* was likewise important. A saccharometer, Richardson insisted, was not a hydrometer: though similar in form, it was crucially different in principle.[46] The only novelty in its material design, in fact, was a sliding tube known as the 'regulator' (Figure 4.1), which could be adjusted to compensate for local variations in the gravity of brewing water.[47] Richardson's contention, however, depended more on the instrument's calibration history. Because it was impossible to machine a perfectly uniform hydrometer stem, and because the rise and fall varied non-linearly with both density and temperature, makers needed to graduate each hydrometer individually, by experiment.[48] The usual approach was to calibrate from a series of reference samples whose specific gravity had been verified using a hydrostatic balance. Although any device could be forced to operate in any gravity region by the addition or removal of weights, an informed buyer would trust it only within the range of this rigorous testing, which, for Martin and his rivals, had always been defined by the distillery. Hydrometers, then, were designed for lighter-than-water spirit samples; the saccharometer was designed for malt sugars. Unlike all his rivals, Richardson repeated, he had designed his saccharometer in the brewhouse itself, with full first-hand knowledge of the brewer's needs.[49]

Equally significant was Richardson's pounds-per-barrel scale, which was certainly novel: Clarke's wort hydrometer gave arbitrary values, while Martin had simply extended his distillery scaling. By referring to the standard beer barrel, Richardson made the valuation concrete for his audience. The proposition 'my wort is 30 pounds per barrel heavier than water' was, he said, intelligible and useful to 'the most perfect son of dullness'.[50] Not, of course, that his readers were dull; but they were, he implied, practical souls seeking clear indications. The approach did, indeed, find favour, not only among brewers, but with maltsters such as John Reynoldson, of Newark-on-Trent, who published a treatise of his own in 1808. The rivals to Richardson's device, said Reynoldson, were 'unintelligible' except by translation into his terms.[51]

What all this tells us is that the saccharometer would not sell itself. Richardson needed widespread distribution of the text of the *Statical Estimates* to affirm that what he was promoting was new, stable, philosophically valid and appropriate to his market. This was actually far more important than the gravimetric accuracy of the device. Richardson himself said almost nothing about his instrument's reliability, referring any concerned readers to an appendix containing the anonymous London correspondent's letter. His goal was rather to make the saccharometric definition of value seem so self-evident that it did not need any external justification. In this, as the next section describes, Richardson faced one further major challenge: the threat of conflict with pre-existing definitions which seemed, on the face of it, equally evident.

Extract, Spirit and Inebriating Airs: Competing Definitions of the Value of Beer

Although most brewers reckoned the value of beer by the malt used to make it, they generally accepted that its strength could also be quantified in terms of the spirit it contained. Distilling beer to determine its alcohol content, though not routine among either Excisemen or brewers, was not unknown: Humphrey Jackson, for instance, had claimed a decline in quality of porter on distillation evidence in the 1750s.[52] There was no immediate conflict between the malt-extract and spirit definitions of strength: a rich wort before fermentation implied a high-alcohol beer afterwards, and Richardson, when he appointed malt-sugar extract the sole determinant of value, took it for granted that his audience also sought a good yield of spirit.[53] Indeed, the saccharometer seemed to offer a convenient short cut to measuring spirit content without distillation. Richardson saw fermentation as a uniform process in which each particle of malt sugar was converted to a single particle of spirit: the weight of spirit gained should be proportional to the weight of malt extract lost. This meant that a quantity exactly proportional to the spirit content could be found as the difference between the original gravity of the wort and the final gravity of the fermented beer.[54]

When Richardson experimented on beers of different kinds and apparent strengths, however, he found a serious problem. The two measures (original wort gravity for the malt extract, fall in gravity for the spirit) bore no proportion at all to each other. Particularly alarming was the comparison between samples of Burton ale – a heavy, dark drink, universally understood as 'strong' and priced accordingly – and ordinary porter. Comparing original gravities gave the expected result, the Burton's being twice the porter's; but the porter's fermentation had been pushed much further, so that the two drinks had almost the same alcohol content, a finding Richardson confirmed by distillation. This told badly against the credibility of the scheme: everyone knew that Burton ale was more inebriating, and this underpinned its market value. There had been no problem so long as the differing definitions of strength had not been routinely quantified, but Richardson's whole approach relied on quantifications which now seemed to contradict each other. He briefly wondered whether his whole approach might be mistaken.[55]

By the time he published the *Statical Estimates*, however, Richardson had hit upon an ingenious solution. The link between economic value and inebriating potential could not safely be challenged; nor could the proportion between spirit content and the fall in saccharometric gravity, which followed from physical postulates he had been invoking since the *Theoretic Hints*.[56] He therefore broke down the intermediate proportionality, pronouncing: '*the apparent* STRENGTH *of malt-liquors, or that* INEBRIATING EFFECT *which they produce upon the animal frame, does not entirely consist of* SPIRIT'.[57]

The idea that intoxication might be due to something other than measurable alcohol was not new. Physicians were often at pains to distinguish the long-term effects of different drinks, and it was obvious to most drinkers that gin and beer, for instance, produced different kinds of intoxication.[58] The dichotomy had been selectively stoked across the first half of the eighteenth century, as anti-gin campaigners emphasized the transient rages, moral collapse and ruined digestion of spirit users as against the well-nourished merriment of a traditional (and supposedly moderate) beer-drinking culture.[59] The distillery lobby had drawn on chemical theory in its defence: since 'it is the spirit alone in the Beer and Ale that intoxicates', there was no rational basis to condemn one drink whilst favouring another.[60] A minority of chemists, however, supported alternative conceptions. In France, Guillaume-François Rouelle (1703–70) had argued that the distillation process did not merely extract alcohol but chemically produced it, as was the case for the distilled oil of almonds or mustard: on this view, beer and wine contained a different inebriating principle whose nature was otherwise unknown.[61] Richardson accepted that beer contained spirit, but postulated that it also contained another principle which modified the inebriating effect, and was particularly prevalent in high-gravity beers. He found a prime candidate in contemporary philosophical studies.

'Fixed air', as the gas produced by fermentation was known, was a key focus for chemical and medical research across the middle years of the eighteenth century, often for its possible role as an antiseptic, preservative or cure for scurvy, as in David MacBride's proposal for shipboard brewing (p. 69). Joseph Priestley, in developing the impregnation process which created artificial mineral water, used a commercial brewhouse in Leeds as a convenient source for large volumes of the gas.[62] Thomas Henry, over in Manchester, believed fixed air could be not only the effect but the cause of fermentation, serving as a substitute for yeast.[63] Such authors' work was examined in a commentary of 1779 by the natural philosopher and physician Matthew Dobson, whose brother was soon to become Richardson's partner in Hull.[64]

It is not surprising, therefore, that Richardson speculated in the *Statical Estimates* on fixed air as an agent of intoxication. He cited the 'transient exhilaration' caused by drinking impregnated water, even in the absence of spirit, which was often reported in contemporary accounts of spring waters.[65] Might fixed air and spirit together produce a more lasting inebriation?[66] If so, here was a plausible solution to the riddle of the Burton ale and porter. The method used to ferment porter was very rapid and forceful, driving off most of the fixed air; the Burton method retained the fixed air, but yielded proportionally less spirit. The brewer's skill, therefore, lay in achieving the particular balance of fixed air and spirit customarily associated with each particular drink.[67] Were there a way of measuring fixed air with the same quantitative certainty as for spirit, Richardson said, he would be able to find the optimum trade-off between the two intoxicating principles. He admitted that he had so far only 'the vague discrimination of the senses' as a guide; in later years, he offered some preliminary computations in a paper to the literary society at Hull.[68]

Richardson's novel interpretation did not take hold. Just as the prospect of fixed air as an antiseptic or scurvy remedy faded out of medical discourse in the nineteenth century, so its possible intoxicating qualities, though never firmly denied, drew little attention from later writers.[69] At the time of his first publication, however, the conjecture bought off a significant objection and provided a useful breathing space for Richardson's scheme to secure acceptance. By around 1800, the saccharometric malt-extract definition of value was well enough established to be able to survive in spite of the unresolved compatibility question. While distillers, distillery Excise assessors and chemical philosophers continued to interpret drink primarily in terms of its spirit content, beer-brewers made little use of spirit concepts, identifying 'strength' with original wort gravity expressed in pounds per barrel. This situation occasionally caused some tensions, as later chapters will show, but was largely stable across the nineteenth century and into the twentieth.

Overall, Richardson's well-publicized claims met with little targeted objection. The field of philosophical instructors was temporarily clear, following the retirement of Humphrey Jackson; the promoters of rival instruments lacked the brewhouse connections to challenge the more detailed arguments in his treatise. The *Statical Estimates* did, however, draw a pointed response from one fellow brewer, whose unusual attack allows us better to understand the fragility of Richardson's position.

Philosophy without Publicity: The Challenge of James Baverstock

James Baverstock (1741–1815) was born to the brewery, following his father into business in the Hampshire town of Alton. In 1785, the year after the *Statical Estimates*, he issued a pamphlet entitled *Hydrometrical Observations and Experiments in the Brewery*, dedicated to the Society of Arts, claiming that he had anticipated Richardson's supposed innovations by well over a decade. Curiously, Baverstock's account of his engagement corresponded at almost every step with Richardson's: the early adoption of thermometers; the response to Benjamin Martin's advertisement for the hydrometer; Martin's admission of failure; the brewhouse intuition that the device must be applied to worts as well as finished beer. Whereas Richardson worked the saccharometer into his public schemes, however, Baverstock approached the most eminent brewers of London privately. He seemingly found only one supporter, Henry Thrale of Southwark, who was already looking to chemistry to revive his fortunes through the projects of Humphrey Jackson (chapter 3), and who soon incorporated the hydrometer into his own practice.[70]

This narrative left no room for an original contribution from Richardson. Baverstock dated his own crucial discussion with Martin to 1770, years before Richardson's first public appearance as a consultant. Thereafter, said Baverstock, wort hydrometry had remained isolated in his own brewhouse (and, presumably, at Thrale's) until around 1780, when 'hydrometers became nearly as generally used as the thermometers' owing to 'the activity of another maker, who had got the hint, and went among the Brewers, in London, to sell them'.[71] This maker was probably John Dicas, whose reference to worts in his 1780 patent was noted above, or conceivably Matthew Quin, whose hydrometric improvements earned him a silver medal from the Society of Arts in 1781.[72] Since that time, Baverstock later added, 'a system of practice' had '*been gradually forming*' among the better-educated brewers, without recourse to formal instruction.[73] Richardson had argued that the philosophical trend in brewing made his guidance crucial: Baverstock's position was that it made it superfluous. Baverstock's annoyance at Richardson's public priority claim may well have been his sole motivation for following him into print: he had no rival consultancy to promote, and apparently tended to turn down requests for instruction.[74]

Hydrometrical Observations, then, represents an atypical solution to the paradox of the trade writer. Baverstock's line was that he had taken the common-sensical route of applying his beneficial discoveries to his own affairs. He had not published, but had given his knowledge freely to a fellow brewer who displayed the appropriate interest, and had seen it reproduced by others, apparently independently. The question of trade secrecy did not arise because there was, simply, nothing publishable about the matter – until Richardson's attempt to enclose and mystify the process by publication demanded a remedy in kind. Even then, Baverstock confined the controversy to a short appendix, and stressed that he largely agreed with Richardson's findings. The main text of the *Hydrometrical Observations* stuck largely to experimental observations; by dedicating the pamphlet to the Society of Arts, Baverstock signalled that his intended audience for these was the practically minded, yet philosophically curious gentleman. Brewers did not need the particulars of his findings, except insofar as they, too, happened to be philosophically curious.

Behind its polite construction, however, the *Hydrometrical Observations* was calculated to block Richardson's putative monopoly, chiefly by establishing that the saccharometer was nothing more or less than a hydrometer. Baverstock criticized the Richardson–Troughton instrument's regulator – its 'only novelty' – as prone to slippage, wear and possibly leakage, all of which would affect the reading.[75] Otherwise, he raised no objections, but professed impartiality between the saccharometer and the various hydrometers whose makers were now courting the brewery. The only notable distinction, he said, was in the units in which they were scaled, and these were simple enough to convert arithmetically: he had tested five different models, and found their readings equivalent 'to great or fully sufficient exactness'.[76]

Richardson, therefore, found his second serious opponent to be a very different creature from Humphrey Jackson: an undeniable brewery insider who sought not to steal his promising monopoly but to dispel it, purportedly for the general good, claiming priority based on an account which unashamedly involved a sixteen-year silence. Bemused, Richardson resorted to what he knew:

the unsubtle art of the pamphleteer. In *Remarks on a Pamphlet Entitled Hydrometrical Observations*, Richardson displayed all the conventional sarcasm, mock pedantry and line-by-line dissection he had deployed against Jackson ten years earlier. This time, however, his adversary could not be faulted for ignorance of the brewhouse, and so Richardson flung himself onto the opposite tack. Baverstock was simply not up to the job of writing, and had most unwisely stepped 'so far out of his brewhouse at Alton, as to be found wandering in the literary labyrinth of Pater-noster-row'. His experiments were hopelessly inexact, and his theory as misconceived: 'with a single twirl of the finger', wrote Richardson, Baverstock had thrown 'neck and heels out of the brewhouse window all the philosophy I had introduced at the door'.[77]

In fact, Baverstock had exercised conventional modesty in disclaiming any ability to meet philosophical standards, his point being that very precise measurement was not necessary to the work of a 'practical brewer'. Richardson's response was withering: 'Bravo! bravo! Mr. B. – down with the philosophers!'[78] If Baverstock's self-assessment was sincere, then it was incoherent to address his brewhouse guesswork to the 'philosophical gentlemen' of the Society of Arts, who could have no use for it: what was needed, and what Richardson supplied, was the converse approach of communicating genuine philosophy to an audience of brewers.[79] Further, '[i]f the information of our common brethren of the white apron was his only inducement', why did Baverstock keep them in ignorance for fully sixteen years?[80] Having built a career, from the time of the *Theoretic Hints*, on the principle that selective revelation on commercial terms was the legitimate and only sound basis of useful consultancy, Richardson unsurprisingly painted Baverstock's professed commitment to openness as suspect.

Richardson also set up a distinction between Baverstock's insider status and his own. Whilst caricaturing Baverstock as an insular product of the family brewhouse at Alton, he pointed out that he had received the *Hydrometrical Observations* during a visit to Russia: breadth of knowledge of practices from across the inland and export trades was needed to achieve the perspective necessary for tracing causes and effects.[81] Baverstock showed his ingenuousness, moreover, in accepting modified distillery hydrometers, whose readings, for denser-than-water fluids, were not linear with specific gravity: they might serve him locally, but bore no constant relation to any wider definition of value, a problem solved by Richardson's attention to the calibration of the pounds-per-barrel scale.[82] As to the charge that Richardson's methods were overcomplicated, Richardson retorted that he was instructing 'a little pupil in Suffolk, green from the boarding school' in managing his father's brewhouse, whose ready mastery of Richardson's philosophically exact scheme showed up Baverstock's limitations.[83]

Always chiefly concerned with his brewery audience, Richardson had taken his eye off the likely effects of such rampant invective in the reviewing press.

John Noorthouck, who had treated Richardson's *Statical Estimates* favourably in the *Monthly Review*, greeted his *Remarks* with dismay:

> however superior he may be, or suppose himself to be, to Mr. Baverstock in the art of brewing philosophically, [Richardson] triumphs over his competitor in a flippant style of contemptuous language, unusual among true philosophers; who proceed in experiments with cautious steps, lest they should stumble upon error; and they speak with modesty, from a conviction that even the highest acquisitions of knowledge are still defective.[84]

It was not only the want of modesty which proved off-putting. Reviewers were generally happy to accept Richardson's self-fashioning as a brewing philosopher, but could not swallow the insistence, which was increasingly integral to his overall scheme, that it was both possible and valuable to include chemical and statical philosophy in the instruction of brewers in general. The usual judgment of the reviewers (who generally claimed competence in contemporary philosophy, and cheerful ignorance of practical brewing) was that simpler prescriptions, such as Baverstock had provided, were the appropriate tool for the job.[85] Richardson also did himself no favours by presenting his *Remarks* as a letter to Samuel More, Secretary of the Society of Arts, on the strength of a mutual acquaintance. More, it seems, knew Baverstock in person, their friendship being indicated (and perhaps fostered) by Baverstock's making More a present of a cask of ale in 1785. More endorsed Baverstock's approach of 'paying no further regard to what [opponents] may say; for as one of them, I believe, is professedly a pamphleteer, they will be happy in any opportunity of exercising their vocation'.[86]

Privately, in fact, Baverstock considered the possibility of imitating the Richardson–Troughton partnership. Around the time of the *Hydrometrical Observations* publication, he was negotiating to commission an instrument 'on a new principle, which, I think, bids fair to exceed all others in accuracy' from Gabriel Wright, a former employee of Martin's. Baverstock was happy to direct an inquisitive sugar-planter to Wright for a copy of the instrument, but stressed he would not be so 'indiscriminately communicative' to his fellow brewers. Like all his peers, Baverstock treated openness as a resource to be employed selectively.[87]

In the end, however, he stuck to his initial path. In 1786 Baverstock moved to Windsor, Berkshire, forming a brewing partnership with Richard and John Ramsbottom, who sponsored his election to the Society of Arts in 1787.[88] Trade prospered under Baverstock's management, and by 1801 production had risen above 30,000 barrels, on a par with the largest London ale-brewers. The firm had established 'Windsor ale' as a prized niche commodity, and its owners were perhaps the most conspicuous leaders of the 'country' trade in England around 1800.[89] Baverstock's success, which he attributed to the hydrometer, removed any need to consider philosophical instruments or consultancy as a source of income.

The Spread of Saccharometry

Ultimately, Richardson gained more influence than he might have done, yet less than he sought. Saccharometric valuation according to his directions undoubtedly caught on: its spread is hard to map precisely, given the limited survival of records and the wide variation in production cultures, but authors in the first decade of the nineteenth century give the impression that it was then general among 'public breweries'.[90] Yet Richardson failed to monopolize the saccharometer. Unsurprisingly, the established hydrometer-makers responded quickly to his challenge, providing tables to translate their instruments' readings to pounds per barrel. In 1789, Matthew Quin produced a simplified 'Universal Hydrometer', covering both the spirit and wort ranges with only four weights, which gained him a second bounty from the Society of Arts.[91] Though Baverstock had decided not to enter the market directly, other brewers were less reticent: the consultant George Blake, whose 'practical' persona was inspired by Richardson (chapter 3), and Thomas Harris, who directed brewing at the major London firm of Combe's, both marketed instruments created to their own specifications.[92]

As the market grew crowded, comparing and contrasting became a fact of life. John Martineau of Whitbread's was performing regular saccharometric experiments with Quin's device by 1806: he had tried, he said, 'every instrument that has ever been invented', and greatly preferred Quin's to Richardson's, which he believed registered some 5 per cent too low.[93] Thomas Fowell Buxton of Truman's, comparing strengths with the clerk at the Golden Lane Brewery, recorded his beer as '20lb by Richardsons Instruments or 20¾ by ours'.[94] Brewery treatises after 1800 recommend a wide variety of new and established devices – unsurprisingly, given that most were published at least partly to advertise apparatus, again following Richardson's lead.[95] Several challengers co-opted the name 'saccharometer', over which Richardson had no formal control.

The chief beneficiaries, in fact, were the successors of John Clarke, who had continued to market his 1740s brewery instrument alongside the official spirits hydrometer. Clarke's lucrative trade had passed by marriage to John Dring, who made some changes to the pattern of the brewery hydrometer around 1779. Following the success of the *Statical Estimates*, it seems that Dring simply rescaled this device in pounds per barrel: rechristened the Dring and Fage saccharometer, it dominated the trade in the early nineteenth century.[96] The Troughtons, whose instrument-making concerns lay increasingly in more prestigious fields such as astronomy, had little to gain from challenging this takeover. Although one author still recommended the Richardson–Troughton pattern in the 1830s, it probably fell out of common use soon after Richardson's death in 1815.[97]

Nonetheless, Richardson's theories and priorities survived. Significantly, although they appealed far more to commercial brewers than to philosophers, their survival was almost entirely due to Richardson's 'scientific' publications. In 1788, he pulled the *Theoretic Hints* and *Statical Estimates* into a single volume, *Philosophical Principles of the Science of Brewing*, with new editions in 1798 and 1805. Material from these texts was excerpted in the *Encyclopaedia Britannica* and quoted verbatim, at great length, in Richard Shannon's lavish treatise of 1805, to which many of England's leading brewers subscribed.[98] William Black, writing half a century after the *Statical Estimates*, considered Richardson's work so thorough that he 'need say little' of it, 'presuming that few brewers are without a copy'.[99] The *Statical Estimates* also had some influence in the German lands, a translation having appeared in Berlin in 1788.[100]

Just as James Baverstock had feared, Richardson became entrenched, in literate brewery culture, as the principal innovator in the measurement of beer strength: even those accounts which criticized Richardson's instrument acknowledged his theoretical contributions.[101] Richardson himself, meanwhile, prospered as a consultant and in his own brewing, living into a wealthy and respected retirement. A son, Philetus Richardson (1776/7–1838), was taken on by the great London brewing firm of Whitbread in 1805, becoming head porter brewer. The younger Richardson introduced various innovations, chiefly in temperature regulation: these were publicized in an encyclopedia account of 1819, which became a key source for depicting large porter breweries as icons of technological progress.[102]

What survived most clearly from the elder Richardson's scheme was the innovation which spoke most closely to the established needs of brewers: the definition of value in terms of the weight of extract, measured in pounds per barrel. Chemical philosophers, who usually measured specific gravity in parts per thousand, sometimes objected to the 'brewer's pound' as liable to cause confusion. The non-additive increase in volume when malt sugars dissolved was such that an actual barrel of wort, registering one 'pound per barrel' denser than water on the saccharometer, would yield approximately two and a half pounds of malt extract when evaporated to dryness. One version of the Dicas instrument, indeed, had a scaling which was claimed to correct for this effect.[103] Most brewers, however, stuck firmly to the brewer's pound: the distinction from dry weight, though it apparently still gave trouble to 'brewery pupils' a century later, was clear enough to those who had mastered the art.[104] Besides, as one author observed in 1829, it was no part of the brewer's business to boil his worts to powder. The quantity expressed by the measure was 'accurately true and obvious' to his daily practical experience.[105]

Extract-based valuation was particularly influential in comparing different varieties of malt. It had long been accepted that the hot-dried brown malts used in London porter yielded somewhat less malt sugar than paler malts, and accordingly they were somewhat cheaper.[106] Saccharometric investigation revealed that the difference in extract was far greater than the price differential suggested: pale malts were better value.[107] Richardson saw this finding as an important justification of the saccharometer's role, and he continued to compile extract results for each year's malts to 1797. Curiously, Richardson's figures do not unambiguously support his case.[108] The suggestion was followed up, however, by the chemist Thomas Thomson, who in 1805 and 1806 investigated extract values for the English and Scottish Excise. As we will see in chapter 6, Thomson rejected Richardson's and all other internal brewery theory as chemically illiterate, and yet he strongly affirmed the inferiority of brown malt: though costing seven-eighths as much as pale, it was perhaps 'on an average but one fifth' as extractible.[109] Commercial brewers in general took note, switching increasingly to paler malts. As demand for brown malt fell, the price differential grew, so that price became, up to a point, a function of saccharometric value.[110]

This broader change, we should note, was peculiarly a consequence of Richardson's scheme. Clarke, Martin and other makers had promoted hydrostatic investigation directly to brewers at various times since the 1730s; so had Baverstock, who enjoyed the advantage of approaching the brewers as a peer. All, however, had worked on the assumption that each brewer would apply the device privately: none had linked it to any general system of valuation. It was Richardson who spotted how the universalism of philosophy could be used to play on the hopes and fears of his audience. Reducing worts to their saccharometrically determined gravities gave the prospect of efficiency savings and the emulation of rivals' products, but also the concern that rivals might have similar thoughts. To Richardson, this scheme was in many ways secondary to his attempted monopoly on instruments: without the saccharometer, there would have been no *Statical Estimates*. In the event, however, it was only at the conceptual level that Richardson achieved lasting dominance.

Richardson's 'science of brewing' had significant unforeseen consequences, feeding into an ongoing controversy which I will discuss in the next chapter. To take the saccharometric project literally was to reduce the value of malt wort to the density of the sugars it contained. Did it matter, therefore, whether the sugars came from malt at all? Contemporary chemistry suggested not; most drinkers, and a belligerent daily press, felt otherwise. In the years around 1800, any invocation of 'chemistry' in connection with the brewhouse was dangerous, provoking immediate associations with a retreat from traditional purity towards adulterants which were fraudulent and probably toxic. In such a climate, any author professing to offer philosophically grounded improvements to the brewing process had to tread very carefully indeed.

5 CHEMISTS, DRUGGISTS AND BEER DOCTORS

> The new mode of compounding porter ... is now carried to the highest pitch of perfection. The compounders are saved all that trouble, which brewers were formerly put to, of keeping their casks and butts in cellars for a length of time, during which they had to wait until it became seasoned, keeping so much capital idle and locked up. All the seasoning stuff is now ready made to their hands, by a set of chemical gentlemen, whose trade it is to prepare it; as it is for others to take off the stale and sour beer; and by a process peculiar to themselves to manufacture it afresh ... This compound is taken and mixed with another liquid, recently made from a small quantity of malt and hops; and from the happy union of these ingredients that once famous liquor porter is now made.
>
> J. Malcolm[1]

In 1818, a House of Commons Committee heard evidence of suspicious goings-on at Barclay Perkins, the greatest of London's huge industrial breweries. A carpenter working for the young managing partner, Frederick Perkins, who had private rooms alongside the brewhouse, had noticed a still, bottles and containers of various chemicals, and had tipped off the officers of the Excise. Called before the Committee, Perkins explained that the apparatus had no connection with his business. It belonged to his hobby of chemical experimentation: he had been distilling rose water and attempting to repeat Humphry Davy's experiments, 'as published in the *Philosophical Transactions*' and as displayed in public lectures at the Royal Institution.[2]

Perkins's reply invoked all the standard features of respectable chemistry: untouched by trade; amateur yet systematic; guaranteed by association with an individual, a journal and an institution of high public standing. The problem was that where brewers were concerned, a very different set of associations had a powerful hold on the popular imagination. 'Chemical' beer was, by implication, drugged or doctored, undoubtedly fraudulent and probably poisonous. This chapter charts the development of this perception, which proved to be the single greatest challenge to the public credibility of nineteenth-century brewing science.

The problem's origins lay in popular awareness that beer was routinely adulterated. In chapter 4, we saw how uniform rates of beer duty fostered tax evasion by mixing strong beer with small, or simply watering it down, to increase its bulk. This operation left no trace, so prosecution was seldom possible unless a servant or partner could be persuaded to inform against the offenders: it could often be managed quietly in the publican's cellar or, in reported cases of collusion between brewers and Excisemen, in the brewhouse itself.[3] In years of poor harvest, when the price of malt and hops rose tremendously, watering was probably more the rule than the exception.[4] Drinkers' fears, however, focused not on watering itself, but on attempts to mask the resulting weakness using secret ingredients.

Fears of what might lurk in a familiar brew have a powerful cultural resonance, and provoked outrage and fascinated disgust long before the establishment of systematic chemistry. To understand the continuity, we might invoke the anthropologist Mary Douglas's observations on how definitions of what is 'dirty' or 'polluted' vary across cultures. Dirt, Douglas proposed, is not dirty in its own right: it becomes so by being put in places where, according to the presiding moral system, it has no business to be.[5] Dirtiness in this sense is central to the grotesque comic effect of John Skelton's sixteenth-century poem, 'The tunnyng of Elynour Rummyng'. The titular alewife's character is reflected in her brewing:

> And somtyme she blennes
> The donge of her hennes
> And the ale together;
> And sayeth, Gossyp, come hyther,
> This ale shal be thycker,
> And flowre the more quicker;
> For I may tell you,
> I lerned it of a Jewe,
> Whan I began to brewe.[6]

Chicken dung, in itself, is a blameless product of husbandry; added to the fermenting tun, it becomes not only unnatural but positively un-Christian, superficially excellent but secretly corrupt. In Skelton's time, the skilled countrywoman was the typical producer across the ale-drinking areas of Britain; her ability to practice secrecy, revealing herself only to the occasional trusted (female) confidante, invited fear.[7] The strong gender dimension in such fear, however, did not prevent its resurfacing in revised form, attached to the exclusively male large-scale producers who became dominant in large towns and cities across the eighteenth century. The common brewers might be 'gentlemen'; but they were tradesmen, too, and their systems of enclosed apprenticeship and trade secrecy often left outsiders wondering what exactly went on behind the brewhouse walls. Whereas Jeffrey Boys (chapter 1), in 1700, had been confident that the common brewers brewed ale in much the same way he did (if less competently), many urban markets were gradually dominated by porter-brewing from the 1720s: the London porter method was unknown to domestic and publican brewers, and subject to much speculation.[8] Where dung had fitted the bill for alewives, the 'matter out of place' suggested by the big common brewers' urban–industrial setting involved chemical manufacturers and pharmaceutical drugs.

Anxiety about the possibility of fair dealing in an increasingly anonymous urban society was summed up in Matthew Bramble's weary contrast between 'country comforts' and 'town grievances' in Tobias Smollett's *Expedition of Humphry Clinker* (1771). London, says Bramble, has turned away from 'what is genuine and excellent in it's [*sic*] own nature': its food and drink, putrefied over long journeys to market or concocted locally in dingy cellars, is sour, watered, contaminated with mud and human excrement; to hide these failings, it is unnaturally dyed and polished.[9] Smollett's account, though comic in tone, was based on various allegations which had circulated in all seriousness in newspapers and pamphlets. Extreme grain prices in 1757, for instance, had fostered stories that the millers or bakers were bulking out their product with alum, chalk and the ash of bones – including not only animal bones, but those which, in the words of one ghoulish pamphlet, 'ought to have been at rest'. The result of such unnatural additions, supposedly, was widespread fatal poisoning.[10]

The implications of such public representations for brewers and beer are traced through the first three sections of this chapter. Poison tales were a gift to satirists; accounts based on rumour and invention cross-fertilized with serious attempts to describe chemical additives, and with investigations into the alleged monopolies of major brewers which focused public attention on the moral character of the trade. To some campaigners, industrial beer-makers were no longer brewers at all, but druggists vending an artificial and toxic compound. Popular sentiment, and a minority of brewers, began to favour a strict purity definition which enshrined malt and hops as the *only* legitimate ingredients, a stricture which acquired the force of law in 1802.

In the remainder of the chapter, I discuss the consequences of these developments for the attempt to formalize the chemistry of the materials for brewing. A porous boundary separated trading druggists from gentlemanly chemical philosophers: the image of the malignant poison-peddler was quite strong enough to destroy a hard-won philosophical reputation, as I show by revisiting the case of Humphrey Jackson, ultimately branded a quack who had left the national drink 'ruined for ever'. On the other hand, a securely respectable chemist could sometimes banish drug-shop connotations entirely, and brewers were quick to recruit consultant lecturers and analysts, including Humphry Davy, as expert witnesses in test cases concerning the legality of ingredients.

Attempting to exploit and develop chemical authority in print, however, carried its own dangers. I conclude by exploring the multiple readings of the *Treatise on Adulterations* published in 1820 by Friedrich Accum, a prominent analyst, retailer and lecturer. Accum's strategy was to argue that chemical knowledge, far from promoting adulteration, provided the means to prevent and detect it; his sensational style quickly brought him celebrity status as a public-minded crusader. The systematic ambiguity of contemporary literature, however, meant that Accum could simultaneously be painted as a disingenuous drug-peddler: in public discourse, all discussion of suspect materials was suspect in itself.

The Politics of Adulteration

Many leading members of the brewing and allied trades were active in Parliament, and it had long been a standard satirical device to symbolize noxious politics by a noxious product.[11] In 1770, for instance, the two Members of Parliament for Southwark – a constituency long dominated by the industry – were Henry Thrale, the porter-brewer discussed in chapter 3, and Joseph Mawbey, a Wilkite Radical and prominent distiller. So lethal were Mawbey's wares, suggested one squib, that Southwark would have become an uninhabited rotten borough like Old Sarum, but for the antidote effect of the more moderate Thrale's wholesome drink.[12] Thrale was decidedly lucky that his interest in chemical innovations was not publicly known in his lifetime. Poisoning allegations, though vague and often purely facetious at first, became increasingly serious as the particulars of alleged drugging began to circulate in print.

Particularly influential was a sixpenny pamphlet, issued sometime around 1790 in the name of 'Samuel Child, Brewer', purporting to explain how householders could brew commercial drinks for themselves by imitating the 'deception' practised in the trade. Child's recipe for porter included cocculus indicus, or Indian berry, the fruit of a climbing plant imported from various parts of Asia and classed by physicians as a narcotic poison. Other ingredients were treacle, as a malt substitute, preparations of burnt and boiled sugar for bitterness and colour, and flavourings including liquorice, capsicum, ginger, linseed and cinnamon. Finally, 'heading', a mixture of alum and copperas (ferrous sulphate), was added to give the drink a convincing frothy head. These ingredients, said Child, 'however much they may surprize', were essential to mimicking commercial porter.[13] The pamphlet was widely read, running through eleven editions to 1810 and attracting hack imitation.[14] Child's list of additives became a standard resource for satirical attacks on brewers across the early nineteenth century.[15]

Most of the additives mentioned were, in fact, generally accepted as non-toxic, being familiar foodstuffs or pharmaceutical remedies. As applied to beer, however, they were read as part of a scheme whose overall effect was to engender poisoning. The inclusion of cocculus indicus, if nothing else, confirmed this: beer laced with cocculus, which supposedly simulated inebriation, had been notorious since the 1730s ('I once drank half a Pint of it in a Forenoon, and it gave me twenty four Vomits and fifty Stools that Day').[16] Displaced from their standard uses, liquorice, capsicum and so forth took on a similar aspect. So did quassia bark, widely imported from South America as a medicine against fevers, but also intensely bitter and useful as a cheap hop substitute. Though not mentioned by Child, quassia was apparently added to the customary litany of additives around the mid-1790s through organized protest by the hop dealers of Kent, whose campaign turned the known curative properties of the substance against it: 'will not any medical decoction ... from constant use in beverage, become a poison to the patient it had in a degree cured?'[17]

James Gillray's cartoons of the first decade of the nineteenth century repeatedly show parliamentary brewers encased or assailed by barrels prominently labelled 'Quassia' or 'Cocus [*sic*] Indicus'.[18] The most elaborate play on this conceit is the 'Triumph of Quassia' of 1806 (Figure 5.1), inspired by an unpopular proposal for a tax on home brewing from Lord Henry Petty, Chancellor of the Exchequer in the recently installed Ministry of All the Talents.[19] Petty and the two senior Whigs in the administration, Grenville and Fox, ride a brewers' dray horse at the head of a parade of three brewer MPs with strong Whig sympathies: George Barclay, the second Samuel Whitbread and Harvey Combe, representing three of the top five porter breweries by output. Barclay has a book of 'Receipts to make a Cauliflour Head', recalling Child's pamphlet: his product needs artificial froth because it is adulterated. While honest home brewers must suffer the tax, the porter-brewers proudly escort a quassia barrel which confirms them as tax dodgers. Just as the spent grains from mashing are given to pigs, so, as home brewing is priced out of existence, the profits of increased trade will pass to the 'New Piggery' and its cronies in Whiggery.

The allusions in such images could be complex. The black figure seated above the barrel is presumably Quassi, the source of the plant's name: born in West Africa and shipped as a slave to Surinam, Quassi was reported to have discovered the medical properties of the wood around 1730, later gaining his freedom and corresponding with European naturalists.[20] The reference is surely to Whitbread's leading role in the campaign to abolish the slave trade, and the most obvious reading is anti-abolitionist: quassia, like Quassi, is culturally alien, and its free circulation is to be feared. An extensive list of ailments including apoplexy, consumption and scurvy radiates from Quassi's quart-pot like a halo.

Adulteration also had more local political meanings. As the industrial porter-brewers achieved ever-greater output, their product replaced that of the traditional publican-brewers, and also of the many smaller common breweries they acquired and closed. This concentration of the trade was tacitly promoted by the Excise, since it made the collection of duties easier. As they began to act by committee to enforce general rises in the price of beer, the dominant brewers came to be seen as a cartel. Public objections came from an unusual coalition of publicans, clergymen and free-trade campaigners, who alleged that a lack of competition and effective regulation fostered not only price-fixing but brothel-keeping, public drunkenness and routine toxic adulteration. Similarly, they claimed, brewers in country towns were monopolizing the supply of the local pubs, leaving the retailers no choice but to sell poor-quality adulterated beer.[21]

Figure 5.1: James Gillray, 'The Triumph of Quassia' (1806). Left to right: George Barclay; Harvey Christian Combe; Quassi (?); Samuel Whitbread II; Lord Henry Petty; William, Baron Grenville; and Charles James Fox. Reproduced courtesy of the Wellcome Library, London.

These complaints were summed up in a major petition to Parliament organized by John Barber Beaumont (1774–1841), an insurer and property developer who had been refused licenses on pubs intended to serve his newly built housing. The greatest brewers, Beaumont asserted, had carved up the London trade between them by systematically corrupting the magistrates responsible for granting licenses. The petition resulted in an 1817 Commons enquiry which put numerous specific allegations onto the public record: Truman Hanbury, in particular, was said to have a stranglehold on the trade around Brick Lane through Joseph Merceron, a local kingpin and notoriously corrupt magistrate who was jailed soon after the hearings.[22] In 1818, a follow-up investigation focused on the allegations of price-fixing and the 'inferior quality of Beer', bringing many successful prosecutions for the use or possession of banned substances to public attention.[23]

Shrewdly, the major brewers did not downplay the abuses, but deflected the blame onto their smaller competitors and the remaining publican-brewers. The 1818 Committee itself included two of the leading brewer MPs of the day, Charles Barclay and Charles Calvert, who were able to steer the framing of the investigation to sharply distinguish the eleven greatest breweries, with recorded annual production above 40,000 barrels, from the rest.[24] The major brewers' case was that they had neither motive nor opportunity for the use of drugs. Their enormous efficiencies of scale meant they had no need to resort to unlawful measures; their continuing success was also a function of their reputation for honest dealing, and to risk the publicity of a prosecution was unthinkable. Moreover, it would be impossible to hide drugs in the necessary quantities. These arguments were, unsurprisingly, endorsed in the 1818 Committee's report: while the greatest brewers' power might not be in the public interest, it concluded, there was no case to answer regarding the material and moral purity of their beer.[25]

In reality, as private correspondence indicates, some members of the 'great eleven' at least experimented with additives.[26] Their public protestations were less than universally convincing. One former Exciseman had testified that evading detection would be far easier in the large brewhouses than anywhere else, citing the availability of workers to keep lookout, the rambling networks of outlying storehouses and the systematic organization which could allow even bulky adulterants to be shuttled from outside the brewhouse into a working vat before the Exciseman had a chance to notice: he had known twenty barrels of molasses to be 'got rid of' within five minutes.[27] Great brewers such as Whitbread, moreover, had built their fortunes and reputations on the ability to abolish traditional limits, sponsoring steam-powered automation and unprecedented feats of civil engineering.[28] The question of how far these brewers also sponsored chemical innovations was a source of unquenchable controversy.

From 'Malt and Hops' to the 'Modern Druggist'

The suppliers of drugs and other additives, on the other hand, initially felt quite safe in encouraging the trade. The law was framed to prohibit use or storage by brewers, rather than sale or promotion; outside the brewhouse, many of the ingredients did not provoke suspicion, having authorized uses in the distillery or elsewhere. The enterprise was widely reported to be on a systematic footing, as 'brewers' druggists' sent travelling representatives around the brewers and publicans with lists of the available ingredients.[29]

In 1798, the Board of Excise was concerned to note the emergence of a 'certain description of Persons commonly called Beer Doctors', who collected 'useless' (sour) beer from publicans, combined it with 'other Ingredients', re-fermented it and sold it as fresh. Because beer doctors did not, in the literal sense, brew, their trade was not subject to duty, and it was not evident that any of the laws governing commercial brewers applied to them. By 1800, the Excise was reporting frantically to the Treasury that a single beer doctor had sent out 'no less than four hundred Barrels' on a single day.[30] The loophole was closed in 1802 by an Act of Parliament which also raised the beer duty from 8*s*. to 10*s*. per barrel and almost doubled the malt tax. Besides forbidding any trader whatsoever to 'mix, compound, fabricate, manufacture, or prepare ... any Liquor to imitate or resemble, or to be mixed with or used as Beer or Ale', the measure made it explicitly illegal to use in beer any 'Material or Ingredient whatever, (except Malt and Hops)'.[31]

The 1802 Act gave legal force to the established popular assertion that moral probity meant 'Malt and hops only'. This easily digested rallying cry was adopted strategically by firms such as the Golden Lane Genuine Beer Company, set up on the rising tide of anti-monopoly rhetoric in 1804. Effectively a joint-stock project, capitalized by hundreds of publicans and other investors, the Golden Lane enjoyed enormous early success by undercutting its price-fixing rivals while stressing – through devices such as the use of the initials 'M. H.' as an identifying mark on its casks – its commitment to malt-and-hops purity.[32] Similar guarantees came from smaller brewers such as John Walsby of Battersea, who dealt mainly with private families and export customers at the upper end of the retail price range: Walsby's advertisements reproduced affidavits of purity sworn before the Lord Mayor of London.[33]

Such boasts could easily backfire. On the eve of Peterloo, Manchester was placarded with claims that Henry Hunt, the ex-brewer turned Radical 'Orator', had proudly advertised beer free from '*any other ingredient whatever*' in 1807, only to be convicted of possessing adulterants the same year. (Inevitably, Hunt was accused of 'infusing poison into the *Constitution*' of his clients.)[34] A cartoon of 1810 shows a framed notice, proclaiming 'Genuine Beer From Malt & Hops Only', with druggists' bills for quassia and wormwood, another unlawful bitter, tucked behind it.[35] Accusations of hypocrisy could have serious consequences for a brewer's reputation, as we will see later.

The rejection of malt-and-hops purity had a rhetorical life of its own. Complainants since Smollett had invoked the spectre of beer not merely adulterated, but absolutely 'guiltless of hops and malt'.[36] In 'The Triumph of Quassia', for instance, George Barclay's placard proclaims the complete abolition of the two core ingredients. This was dramatic licence: total substitution would have produced a markedly different taste, and there is no evidence that any brewer achieved it.[37] Many consumers, however, would have believed the Kent hop-planters who complained in 1796 that the chief ingredients in beer were now molasses, cocculus, quassia and other drugs, with 'little malt, and less hops' retained to conceal the deception.[38] Drastic fluctuations in malt prices around 1800 stoked fears that the urban brewers, too powerful to fear either public sentiment or the law, were turning decisively away from the traditional materials of agriculture in favour of drugs.[39]

Part of the concern lay in chemists' extensive familiarity with techniques of extraction and concentration, which offered a counter to the argument that detection would be unavoidable in large-scale brewing. Cocculus indicus, suggested one Exciseman, could be extracted and refined to a crystalline solid so potent that a single man might hide a quantity 'sufficient for any brewery in England'.[40] Such intense concentration might also be useful in managing knowledge of the abuses. One lurid pamphlet narrates how 'the head-brewer ... in the dread moment of the poisonous immersion, turns all the usual assistants out of the brewhouse, and does the deed of ruin alone'.[41] This tactic minimized the opportunities for disgruntled ex-workmen to inform the Excise, and meant the remaining staff could not be obliged to explain the incident under oath.

Fears were heightened, too, by unfamiliarity with most of the druggists' stock-in-trade. The social reformer William Cobbett's equation between urban living and the collapse of trust and decency strongly recalled the plaints of Matthew Bramble: a central theme of *Cottage Economy* (1822) was that householders must revive home brewing along traditional lines to know what they were drinking. Describing a porter recipe which called for liquorice, capsicum and salt of tartar, Cobbett exploded:

> What a curious composition! What a mess of drugs! But, if the brewers *openly avow* this, what have we to expect from the *secret practices* of them and the *retailers* of the article! When we know, that *beer-doctor* and *brewers'-druggist* are professions, practised as openly as those of *bug-man* and *rat-killer*, are we simple enough to suppose that the above-named are the *only* drugs that people swallow in these potions, which they call *pots of beer*?[42]

One answer to this question had been widely publicized in 1807, when a paragraph in the *Morning Post* asserted that 'the principal part' of all porter was made up of pharmaceutical poisons. The familiar cocculus indicus was joined by opium, deadly nightshade, thorn-apple, houndstongue, nux vomica (whose active principle was strychnine) and tobacco; the porter-brewers, allegedly, had single-handedly caused an 'alarming increase of paralytic complaints, and sudden deaths in the Metropolis'.[43] Recalling the most melodramatic contributions to the 1750s bread scare, the text was probably a straightforward piece of mischief-making: of the alleged ingredients, only cocculus, and occasionally nux vomica, appear in the records of seizures and convictions, and it is hard to imagine most of the others as useful substitutes. It was a serious business, however, for the brewers of London, who moved for criminal libel actions until placated by a penitent editor.[44]

The story proliferated, nonetheless, to become a source of grim amusement. Isaac Cruickshank portrayed a porter-brewer (Figure 5.2) gleefully introducing the demon brood of poisons to his vat, while his two legitimate children – malt and hops – sit neglected. The operation is imprudently performed in view of an open window, through which peers an astonished John Bull: 'be I to take all those Fellows in my guts', he laments, 'why I shall ne'er want any more Physick'. The porter-brewer, like the beer-doctors, is really no brewer at all: he is a 'Modern Druggist', and, like the stock figure of the quack apothecary, he will as likely kill as cure.

Purity in Principle and Practice

The sharp dichotomy between 'traditional' malt-and-hops purity and the synthetic evils of the drug shop found little support, however, among those with specialist knowledge of brewing. Historically, a wide range of ingredients had been legitimate, many of which had nothing to do with druggists. William Harrison's sixteenth-century process account (pp. 16–17) included unmalted wheat and oats in the grist, with orris root and bayberries for flavour; the published process narratives of the following two centuries recommend non-hop bitters such as wormwood, and colouring materials ranging from horse beans to elderberries.[45]

The definition of malt and hops as the only proper ingredients derived not from tradition, but from legislative efforts to protect, firstly, the barley-growers, and latterly the revenue, by preventing the substitution of these increasingly highly taxed commodities. Though the wording of the law sometimes mentioned the 'unwholsome' nature of materials such as cocculus indicus (banned in 1701), toxicity was never a primary concern: molasses, as a malt substitute, had been banned before cocculus and was subject to harsher penalties.[46] Moreo-

Figure 5.2: Isaac Cruickshank, 'The Porter Brewer and his Family – or the Modern Druggist' (1807), © The Trustees of the British Museum.

ver, traditional additives remained in common use among private householders: private brewing had never been taxed, a position which was secured by the defeat of Petty's 1807 proposals.[47]

Some brewers, therefore, publicly supported 'wholesome' additives as beneficial and necessary. When the price of malt could double in a year, argued one country brewer in a pamphlet of 1802, the only alternative to substitution was 'the shutting up of his brewhouse'.[48] Substitutes for the corn used in bread were welcomed in times of scarcity: why not the same for beer? Brewers also defended an assortment of long-established additives which had never been understood as malt or hop substitutes, but were intended to correct, or prevent, perceived technical deficiencies. William Ellis had recommended 'Balls' of his own devising, containing isinglass, molasses, alabaster and spices, for 'fining, relishing and improving' the beer.[49] Isinglass had been known to brewers since the seventeenth century, and was deemed indispensable as a fining agent for porter by the 1750s.[50]

Brewers' rejection of malt-and-hops purity did not necessarily imply acceptance of the druggists' overtures. James Best of Chatham was content to use liquorice and grains of paradise, a harmless but unlawful African spice, in the 1770s; he was nonetheless disgusted to hear of a Lowestoft brewer profiting by the advice of 'a low, dirty fellow from London', who had taught him to use 'certain Ingredients' to undercut his rivals. These included molasses – for which the miscreant had been detected by the Excise, escaping with a light fine owing to personal influence – and an unidentified 'fine, brown powder very heavey [*sic*]', introduced at the fermentation stage. Best's informant told him that the beer fermented with excessive force in the cask, while the yeast was 'almost like Clay' and rejected by the town's bakers, who would normally have bought it as a raising agent.[51] Such innovations could easily appear as a threat to the skills and conventions on which an established brewer's livelihood depended.

The legislature, meanwhile, repeatedly varied the list of banned substances in response to a variety of factors, including the brewers' pleas: cane sugar was briefly permitted as a substitute for malt during the dire shortages of 1800 and 1801.[52] Non-substitute materials which did not affect the revenue, such as isinglass and heading, were typically ignored until the strict Act of 1802. Even this measure, however, did not necessarily interfere with their use. The practicalities of regulation were often less to do with the material nature of the ingredients, than with the circumstances and connotations of their use. In formally prohibiting all ingredients *whatsoever*, except malt and hops, the 1802 Act hardly encouraged literal enforcement. Water, by common sense, was 'no doubt understood' to be permitted.[53] So were yeast and finings, provided their use followed the established pattern of brewery practice under Excise supervision, which had settled into common law.[54] In managing the grey area, there was often conflict between the Treasury, which ultimately received the taxation revenue, and the

Commissioners of Excise who collected it. The Treasury tended to tolerate new practices so long as its income was secure, and sometimes acted to bring them into the hands of licensed monopolies, preferring an easily monitored licit trade to unenforceable prohibition. The Excise, more directly acquainted with the brewhouse environment, scrutinized every change in terms of its potential to conceal frauds and disrupt the bureaucracy of collection.

This tension is illustrated by controversies over the status of artificial colouring under the 1802 Act. The dark colour of London porter – and of the imitations that had become established in many provincial towns – had originally derived from high-dried brown malt; from the 1780s, however, the proportion of pale malt had increased sharply, as John Richardson's demonstration of its higher fermentable yield (pp. 105–6) became widely known, and brewers resorted increasingly to artificial colour to maintain the traditional appearance.[55] The customary colouring material was highly roasted or burnt cane sugar, although a similar preparation made from evaporated malt extract could be used instead: the name 'essentia bina' was apparently used for both. Essentia had been explicitly banned since 1701, as an aid to 'evil practices': though generally considered non-toxic, it was historically the first resort in hiding the fact that beer had been watered down.[56] The brewers' claim that recent technical improvements had reduced the coloration of their product *without* reducing the strength, hence the need to legitimize additives for correction, was unsurprisingly unappealing to the Excise.

The sweeping restrictions of 1802 appeared as an opportunity to Matthew Wood (1768–1843), a former druggist's traveller now established as a wholesaler in Cripplegate. Wood enjoyed a large trade with brewers, and was soon to recruit a hop dealer, Edward Wigan, as a partner. Wood reasoned that the 'Malt and Hops' provision legitimated colouring made from malt extract, and, as the Act came into force, patented a process to produce it.[57] His specification included little more than a general description of the customary (long-prohibited) method, but Wood had a personal assurance from Nicholas Vansittart, Secretary to the Treasury, that his particular practice would be deemed legal. This should have amounted to an unusually watertight monopoly: any rival colour-maker's process would be deemed to infringe either the patent, if it resembled Wood's, or the Act, if it did not.

If this arrangement suited Vansittart, however, it did not suit the Commissioners of Excise. Taking the line that 'Malt' encompassed only products customarily made by maltsters, they systematically prosecuted brewers who had unsuspectingly bought Wood's colouring. When the brewers protested that the colouring was harmless to health and revenue, the Commissioners retorted that this was beside the point: leave to add *anything* to beer would be used as cover for more harmful additives, as past cases showed. The brewers, they said, should attend to the colour of their beer themselves: placing it in the hands of a druggist would 'open the widest Door for introducing Ingredients forbidden by Law as well as of the most pernicious Effect'. Raids at Wood and Wigan's in August 1806 served to confirm this point: the Excisemen found thirty bags of cocculus and quantities of Spanish liquorice, grains of paradise and chilli peppers. Wood and Wigan pleaded that the Excise was wilfully misrepresenting the nature of their business: they dealt largely with apothecaries, grocers and dry salters, who had legitimate uses for the materials seized.[58] The Excise achieved its purpose, however, impounding the utensils for preparing the colouring and breaking up the trade.

Wood, nonetheless, continued to deal in other commodities to the brewers. As a Radical Whig who served as Sheriff of London (1809–10), Lord Mayor (1815–17) and thereafter as an MP, he operated under constant attack from political enemies who regularly drew attention to his trade with brewers said to include Whitbread and Combe.[59] When one of the huge porter vats at Henry Meux's brewery in St Giles collapsed in 1814, causing a flood of beer, major structural damage and eight fatalities, a facetious poem reported it to have 'exploded of the drug-gripes', following a constant diet of Wood's 'nauseous Physic'.[60]

In its practical operation, then, formal regulation was approximately in tune with popular sentiment: innovations involving druggists and patentee projectors were presumptively fraudulent and probably poisonous. The legislature, however, was by no means wedded to the strict purity definition. By 1811, the Treasury was seeking the re-legalization of sugar-based colouring, so as to encourage the brewers to use less brown and more pale malt, experience having shown that the much slower pale-drying process was 'more open to the Observation and vigilant Attention' of the Exciseman.[61] The move was also influenced by pressure from the West Indian sugar interests, which eventually inspired a general revived permission of sugar from 1812.[62] Cane sugar, though not malt, was relatively unproblematic, being evidently agricultural and a product of Empire; Wood's colouring, though made of nothing but malt, was rejected chiefly because it was an industrial contrivance. The challenges that Wood faced, both legal and reputational, were faced to varying degrees by all the innovators who tried to modify the materials for brewing, and strongly constrained the opportunities available to those who claimed philosophical authority.

The Challenges of Chemical Instruction

Philosophers had a long-standing interest in brewing from unfamiliar substances. Samuel Hartlib's correspondents in the 1650s had debated the possibility of making drinks 'not inferiour to *Wine*' by fermentation of peas, oats and other crops. Already, departure from custom was attracting scepticism: one of the group decried the 'Mountebank boastings of *brewing without malt, and of drawing great store of Aqua-vitae out of the imaginary Beer*'.[63] The investigation was nonetheless carried into the early Royal Society by Hartlib's disciple Lord Brereton and others; their work was an influence on agricultural improvers such as John Mortimer, who in turn influenced the brewing experimentalists of the eighteenth century.[64] In categorizing fermentable substances analytically, many saw 'sugar' as a uniform principle: questions of legality aside, it did not matter whether it came from malt, molasses or other materials, and determining the most cost-effective source was an obvious project for a commercial chemist. Similarly, the manipulation of the properties of drink – forwarding or checking fermentation, dispelling cloudiness, correcting acidity or colour – was an established focus for philosophical enquiry and experiment.[65]

Chemical philosophy, then, tended to reject the malt-and-hops purity which popular sentiment increasingly demanded, and the law intermittently dictated in commercial beer. This was unproblematic when writers confined themselves to interpretation, and did not invoke chemical authority to promote reforms of the brewing process. Michael Combrune, as we have seen (chapter 2), focused on chemistry as a means to make existing drinks more reliably; he noted the existence of cocculus and other additives, but largely condemned them.[66] The instructors and authors who followed him, however, needed to demonstrate useful innovations in practice to gain an audience. Their reliance on strategically withholding key details raised awkward questions among brewers and non-brewers alike.

We can get a good sense of the tensions involved by revisiting the case of Humphrey Jackson, the manufacturing chemist discussed in chapter 3. Jackson, in fact, had published several condemnations of toxic additives from the time of the 1750s bread scare. Responding to the hysterical pamphlets which had carried the 'bone' allegations, Jackson had made shrewd use of modest language and analytical authority to acquit the bakers of blame while presenting evidence of harmful substances in other foodstuffs, including beer.[67] His warnings, which included descriptions of chemical tests for the presence of copperas and oil of vitriol, reappeared in both his public and private writings for brewers.[68]

At the same time, Jackson was content to recommend any illegal addition he considered harmless. His 1762 prospectus, distributed privately to brewers, promised guidance on 'the best Methods of preparing an effectual wholesome colouring Matter for Beer, with and without Sugar'. In work published at large, his wording was more cautious. Pronouncing the management of colour 'not a little arduous, since we are debarred the Use of Sweets', he wrote as follows:

> as the Description of the Properties, which a true colouring Matter for Beer ought to possess, may afford Hints to others interested in Pursuits of this Nature, we give them as follows: *Cheap, wholesome, unfermentable, nearly inodorous, unalterable by long Time, commiscible with Fining, indiscoverable by Taste, and imparting no yellow Tinge to the Head, or Spume of Beer*. That such a colorific Liquor can be prepared ... we can warrant by Experience.[69]

At one level, this was typical trades-improvement language which could be justified as part of an enquiry into principles. Jackson clearly intended, however, that brewers should read between the lines to find a proposal for private instruction in a banned practice.

This dissembling strategy was risky, since it fostered the suspicion that Jackson's objections to 'unwholesome' ingredients were likewise merely a matter of public show. This was one of the many thrusts against Jackson's moral character in John Richardson's pamphlet of 1775 (chapter 3). Jackson's professed ability to mimic the properties of local beers, Richardson insinuated, rested less on the analysis of malt and hops than on '*mundungus* [tobacco], *wormwood*, and *molosses*', while a private proposal to instruct on '*inebriating additions*' spoke for itself.[70] We may note that Richardson himself transmitted privately, under promise of secrecy, directions for the use of vinegar to simulate the effect of age, aloes and quassia to simulate hop bittering, orange pea for flavour and salt of steel for heading.[71] Jackson's status as a trading chemist and a non-brewer did not necessarily make him keener to promote additives than his brewery rival, but popular convention suggested otherwise, and Richardson played up the associations for all they were worth.

The final collapse of Jackson's reputation, however, had more to do with his perceived trespass against the non-brewers' concept of malt-and-hops purity, stemming from the failure of his patron Thrale's production in 1772 (pp. 68–71). Few details of this incident were ever reported; when Hester Piozzi published her collection of Samuel Johnson's letters in 1788, however, she noted in passing that

> H—ph—y J—ck—n was a pretender to chymistry, who obtaining much of Mr. T[hrale]'s confidence, used it to his own emolument only, not his friend's, who suffered exceedingly from the experiments made by him at the brewhouse, in consequence of a notion that he had some secret to preserve wood from decay, and brew at a smaller expence than was possible with malt or hops.[72]

What did Jackson's experiments really involve? From the content of his prospectuses, it seems likely that he was experimenting with non-traditional malting practices, or attempting to substitute part of the malt with unmalted barley.[73] Another possibility is that Jackson's wort concentrate scheme (which used no *fresh* malt or hops, but only extract) was somehow involved.

Piozzi's charge, however, was that Jackson had tried to remove the two core ingredients from the process entirely and substitute artificial materials. Thrale, as a man born to the trade, should have known that this could only end in disaster; Jackson, apparently, had beguiled him into believing the impossible through quackish displays of chemical showmanship.[74] This narrative quickly established itself as an origin story for abuses in general. One account of 1795 stated that in the days of cheap malt and hops, deleterious additives had been known at 'but one brewhouse in London', spreading to others as prices rose. A later elaboration identifies 'the rival of the celebrated Whitbread', by implication Thrale, as the rotten apple.[75]

As this account circulated in the years after Jackson's 1801 death, the guardians of chemical authority took steps to affirm that Jackson was, indeed, a mere 'pretender', and not a chemical philosopher at all. Given his apothecary's background, focus on commercial projects and the lack of a systematic treatise bearing his name, it was easy enough to portray him as a mercenary druggist who had somehow wormed his way into the Royal Society with a borrowed display of learning. This characterization first appeared in the 1809 abridgement series of the *Philosophical Transactions*, when the editors took the unusual step of qualifying the inclusion of Jackson's isinglass work with a biographical note relating how,

> speculating on schemes how at once to make a great fortune, [Jackson] fell on that of brewing porter by certain drugs substituted as materials instead of malt and hops. With these he set up as a general instructor of the brewers ... by giving private lessons in the art, at an enormous premium. This art it seems they have, in most instances, practised ever since in so extensive a manner, as to have produced a general complaint, that the ancient national malt liquor is miserably degenerated, with universal execrations on the memory of the man who could be so wicked as to introduce a practice, in consequence of which the natural beverage of the country has been ruined for ever.[76]

Jackson's moral standing was further undermined by reference to his role later in life as a 'trading justice' or magistrate, extorting fines for trivial offences.[77] This account was carried through into Thomas Thomson's *History of the Royal Society*, which cast doubt on the originality of the isinglass research, and to Thomson's entry on 'Brewing' for the *Encyclopaedia Britannica*.[78]

To draw a boundary between good chemists and bad druggists was to accept that some kinds of chemical intervention were legitimate. This was the position of the Boards of Excise: as we will see in chapter 6, they relied increasingly on chemical experts, including Thomson, for authoritative determinations of the value of materials. In somewhat similar fashion, 'scientific' brewer-authors such as Richardson distinguished the philosophically informed practicality of their own work from the mercenary opportunism of their non-brewing rivals. To sell *knowledge* in secret was legitimate, according to Richardson; Jackson, however, refused to communicate the method for his beer-colouring at all, instead selling it ready-made, an arrangement which could be in neither the brewer's nor the drinker's interest.[79]

Such distinctions did not necessarily carry any weight in public discourse. Many drinkers would have agreed with the ex-brewer who in 1807 wrote of 'scientific and chemical brewing' as a byword for toxic adulteration, lamenting the disappearance of a culture founded on 'plain rules', brewhouse experience and the generous use of malt.[80] It was not the beer-drinking public, however, who determined what was permitted. Regulation in practice was determined by a pattern of case law which could, on occasion, be strongly directed by appeals to chemical authority.

Expert Authority and Experiment

In May 1809, Excise officials detected half a dozen casks of a mysterious and malodorous substance being transferred from the warehouse of James Butcher, a yeast-dealer, to the premises of the nearby Golden Lane Genuine Beer Company. The proprietors turned out to have engaged Butcher to buy up the skins discarded by fishmongers and dissolve them in stale beer, to make an experimental substitute for isinglass. The Company was duly prosecuted under the Act of 1802. The Solicitor General went to town on the peculiarly disgusting nature of the innovation:

> Keep your stinking fish to yourselves ... If [the Company] contend they have a right, by law, to fine their beer with the offals of the fish market, because it will fine it, I will not enter into a chemical discussion whether it will fine it or not ... it is enough for me to say the law upon the subject will not let you put stinking fish into your beer ... I will not hold my nose, I won't drink it, because the law says I am not to have any of this stinking stuff put into my beer.[81]

Public knowledge of the seizure – following which the seized casks stood openly in the Excise Office yard, their stench increasingly apparent – was particularly embarrassing for the Golden Lane Company because of its founding promise of a 'genuine', unadulterated product; the episode followed earlier, unproven rumours of 'pernicious ingredients'.[82] Through the pages of the scabrous *Satirist*, scrutiny focused on the Company's founder, William Henry Robert Brown, as a shady projector involved in a worryingly wide variety of schemes.[83]

Given the Golden Lane's price-cutting tactics, we might suspect that the prosecution was sponsored by its established competitors, who had already used their influence in Parliament to intensify problems surrounding the firm's complicated corporate status.[84] In fact, the Excise was equally willing to target the old-established brewers. A similar raid followed at Calvert's of Thames Street, where officers alleged that pieces of skate had been substituted for the customary isinglass. The Calverts evidently had some pull with the Treasury, which responded frostily to this display of Excise zeal; the Solicitor of Excise, however, pointed out that Calvert's had previously been prosecuted for making molasses-based colouring in secret. Ultimately, the action was dropped, although the damage had already been done. Calvert's fining-house stood in a highly public location, near Allhallows Stairs and the busy pathway running alongside the Thames west of London Bridge. The flurry of Excise activity provoked much gossip and at least one practical gesture, in the form of a trail of dead fish strewn between brewery and street.[85]

In such circumstances, however much its competitors despised the Golden Lane, their common interest was threatened by any public prosecution of a major brewery for illicit additives. Moreover, it appears that the trade in general had a shared interest in the fish-skin process. Its promoter was the engineer William Murdock (1754–1839), chief technical supervisor to Boulton and Watt: Murdock had overseen the installation of the first steam engines at several large breweries, and was around this time experimenting with a wide variety of chemical and mechanical inventions. He claimed personally to have devised the fish-skin process, which – presumably not trusting to his chances of enforcing a patent – he disclosed to a consortium of brewers, under conditions of secrecy, for a one-off payment of £2,000.[86]

Called for the defence at the Golden Lane trial, Murdock testified that he used the substitute in his own family, and that it was 'exactly the same thing' as isinglass, not only functionally but materially. Another witness, William Yeates, who had seen the finings soon after seizure, testified likewise that they were unexceptional and wholesome. Whereas the Excise case, as usual, focused on circumstances and opportunities, the defence argued purely from material composition: isinglass and the substitute being equivalent, it could make no sense to sanction one while hounding after the other. The prosecution responded by questioning the witnesses' authority. Under cross-examination, Murdock stated that he was neither a physician nor a 'professed Chymist'; Yeates, that he was an ex-brewer who now dealt in wine. In pointing to the absence of chemical expertise, however, the prosecution played straight into the hands of the defence, which was about to produce its trump card.

Though not yet the grand old man of science, Humphry Davy (1778–1829) was known throughout the international philosophical community as a formidable research talent by 1809; he was also, as Professor of Chemistry at the Royal Institution, at the height of his fame as a dazzling public lecturer. In substance, Davy's evidence simply confirmed that of Murdock and Yeates: true isinglass and the substitute were materially one and the same; neither was harmful in normal use; such substances should not be considered additives, because they sank to the bottom of the beer and precipitated out. Davy had made no independent trials, had not been briefed on the particularities of the brewhouse and was unable to answer to specifics under cross-examination. Nonetheless, his confident affirmations (or perhaps the simple fact of his status) apparently made a powerful impression on the presiding judge, Archibald Macdonald, who waved away the defence's intention to call another chemical expert: 'After Mr Davy you need not trouble yourself'.[87]

In vain, the Solicitor General pleaded that a dismissal would mean violating a key Excise principle: that 'the conduct of those in the trade should be universally known'. If beer was made 'of any thing the ingenuity of Chemists can think of provided the fineing will at last get to the bottom', innovation would drift beyond the safe cognizance of the authorities.[88] In finding for the defendants, however, Macdonald considered Davy's word authority enough. It was unreasonable, he said, to object to practices on the grounds that they had never been used elsewhere, or were not customary: for an innovation 'spoken of by men of science as a very advantageous discovery', there must be legal protection for experiment.[89] Ultimately, in 1817, the letter of the law on finings was brought into line with this view. The 1802 purity definition was replaced by a ban on substances intended 'as a Substitute for Malt or Hops', indirectly permitting all fining agents.[90]

Davy, like most of the prominent chemists of his generation, drew a significant professional income from analysis and expert testimony, and the Golden Lane's backers presumably paid handsomely to get him onto the witness stand. His public reputation, however, was firmly that of the disinterested man of science. Davy stood as a symbol – perhaps *the* symbol – of polite, authoritative public chemistry, imbuing the innovation with a set of associations quite different from those of the drug shop. Though the fish-skin process was quite obviously a commercial proposition, the combination of Murdock and Davy allowed it to be seen as ultimately disinterested, part of an agenda of import substitution and rational improvement which owed something to the template laid down by Peter Shaw and the Society of Arts.

The Excise objection retained some of its force, however. In 1813, Meux Reid of Liquorpond Street – then second only to Barclay Perkins in production – was convicted of receiving casks containing a solution of salt of tartar, prescribed by a London chemist, Daniel Wheeler, as a remedy for persistent problems of acidity in the beer. Meux had originally engaged Wheeler to supply the colouring legalized in 1811: this was also supplied as a casked liquid, and carried by the same cart driver. Barber Beaumont made particular play of this incident at the 1818 hearings, since the intention to conceal was obvious on both sides. Wheeler had sent the illicit consignments out extremely early in the morning to avoid scrutiny, while Meux's employees had marked the casks to suggest they contained beer returned as sour.[91] Such incidents inevitably raised questions about what else might be changing hands: one later story had it that the druggists had taken to packing cocculus indicus in soda barrels, with a thin layer of soda to defeat casual inspection.[92] The legislation which came into force in 1817, while it abolished the purity requirement, also restored the ban on sugar colouring.

The dividing line between the legitimately protected improvements of respectable chemists and the hidden swindles of druggists or projectors was open to constant renegotiation. Daniel Wheeler, certainly, did not see his involvement in the 1813 offence as any barrier to a subsequent public career as a technical innovator. Interviewed by the 1818 committee, he emphasized that salt of tartar was a harmless substance in common domestic use, and that he knew of no 'brewers druggists'; he was by profession a chemist, and now patentee of a process for making colouring from malt.[93] Coinciding with the arrival of the 1817 regulations, Wheeler's project was presumably inspired by Matthew Wood's attempt to turn the 1802 restrictions into a production monopoly. The partnership Wheeler formed to work this patent gives an intriguing glimpse into the company such a commercial chemist might keep. It included Thomas Meux, formerly his chief contact at Meux Reid's, who had recently left the brewery; Simon Cock, Secretary of the African Trading Company; and 'G. Imlay' – presumably Gilbert Imlay, the adventuresome American philosopher, land-jobber and blockade-runner, best known to posterity as Mary Wollstonecraft's faithless lover. Imlay's projects had included beer-colouring since at least 1812.[94]

Wheeler seems to have escaped further obstruction by the Excise, but his activities faced other hazards. Patenting, while it avoided charges of secretism, could lead to inconvenient scrutiny of the contended innovation. Wheeler's specification had claimed the slow-roasting of malt to above 400°F as a novelty: he indicated that a coffee-roasting machine would do the job, but sought to extend his protection to all 'other contrivances'. What became known as 'patent malt' was immediately imitated by Joseph Malins, an established coffee roaster, provoking an infringement action in which, as was increasingly common, consultant chemists featured as witnesses on both sides.

Malins engaged Henry Ronalds, of the Society of Apothecaries, and a 'Mr Hume' – probably Joseph Hume (1756–1846), a commercial chemist of Long Acre (who was to make a literally explosive contribution to a subsequent case, accidentally discharging a box of percussive compound in court and setting himself on fire).[95] To show that roasting above the specified temperature had been commonplace before the patent, Malins's chemists constructed a chain of witnessing involving observations on samples of alloys, of reliable melting point, thrown into malt kilns and 'fused among the malt'. For their part, Wheeler and partners engaged the young Michael Faraday (1791–1867), then working as chemical assistant to the Royal Institution, but to no avail.[96] The specification was ruled incoherent, and the patent repealed in 1819, whereupon Malins had himself listed in London directories as an 'Anti-patent Malt-maker'.[97] Wheeler was bankrupt the same year. With the prospect of a druggist's monopoly removed, the process garnered little further public controversy. It passed largely to maltsters, and was restricted to an oligopoly of 'malt-roasters' following legislation of 1842.[98]

Overall, for those who sought to secure an additive as legitimate, the most important consideration was to prevent any public incubation of a 'matter out of place' narrative. Evidence of commercial self-interest was not problematic in its own right, and could even be applauded: brewing, after all, was now chiefly undertaken on a commercial basis. It was crucial, however, that the enterprise carry no whiff of drug shops, shadowy projectors touting multifarious projects or counter-intuitive operations without an obvious brewery precedent. This was seldom easily achieved. The image of the lethal druggist, in particular, could manifest in the most unexpected of circumstances, turning even against those who had called it into being.

Friedrich Accum and the Ambiguities of Brewery Literature

Friedrich Accum (1769–1838) was in many ways the quintessential public chemist of the early nineteenth century. A background as a retail pharmacist in the service of the Anglo-Hanoverian Brande family, apothecaries to King George III, brought him into contact with chemists such as William Nicholson, whose influential *Journal* published his early research.[99] In 1800, Accum set up shop on his own account in Soho, retailing chemicals and apparatus whilst developing a career as a commercial analyst, instrument-maker and lecturer. He gained a reputation both for his private training, catering for professional men seeking practical laboratory skills, and for public lectures aimed at fashionable audiences, which brought him aristocratic patrons. Accum was soon perceived – in the admiring words of Benjamin Silliman, whom he instructed – as London's 'pet chemist'.[100]

Like his most successful predecessors in public chemistry, Accum keenly appreciated the value of book publication as a promotional device. His concern to appeal to a broad audience led to a focus on domestic chemistry, and in particular on food and drink. His most successful book by far was *A Treatise on Adulterations of Food, and Culinary Poisons*, first published in January 1820 as a nine-shilling duodecimo. The cover bore an unforgettably grisly motif of serpents, darts and a spider devouring a fly, below a biblical quotation: 'There is Death in the Pot'.[101] The sensational text brought together numerous poisoning and fraud narratives which had swirled through newspapers and pamphlets, with varying degrees of documentary support, from the mid-eighteenth century onwards. Beer, water, bread, wine, coffee, tea, pepper: all, said Accum, were routinely adulterated, often with toxic consequences. Only with appropriate chemical testing could the fraud be detected.

The *Treatise* is often considered to mark a turning point in public attention to adulteration and its detection.[102] Its lurid tone and packaging marked a drastic departure from the conventions of polite chemical publication: Accum was aiming unashamedly to provoke popular alarm, and in this he certainly succeeded. The initial edition of a thousand copies sold out within a month; three further editions, an American reprint and a German translation swiftly followed. The book received exceptionally widespread attention, for a chemical work, from the gentlemanly periodicals which informed his intended audience.[103] Accum's status in public chemistry was suddenly eclipsed by accounts depicting him as a moral crusader, with varying degrees of sympathy. Five years on, the reverberations from the *Treatise* remained potent enough that a popular comedian could draw ready laughs from the character of 'Mr Allum, the writing chemist', who delights in repeating that 'All food is bad, and I can show it/I've analized it, and I know it'.[104]

By this point, however, Accum's public career had been abruptly curtailed. He fled England around the beginning of 1821 to avoid standing trial on the extraordinary charge of mutilating volumes in the library of the Royal Institution, where he had once served as a chemical operator. In a preface to the *Treatise*'s second edition, Accum had written elliptically of anonymous 'maledictions' from 'assailants in ambush'; some historians have taken for granted that Accum's revelations had angered powerful manufacturers, who found ways to turn a trivial incident into a storm of negative publicity sufficient to silence his campaign.[105] Whatever the reality behind such threats, however, it is important to appreciate that Accum had been scrupulously careful not to court them.

Though packaged as an unprecedented exposé, the *Treatise* was in fact compiled almost entirely from published material. Most of the discussion was confined to general principles and possibilities; such case histories as did appear mostly lacked names, dates or precise locations, and more often concerned accidental contamination than deliberate adulteration. Beer provided the only significant exceptions. Besides firmly cementing Humphrey Jackson as the godfather of brewers' druggists, Accum filled many pages with the names of brewers and publicans convicted of adulteration, and druggists and grocers convicted of selling adulterants to brewers, taken from the published minutes of the 1818 Committee.[106] Even here, though, Accum took pains to avoid riling the most powerful of potential enemies. Accepting at face value the major brewers' strategic dichotomy, he stressed that

> there are no convictions, in any instance, against any of the eleven great London porter brewers for any illegal practice. The great London brewers, it appears, believe that the publicans alone adulterate the beer. That many of the latter have been convicted of this fraud, the Report of the Board of Excise amply shews [...] The eleven great porter brewers of this metropolis are persons of so high respectability, that there is no ground for the slightest suspicion that they would attempt any illegal practices.[107]

Accum's disclaimer was cautious to the point of inaccuracy, passing silently over the well-reported 1813 Meux Reid conviction for salt of tartar, even though Accum himself had attacked such alkaline additions as 'the worst expedient that the brewer can practise'.[108]

Small brewers and publicans were unsurprisingly displeased. A particularly eloquent response came from John Tuck (b. *c.* 1761; fl. 1820–2), a forceful exponent of the 'practical brewer' tendency (chapter 3) with a somewhat unexpected background. Tuck was anything but a stranger to what John Richardson had called the 'literary labyrinth of Pater-noster-row', where he had kept a successful bookbinder's shop from around 1780, becoming the founding President of the Master Bookbinders' Association and boasting of earning 'as much as a stockbroker'. Around 1810 he left the trade to take on a brewery in Croydon, and was offering himself as a consultant and instructor when he published an anonymous treatise in 1820: the dedication was to John Elliot, one of the leading porter magnates of London, which suggests that Tuck might have served Elliot in an advisory role.[109] Learning of Accum's accusations after the completion of this work, Tuck addressed them at length in a second edition, published under his own name in 1822.

Tuck's *Private Brewer's Guide* strongly belied its title, which was perhaps chosen to solicit as wide an audience as possible. He wrote explicitly for the commercial brewer, underlining the point by including a subscriber list of nearly fifty brewery proprietors. In his preface, Tuck framed a counterblast to Accum's 'violent accusations' against the trade. His core objection was the standard one, as demonstrated by Richardson and George Blake the previous century: Accum's 'reliance on science instead of practical acquirement' had led him astray.[110] Tuck was helped by the fact that Accum had recently published a brewing treatise of his own, which, like most of his book-length publications, was hacked together with limited independent research.[111] Tuck was able to show that Accum's descriptions of various basic brewery operations were erroneous.[112] Echoing Blake's criticisms of John Long, he surmised that the chemist had probably been 'imposed upon' regarding matters he had not the experience to judge.[113]

Tuck also followed earlier brewer-authors in co-opting parts of the philosophical agenda. Brewing, he said, should be managed on the basis of systematically recorded, quantitative data from instruments including the thermometer and saccharometer.[114] Whereas Richardson and Blake had taken the opportunity to write off their opponents as miscellaneous quacks masquerading as serious chemists, however, Tuck was prepared to assert that there was an institutional problem with chemistry itself. Chemists and druggists, in his eyes, acted as a body to protect their common interest, and Accum's *Treatise* was merely the latest stage in an extended scam that had begun with Humphrey Jackson and the beer doctors:

> [W]ho learned Brewers the use of Drugs? The answer must be, *Chemists*; Brewers were tempted, and have since completely discovered the fallacy of the experiment; and now the game is up, one of their own body comes forward to expose the evils they have brought on the Brewery.[115]

This condemnation of Accum as a fellow traveller of brewers' druggists begins to make sense when we appreciate the routine use of strategic ambiguity in contemporary instructional literature. Guides for publicans – who needed to be on the lookout for adulteration by distillers and spirit vendors, but often used such tactics themselves to extend their profits from drinkers – commonly interspersed virulent attacks on particular techniques with directions for performing them, clear and precise enough to allow emulation. Various generic forms of words excused such inclusions. Some authors argued that 'it is proper to mention every thing that may occur'; others would absent themselves from responsibility and 'leave the reader to judge of [their] good and bad qualities'.[116]

Though initially focused on spirits, the publicans' guides increasingly encompassed beer from around 1800, as the content of Samuel Child's abovementioned pamphlet worked its way into the corpus of widely borrowed material, and druggists and instrument-makers began promoting the guides to victuallers and small brewers.[117] Through the customary chain of hack work, the directions passed, in turn, into manuals aimed at the common brewer, whose authors could exploit the presumed multiplicity of their readerships: additions prohibited in the commercial brewery were perfectly legal among private individuals brewing for their own consumption.[118] Meanwhile, the imperative to give as much information as possible, as a safeguard or 'for choice and experiment', remained a useful prop.[119]

Accum's work was inescapably part of this process. In the second edition of the *Treatise*, Accum censured Alexander Morrice for recommending various additives; in the course of doing this, he reproduced Morrice's guidance verbatim.[120] Tuck, who as a brewer must have been fully aware of the possibilities of strategic ambiguity, thus had no hesitation in depositing Accum among the brewers' druggists. Though both Tuck's and Accum's writings were sharply critical of Samuel Child's pamphlet on brewing with cocculus and other additives, Tuck's charges against Accum included 'plagirism' [*sic*] of Child (on brewing in wash tubs, an innovation Tuck considered ludicrous).[121]

The suggestion that Accum had produced a serviceable manual for adulterators, whether ingenuously or not, recurred.[122] In a comic poem of 1826, 'Death (a Dealer) to his London Correspondent', the grim reaper glories in 'extracts of coculus [*sic*], quassia and copperas', playing on the words *ail* and *bier*.[123] The accompanying cartoon, with Death surrounded by adulterative poisons, shows a copy of 'Accum's List' pinned to his wall for ease of reference (Figure 5.3).

Discussing adulteration from any angle was thus a risky business, open to multiple readings: later brewing writers, indeed, played on their refusal to do so as a guarantee of probity.[124] The only safeguard open to the chemist was to establish an unambiguously sound reputation as an analyst motivated by public rather than personal interest. This proved impossible in Accum's case, given his lurid promotional tactics and his pronounced habit of scissors-and-paste borrowing, sometimes without attribution or evident full understanding of his material. The iconic 'Death in the Pot' motto was itself second-hand, as was the general tone of Accum's self-presentation as a threatened crusader.[125]

These tendencies were subjected to public scrutiny in periodicals including the *Quarterly Review*, which convicted Accum of a dual sin in too often factitiously 'filling a page' while needlessly 'frightening his audience with a formidable array of hard words'. The more welcoming *London Literary Gazette* also questioned Accum's refusal to bring forth his 'own experience, rather than the loose statements of newspapers'.[126] Such censures affirmed the position – contrary to the objections of John Tuck – that the multiple meanings of chemistry could be distinguished: it would be to the public benefit, sighed the reviewers, if a *reputable* chemist were to write on brewing in a systematic and disinterested fashion.

DEATH'S REGISTER.

Figure 5.3: 'Death's Register', in R. Dagley (ed.), *Death's Doings* (London: Andrews and Cole, 1826), plate facing p. 297. By courtesy of the University of Liverpool Library.

In chapter 6, I will show how one such chemist, Thomas Thomson, set out to do so. Thomson's aim was to redefine brewing as an industrial process dependent on current chemical literature for its conceptual underpinning and the direction of innovation. This manoeuvre brought the print culture of brewing closer to the professedly disinterested norms of chemical scholarship; though brewer-authors were not content to take a purely reactive role, they eventually found means of collaborating credibly with the new chemical professionals. One consequence of this resolution, however, was a more esoteric literature, presuming a shared knowledge of chemical preliminaries. Such literature could do little to counteract the public's adulteration fears, which remained a constant backdrop to technical development in the brewery for decades to come.

6 PROFESSORS IN THE BREWHOUSE

> As this important art has been, in a great measure, overlooked in the *Encyclopædia Britannica*, and as nothing like a satisfactory account of it is to be found in any book on the subject, which we have seen, we consider it necessary to lay down the principles on which it depends, somewhat in detail.
>
> [T. Thomson][1]

From around the turn of the nineteenth century, the institutions and practices of natural knowledge-making came increasingly under the control of a new breed of professionalizers.[2] Jon Topham has elegantly summarized the shift

> from a logic of discovery, theoretically open to all, to a more restrictive notion of discovery as the preserve of scientific 'genius', and from an open-ended philosophy of 'experience' to a far more restrictive notion of disciplined 'expertise'. Both of these moves were intended to do boundary work, restricting the community active in creating and validating scientific knowledge, and producing a passive public.[3]

Most historians who have looked at this shift have focused on how it marginalized amateur enthusiasts, but the professionalizers' boundary work also had implications for trade communities such as the brewery. This chapter examines the question through the case of Thomas Thomson (1773–1852), an active professionalizer in his various roles as textbook author, journal editor, consultant on industrial topics and, from 1818, Professor of Chemistry at the University of Glasgow, where he sought to build a research school of younger chemists.[4] In the years prior to his professorship, Thomson gained opportunities to shape the direction of brewing chemistry through an accomplishment which was probably unique: direct experience in brewhouse processes without any direct relationship with the trade itself.

The key to Thomson's credibility was consultancy work undertaken for the Boards of Excise in Scotland and England. In the years before 1800, the Treasury and Excise had learnt the value of consulting elite philosophers as a means to defuse controversies about the taxable value of goods of all kinds. Philosophical statements about the composition of matter appeared both disinterested and universal; laboratory methods, often involving demonstrations of unparalleled precision, set standards that were hard to question for anyone without access to similar facilities. The Excise in particular consulted the growing class of professional chemists who, though they often depended on income from such consultancy, vested their authority in lecturing roles at universities or public institutions, dignified, increasingly, with the title 'Professor'.

The challenge, of course, was to overcome the problem which had fatally compromised Humphrey Jackson's credibility (chapter 3): bridging the gap between the knowledge-making regimes of the laboratory and the brewhouse in a manner convincing to the brewers themselves. The first section of this chapter relates how, in 1804, the Scottish Excise addressed the challenge in an uncommonly literal fashion: it took control of established common brewhouses, and sent Thomson and other respected professional chemists to direct production. This unprecedented project demonstrated the extraordinary difficulty of subjecting industrial practice to modes of comparison developed for laboratory use: indeed, assessed by its intended aim of settling valuation controversies, it largely failed. Nonetheless, as the remainder of the chapter shows, its findings served Thomson as a long-term source of scientific authority.

One key effect of the Excise investigation was to direct the chemical conceptualization of beer down paths which would not have arisen from within the brewing trade. Chemists and the revenue authorities tended by default to address drink in distillery terms: spirits had been subject to rigorous hydrostatic assessment since the eighteenth century, and chemical approaches to intoxicating drink in general tended to focus on the finished product's alcohol content. The philosophical brewers discussed in chapters 2, 3 and 4, by contrast, were more concerned with the colour and character of malt, the sugar content of worts and the stability of beer in storage. Thomson presented his analyses as a potential key to wrestling both beer and spirits into a common legislative framework that would safeguard the revenue. Although his proposal was not taken up, its elaboration cemented Thomson's relationship with the Excise, and provided material for his subsequent writing.

Whereas some professional chemists consulted or served as expert witnesses for brewers (chapter 5), Thomson, so far as is known, had no dealings with the trade outside his Excise consultancies. The chief attractions of government work were the obvious ones – reliable payment and opportunities to cultivate further projects – but the arrangement provided another advantage for a man whose livelihood often depended on his pen. Thomson was free to publicize his hands-on brewhouse experience without raising the suspicions which usually attached to trade writing. Around 1816, he used the work as the basis of a new entry for the *Encyclopedia Britannica* which served not only as an influential outlet for Thomson's theories of malting and fermentation, but as a manifesto for his professionalizing agenda.

Thomson made a point of rejecting virtually the whole of the literature written by brewery insiders as intrinsically unphilosophical. Interestingly, later brewer-authors did not object to his strictures; some, indeed, praised his detailed engagement with the subject and sought to emulate his approach. Thomson, however, did not reciprocate such attention: after the *Britannica* piece, he paid no further attention to the theoretical interpretation of brewhouse particulars. In the closing sections, I demonstrate that Thomson's limited familiarity with the brewers' context led him to overlook their ability to debate and systematize knowledge-claims through general journals, pamphlets and longer publications, in a makeshift but effective fashion, from the early years of the nineteenth century. If, by mid-century, the manner in which brewing chemistry was written owed much to Thomson, its conceptual content owed more to the brewers.

The Brewhouse of Experiment

Controversy over the value of drink was perennially strongest in Scotland, where the use of malt was dominated by whisky distillers. The compromises of Union in 1707 had left Scotland and England with independent Excise regimes, and, at least on paper, a relatively lenient scale of duties north of the border: by 1804, after soaring increases to meet the costs of continual warfare, English barley malt paid *4s.4d.* the bushel, as against *3s.8⅛d.* for Scotland. The difference was conventionally explained as an allowance for the greater difficulty of production in the northerly climate and the moderate inferiority of the product, Scottish malt typically selling for around 20 per cent less than English.

The comparison was complicated by the cultivation of bigg, a barley-like grain little known in England, but widespread in colder and more barren regions of Scotland where barley refused to grow. Bigg could be used interchangeably with barley, but gave a poorer extract; its price tended to hover around nine-tenths that of Scots barley, but fluctuated greatly depending on the year's harvest. Landowning proprietors in the north of Scotland persistently argued for the lightest possible taxation of bigg, on the grounds that it took more 'Trouble and Expence' to malt, brew or distil, and that its growth needed to be sustained as the only means of feeding the populace of the far north in particularly bad seasons when the oat crop failed. In 1804, bigg malt paid only *3s.⅛d.* per bushel, just over four-fifths the Scots barley duty, but there were continual rumblings that this value was still too high.[5]

The bigg-growers' complaints were mostly funnelled through George Skene Keith (1752–1823), a Church of Scotland minister, metrologist and agricultural improver with farmland of his own in the bigg country of Aberdeenshire. Keith had performed analyses for the Scottish Excise in the 1790s: prompted by German and Russian research on beet sugar, he had distilled the fermented extracts of a wide range of vegetable materials and measured the yield of spirits.[6] In 1804, he used similar spirit determinations to argue for large cuts in duty. A Commons Select Committee set up to look into the question offered a compromise which failed to satisfy the bigg-growers.[7] The Treasury therefore directed the Scottish Board of Excise to commission an experimental enquiry of its own, hoping to determine beyond question the 'relative Qualities' of malt from English and Scottish barley and from bigg. Unlike Keith's experiments, the enquiry was to encompass both distilling and beer-brewing.

The Board enlisted the holder of the senior chemical chair in Scotland: Thomas Hope (1766–1844), Joseph Black's successor as Professor of Chemistry and Medicine at the University of Edinburgh.[8] Hope was to lead what became a three-man committee alongside Andrew Coventry (1764–1832), founding Professor of Agriculture at the University, and Thomas Thomson, an Edinburgh graduate, then supporting himself as a private tutor in the city and best known as the compiler of a widely used *System of Chemistry*. Coventry was no doubt appointed as an authority on crops, having farmland of his own in Kinross-shire, and was respected as an arbitrator in land disputes; Thomson had a stronger background in industrial applications. State consultancy offered an important income stream to Thomson, and perhaps also to the unsalaried Hope, who otherwise supported himself from teaching fees.[9] The Board offered a handsome 600 guineas to each chemist, for what became roughly a year's work. The intention, of course, was that the combined expertise of the three professors would put the issue of valuation beyond reasonable question. The chemists' analytical authority, however, was only one component – if a particularly important one – in a wider credibility-building project which ultimately cost almost £5,000.[10]

As we saw in chapter 3, the gap between laboratory and brewhouse had exercised writers on brewing since the 1780s. The chemists were evidently aware that small-scale demonstrations under ideal conditions, managed by men without a working knowledge of commercial realities, would fail to convince the trade: they emphasized that the experiments must be on 'a scale nearly as extensive as in ordinary manufactories'.[11] The simplest way to do this was to co-opt, as experimental space, a working brewhouse – or rather, as eventually proved necessary, two working brewhouses in Edinburgh, with a third converted to a model distillery. The workmen brewers were retained in their existing roles; the operations were superintended by Archibald Richardson, an experienced local brewer.[12] Most of the brewings used around 60 or 70 bushels of malt, to produce, with remashing, 10 to 20 barrels of wort: figures belonging to the lower reaches of commercial manufacture, well beyond ordinary domestic production, let alone laboratory simulation. The ales produced, indeed, were sold on the open market at the end of the process (duty free, to the disgust of the Edinburgh brewers).[13]

Mounting such large-scale trials generated significant problems of trustworthiness concerning the materials and methods. These were addressed, as far as possible, through the established Excise approaches of hierarchy, interchangeability and documentation. Senior Excise officials were sent to the London and Scottish markets to buy barley from multiple dealers. Their instructions were to see that the grain was bagged up in full view and closed with a unique seal, to shepherd it in person to Leith, supervise the unloading, and make oath that nothing had been interfered with during the journey. All this was relatively straightforward for barley: the trade was concentrated in a few major hubs, such as the London exchange at Mark Lane, and its practices were already strongly shaped by the bureaucracy of Excise supervision. The market for bigg, by contrast, remained geographically diffuse and prone to local customs, which included mixing grain from more than one source before sale. The Board sought to enforce comparability by assigning two senior Excisemen to scour the bigg-growing areas in company with an 'experienced Farmer' hired as consultant, if possible observing the threshing process directly.[14]

The professors, however, soon found that comparing barley and bigg malt entailed complications unknown in the laboratory. To begin with, should the materials be compared by weight or by volume? Chemical precedent favoured weighing, but brewers, distillers and legislators invariably interpreted grain volumetrically, measuring by the bushel. The mass of a filled bushel varied significantly depending on the rate and manner of pouring, and standardizing the procedure was impossible with the plant, workforce and time available. The analysts decided to follow the trade convention and simply ignored the variation in mass, which they deemed to amount to roughly one percentage point either side of the mean.[15]

Secondly, should the trials compare equal volumes of finished malt, or the malt produced from equal volumes of raw grain? These would inevitably differ, as different barley and bigg samples swelled to varying degrees, especially when they were malted according to different systems of management. England and Scotland had distinct customs of malting: should the English and Scottish barley samples all be managed in the same way, or each according to local practice? The analysts ultimately made this question part of the experiment, comparing the consequences of the different approaches, and noting which samples 'were malted by an Englishman' (later indicated, more doubtfully, as one 'who professed to follow the mode practised in England'). Further problems arose in attempting to standardize the malt-drying process, which again affected the volume. The outcome was a comparative table of specific gravities on which the authors discouraged '[t]oo great reliance'.[16] Meanwhile, a third criterion used in the distillery experiments, where the extracts from unmalted grain could be compared directly, gave the preposterous finding that bigg was very nearly as good as Scottish barley, and English barley worse than either.[17]

The investigators had come up against an awkward point of tension between the goals of authenticity and universality. It so happened that in 1804 the barley crop of Scotland, coming into flower several days later than the English, had entirely escaped a snap of bad weather that had badly damaged the English crop; similarly, the bigg harvest had been of untypically high quality. The Excise representatives had asked after the grain of other years, but found it unavailable.[18] Throughout the trials, moreover, supposedly identical operations had produced highly variable results, and there had not been the time or materials to seek consistency: instead, the figures were simply averaged. The professors' conclusions firmly endorsed the high duty on bigg, but it was clear that the imperfect combination of laboratory and brewhouse discipline left their report lacking in authority.

The loudest objections came from George Skene Keith, whose request to witness the experiments of the 'Scientific Gentlemen' had been denied.[19] In a detailed study published in the widely distributed *Farmer's Magazine*, Keith attacked or queried almost every aspect of the investigation, from the definition of terms to the calculations, which he showed to contain arithmetical errors. He did not, however, challenge the principle of deferring to chemical authority – unsurprisingly, since many of the analytical criteria used by Hope, Thomson and Coventry were also his own. Keith carefully acquitted the professors, blaming their operatives and the samples with which the Board had required them to work.[20] In rejecting gifts from 'respectable' farmers, said Keith, the Board had opened itself to the impostures of the commercial market, as when it bought barley 'from four different counties in England, (or at least at Mark-Lane, under the designation, of belonging to four counties)'.[21] In the brewing operations, similarly, excessive fidelity to real-life practice had sacrificed too much of the uniformity needed for experimental certainty.[22]

It was painfully apparent that the Report's conclusions, though invested with meticulous numerical precision, failed to capture the common state of affairs. The bigg question remained controversial in the long term: chemical authority, in this instance, did not point the way to a resolution.[23] For all this, however, Thomas Thomson's contribution to the Report provided the foundation for a significant effort to bring the understanding of beer under chemical control.

Alcohol against Extract

Thomson's contribution was motivated by the desire to set up a valuation scheme for alcoholic drink in general. In the distillery, particularly in England, philosophical authority was well-established as a means of settling valuation questions by the time of the Edinburgh experiments. Spirit duties, as noted in chapter 4, were assessed by flotation hydrometer readings: these were often disputed, not only by distillers and merchants, but by instrument-makers bidding to supplant the official Clarke hydrometer.[24] In 1797, a frustrated Board of Excise had referred the problem to Sir Joseph Banks, President of the Royal Society. Like the well-known enterprises of the 'Banksian Learned Empire' in economic botany and agriculture, this was an opportunity to demonstrate the usefulness of scientific enquiry in the service of the state.[25] Charles Blagden and George Gilpin, respectively Secretary and Clerk of the Society, were directed to provide a new basis for calibration by producing gravity tables of unparalleled precision, and a Royal Society committee was enlisted to select a new official hydrometer pattern in 1802.[26]

Beer-brewers, however, mostly followed John Richardson in valuing beer by the quantity of malt extract dissolved in the wort before fermentation, as measured by the saccharometer in Richardson's units of pounds per barrel (p. 87). Richardson had denied the validity of valuation based on alcohol content alone, believing that the effect of the alcohol was crucially modified by fixed air, which could not be quantified. Although later brewers sidelined this theory, they retained Richardson's doubts about the relevance of alcohol determinations: in their investigations, the unfermented matter, hop characteristics, colour, clarity, acidity or carbonation of the beer might be more important.

To non-brewers, however, it seemed obvious that distilling the product and quantifying the alcohol offered a promising basis for comparison across the whole of the vegetable kingdom. This was the prospect highlighted by George Skene Keith's experiments for the Scottish Excise, which had proclaimed it economically plausible that drink might be made from such improbable materials as carrots and salsify.[27] The potential for a universal scheme appealed to chemical philosophers and administrators of the public revenue alike, and the Edinburgh investigation, with its parallel beer and spirit trials, offered the first major test of the practicalities of extending such a scheme into the brewhouse. Thomson, no doubt, recognized a promising opportunity for a researcher reliant on consultancy income. He took particular charge of the beer-brewing side of the investigation, writing the brewing section of the 1806 Report single-handedly.[28]

Like almost all chemists of his generation, Thomson understood alcoholic fermentation as a uniform process whereby sugar, or 'saccharine matter', decomposed into two constituents: fixed air – increasingly reinterpreted as 'carbonic acid gas' under the oxygen theory of Antoine Lavoisier – and alcohol. These substances were in turn thought to consist of carbon, oxygen and hydrogen in fixed proportions, various possibilities for which were suggested on the basis of combustion and other analysis.[29] Thomson, who was busily revising his *System of Chemistry* to promote John Dalton's atomic theory around the time of the brewhouse experiments, embraced Dalton's conception of the sugar particle as a simple composite of one particle of alcohol and one of carbonic acid, which cleanly disengaged in fermentation.[30] The difficulty was that the malt extracts which actually fermented in the brewhouse were qualitatively quite different from the refined cane sugar which was usually taken to approximate 'pure saccharine matter', and contained a number of apparently non-fermenting substances in varying proportions. Thomson was confident, however, that the variations would prove minor enough to be safely ignored in valuation.[31]

To secure this point, Thomson repeatedly shifted – silently, so far as the Report was concerned – from the brewhouse to the laboratory and back. Taking small-scale samples of wort from each of the sample brewings, he evaporated them to dryness, weighed them precisely, and compared this weight of dry solid extract to the gravity of the wort. '[F]or practical purposes', he concluded, the two quantities were reliably proportionate.[32] The experimenters likewise determined, over multiple trials, the weight of extract 'which actually disappears during fermentation', and found it acceptably proportionate to the quantity of alcohol generated, as measured by distillation.[33] Malt extract, then, was a tolerably uniform substance, insofar as a given quantity of malt sugar, decomposed by fermentation, reliably produced a given quantity of alcohol.

More awkward was the observation that had so puzzled Richardson in the 1780s: the maximum extent of decomposition varied between methods of brewing, and sometimes apparently at random, so that there was no proportion at all between the yield of extract before fermentation and the yield of alcohol afterwards. How could this be, if fermentation was essentially a uniform process involving uniform chemical compounds? The question did not demand an immediate answer: the nature of fermentation was notoriously obscure. But the variation spelled trouble for any attempt to extend a definitive scheme of valuation across both the beer and spirits trades, and the evidence of the Edinburgh trials was particularly unhelpful. The brewing experiments appeared to show that Scottish malt (volume for volume) produced more extract than English; the distilling experiments, that English malt produced more alcohol than Scottish.

The problem, indeed, was sufficient to provoke a split among the three investigators. Hope and Coventry decided the discrepancy could be finessed: their final analysis took a simple arithmetical mean of the extract and alcohol data as the best representative of value.[34] Thomson would have none of this, and characteristically produced a minority report.[35] Extract and alcohol, he said, provided incommensurable criteria; and, since the Excise was committed to alcohol, it would be '*unfair*, in the present state of the question', to accommodate the extract definition.[36] Here was the opposite of John Richardson's stance: the brewers must be *denied* a voice in the operation of the Excise, if it might compromise the coherence of the regime which had grown up around spirits hydrometry. This was not the way to win friends in the brewery, of course, but Thomson's goal was to establish a recurring role as an Excise consultant. In this, he was apparently successful: the Scottish Board gave him remarkable freedom to act in its name in the wake of the 1806 Report.[37]

In the process, Thomson found a significant commercial sideline. He had begun the Edinburgh trials equipped with the conventional Dring and Fage saccharometer, scaled in Richardson's pounds per barrel. The unit, Thomson felt, was deeply misleading: an actual barrel of wort which registered one 'pound per barrel' heavier than water would, when evaporated, yield somewhere in the region of two and a half pounds of dry extract.[38] As we saw in chapter 4, most brewers understood this perfectly well: they had no reason to deal with dry extract, and were happy with a conventional measure giving locally comparable values. To Thomson, however, the exceptionalism of the 'brewer's pound' was a travesty; he also questioned the accuracy of the device. He therefore commissioned a local instrument-maker, Alexander Allan, to produce and retail a saccharometer to his own specification.[39]

Again, Thomson's tactics were the converse of Richardson's. Richardson had begun by promoting the saccharometer to his fellow brewers, seeking a trade consensus which would make its legal adoption inevitable; Thomson began by courting the Scottish Excise, hoping to impose his instrument from the top down. Invoking his brewhouse experience, he apparently convinced the Board that the average Exciseman was not sufficiently astute to take the gravities unaided using the existing equipment, and promoted the Allan pattern as part of a general reform of practice.[40] The Board adopted the device, unofficially but exclusively, in 1805, before the Edinburgh experiments had even been completed.

The calibration of the Allan saccharometer reflected Thomson's universalizing agenda. Its readings were on the specific gravity scale used in spirits determination and philosophical papers, taking the density of distilled water as unity, while the equivalent weight of dry extract could be determined by a sliding rule.[41] Thomson also promoted it to distillers, and in 1816, it was made statutory for Excise testing of distillers' wort and wash in Scotland, meaning it was also in the spirit trade's interest to acquire a copy.[42] Despite his interest in beer-brewing, then, Thomson manifested the focus on spirit which was typical of contemporary chemists. This focus evidently extended beyond his professional life: 1816 was also the year of his marriage to Agnes Colquhoun, daughter of a Stirlingshire distiller.[43]

Thomson was not the only chemist working to extend alcoholometry beyond the distillery. The general approach had seemed to suffer a setback in the 1790s, when a Florentine chemist, Giovanni Fabbroni, revived Rouelle's position that the intoxicating principle in undistilled drinks was not alcohol (p. 99).[44] By 1813, however, Fabbroni's interpretation had been convincingly overturned in papers by William Thomas Brande (1788–1866), a newly recruited Fellow of the Royal Society and rising star among the chemical lecturers of London, and by the most dedicated French alcoholometrist, Joseph-Louis Gay-Lussac.[45] Brande's account of his experiments in the *Philosophical Transactions* acknowledged the 'liberality' of Joseph Banks in providing him with samples of wine required for the determination;[46] we might reasonably speculate that Banks, playing the state servant as ever, had commissioned the investigation in order to safeguard the revenue's options.

To cement the validity of the approach, and his own experimental authority, Brande determined the alcoholic concentration of a wide range of wines and other drinks by distillation, creating the first comparative tables of alcohol by volume. Editorship of the *Quarterly Journal of Science* gave him a ready outlet for publication.[47] Among the most striking samples was beer from Joseph Banks's own brewhouse: using ten bushels of malt to the hogshead, Banks's brewer had achieved a formidable 10.84 per cent by volume, a value which was duly entered in the table under 'Lincolnshire ale'.[48] Friedrich Accum's *Treatise on Adulterations* and *Treatise on the Art of Brewing*, both of 1820, included laboratory assessments of the alcohol in porter and ale.[49] To Accum, the approach was a tool to detect adulteration: if the leading brewers' porter, taken straight from the breweries, averaged 5.25 per cent, and the same beer, from the publicans, averaged only 4.50 per cent, it was obvious where the mischief lay.[50]

Such interpretations took time to become entrenched, however; and the power of philosophical authority as a tool of persuasion had its limits, as the reception of the 1806 Report had demonstrated only too well. The Boards of Excise did not, in this period, show any serious signs either of extending alcoholometric taxation into the brewery or of endorsing the brewers' commitment to extract-based valuation. For want of any uncontroversial alternative, the age-old system of uniformly rated strength bands creaked onwards until 1830, when the beer tax was abolished outright as part of an uncommonly wide-ranging reform involving radical deregulation of the licensing system.[51] Throughout this period, however, Thomson retained his status as a valued consultant to Excise investigations, and meanwhile found another means to turn his brewhouse experience to advantage.

Beer and Britannica

In 1813, work began on a major supplement to the *Encyclopaedia Britannica* under the editorship of the Edinburgh lawyer and man of letters, Macvey Napier (1776–1847). To distinguish the *Britannica* in an increasingly crowded market, Napier broke from the convention of textual recycling, seeking new-minted contributions from the most prestigious authorities in Scotland and England. No natural philosopher himself, Napier relied frequently on the guidance of his scientific authors, and gave them considerable freedom in the extent and scope of their coverage.[52] Thomson was an obvious choice, having served as assistant editor to a previous *Britannica* supplement; now based in London, he had been elected to the Royal Society and was editing a journal of his own, the *Annals of*

Philosophy. Though Napier appointed William Brande to write a general dissertation on chemistry, he gave Thomson most of the specialist chemical entries, particularly those addressing the useful arts. Thomson's letters to Napier show him struggling against chronic ill health and the incessant workload of teaching, the editorship of the *Annals* and the revision of his *System*; yet he persistently took on as much *Britannica* work as he could conceivably manage. The attraction was not only the payment – which was generous – but the opportunity of shaping how his field was presented in what had become the pre-eminent reference work of the age.

Since the supplement was to be issued in instalments, Napier had to commission entries in approximately alphabetical order. Around the beginning of 1816, Thomson staked his claim to a clutch of industrial processes under the letter B. For 'Baking', he compiled an article of around 10,000 words; for 'Brick-making', half as much and for 'Bleaching' somewhat more. His entry on 'Brewing', however, was longer than all three put together, comprising over 20,000 words of text and, unusually, many pages of quantitative data on barley, malt, worts and finished beer. The reason for such special treatment, Thomson explained to Napier, was that 'no account of this manufactory of the least value' had yet been printed.[53]

The brewing entries in early editions of the *Britannica* had been hacked together from standard sources such as Combrune and Richardson. Signalling a clean break, Thomson opened with a 'history' of brewing whose real purpose was to reject every word of the prior literature written by practising brewers. Combrune, for one, had offered a far from 'rational' account: his thermometric diagnostics were fatally flawed, said Thomson, because malt colour depended far more on the rate of heating than on the degree of heat attained. At least Combrune had given his theory freely to the public, however, unlike the 'reprehensible' Richardson:

> If a brewer conceives he knows more of his art than his neighbours, and chooses to keep his knowledge to himself, there is nothing to be said; but if he publish a book upon the subject, and yet persists in his concealment, he deserves no quarter. His book, in such a case, can be looked upon in no other light than as a quack bill to advertise the goodness of his wares.[54]

Richardson was, on the other hand, to be praised for bringing the saccharometer into use: Thomson saw the brewhouse as a valid source of practical innovations, but not of practical philosophers. He did not mention James Baverstock, the brewer-author whose approach was closest to his own. This was perhaps a strategic decision, but more probably a simple reflection of his limited familiarity with the trade literature, which, Thomson felt, did not need or deserve a systematic survey. His strongest invective was reserved for Richard Shannon, a patentee of attemperation equipment and professed 'M. D.', who had published a huge, lavishly produced quarto on brewing and distilling by subscription in 1805. This apparently systematic treatise was in fact built around the text of Shannon's patent specifications, bulked out with extensive borrowings from Richardson and others. Thomson's verdict was that the author 'must have been quite illiterate', his theory compiled from 'scraps and sentences, taken out of chemical books, and tacked together, so as to have no meaning whatever'.[55]

Productions such as Shannon's had been commonplace, in various fields of knowledge, across the eighteenth century; even for the early nineteenth, when respectable compilers increasingly frowned on uncritical recycling, Thomson's judgment was uncommonly fearsome. John Farey, Jr's article for the *Edinburgh Encyclopaedia*, published a few years earlier, had pronounced Shannon's volume an 'excellent' guide to the practices of commercial brewers.[56] The rival *Encyclopaedia Edinensis*, which reached 'Brewing' around the same time as the *Britannica*, noted Shannon's inaccuracies alongside the defects of many other sources (Combrune's old-fashioned Boerhaavianism, Alexander Morrice's unfortunate tolerance of toxic additives), but deferred to each of them as containing information which seemed useful, and was not to be had elsewhere.[57]

Thomson, however, insisted that what Shannon had produced was literally beyond comprehension. It was because of the risk of such monstrosities that there must be a sharp dividing line between the 'treatise' of philosophy and the mere 'book' of commerce. To have 'some practical knowledge of brewing' (the most he was prepared to grant for Shannon) was no basis for credibility. The right to write systematically on such topics must be reserved for chemists – or, to be precise, for experts in quantitative chemical analysis. Thomson appealed to a twofold demonstration of his own credibility on this point.

Firstly, he took pains to establish that he (or rather 'the writer of this article', technically anonymous) had performed extensive analysis at first hand. Unsurprisingly, he relied mainly on the 1806 Report, reworking the quantitative data on volumes, masses and specific gravities into chapters on malting and 'the different kinds of grain used by brewers'. As a result, these sections had a distinctly Scottish flavour. Privately, Thomson confessed to Napier that he did not know the London brewers' practice, and felt that if he 'were to apply to them for information it is very likely that they would mislead'. In 1817, however, his reputation brought him a consultancy – apparently on a relatively small scale – to the English Board of Excise. In this capacity he determined wort gravities at the major London breweries ('information not easily got at'), and added brief notes of average values to the *Britannica* article at the proof stage.[58]

Secondly, Thomson offered a systematic survey of analytical literature written by others, including numerous Continental authorities: Heinrich Einhof on the composition of barley; Johann Friedrich Westrumb on the analysis of yeast; Gay-Lussac, Louis Jacques Thénard, Jöns Jacob Berzelius and Théodore de Saussure on the process of fermentation. Thomson revised his atomic formulae for sugar and alcohol, for instance, in line with Saussure's analyses.[59] Much of the survey was reworked from the *System of Chemistry* and the *Annals of Philosophy*, both of which had played a significant role in disseminating French- and German-language literature to English-speaking audiences.[60] Thomson was proud of this synthesis. His rival Brande, he hinted privately to Napier, had lost touch with the international community through being too much in the thrall of Humphry Davy: Davy's notorious early objection to the atomic theory had its legacy in a hidebound attachment to conventions which ran contrary to 'all the rest of the world'.[61] The same sentiment was probably a factor in Thomson's contempt for the internal brewery literature, none of which had paid any systematic attention to foreign-language journals. His own brewhouse experience, of course, justified him in occasionally correcting his Continental sources: neither Lavoisier nor Thénard, he said, had really grasped the crucial point that sugar was only partly decomposed by fermentation.

The chief novelty of Thomson's contribution, then, lay in his insistence that a general reference work under the heading of 'Brewing' must build on the current literature in analytical chemistry, and *not* on the prior literature on brewing. The article's focus on the composition of materials sets it apart from the comparable entries in rival encyclopedias. John Farey, Jr's piece in the *Edinburgh* had been technical without being philosophical, working mainly from a survey of recent patents; the *Edinensis* contributor (possibly the editor, James Millar) concentrated on the operations of common practice; and the *London Encyclopaedia*, issued in parts from 1826, gave a far more economically and commercially oriented survey. If Thomson's approach was not typical, however, it was also a significant influence on the literature which followed. Purely derivative accounts such as the *London*'s, which drew from the 1817 Police Report, Accum's treatises and assorted surveys and engineers' specifications, drew likewise from Thomson ('perhaps the most philosophical writer on this subject'), and so imbibed some of his language and concerns.[62]

Thomson's sharp separation of philosophical from trade literature was, of course, a tool of his own trade, and this led occasionally to ironic results. The unsigned *Britannica* treatise contains an unabashed puff for the Allan saccharometer: in deeming it 'perhaps the best' available, Thomson blithely mentioned his own name as a recommendation (a surprising hostage to fortune, since his authorship was easily guessed).[63] Further, Thomson was, if anything, more prone to engaging in noisy public controversy with his peers than most of the brewer-authors. Soon after his appointment as Professor of Chemistry at the University of Glasgow in 1818, Thomson joined the impressive roll of enemies of the even more disputatious Andrew Ure, Professor of Natural Philosophy at the nearby Anderson's Institution.[64] Ure had formed an alliance with William Brande, and concocted a policy of deliberately vicious reviewing to boost sales of the *Quarterly Journal*.[65] Thomson was attacked here and in Ure's *Dictionary of Chemistry* of 1821, which named him as the author of the brewing article. Ure lambasted Thomson for a dependence on his French sources which, he suggested, crossed the line from systematizing into plagiarism.[66] Thomson gave as good as he got, extending the feud to Brande's contributions to the *Britannica* supplement.[67]

Nevertheless, Thomson's widely distributed article provided ammunition for those who shared his professionalizing agenda. In 1825, the *Dublin Philosophical Journal*, recently founded by a group of Trinity College dons, took aim at the *Young Brewer's Monitor*, a rather derivative treatise from a self-styled 'brewer of thirty years' practical experience'. The reviewer had fun with the author's high-flown verbiage, while upbraiding him for failing to engage with chemical literature as Thomson demanded. His quantitative analysis of malt and barley, for example, overlooked 'the analysis of Einhoff [*sic*, following Thomson's spelling], in which the estimates are far different from his'. Emphasizing the boundary, the *Dublin* reviewer added that he would much prefer an honest-to-goodness, unscientific craftsman to this would-be philosopher, whether he was genuinely a befuddled brewer in unconvincing 'scientific garb', or – as seemed more likely – an opportunistic hack (and, in his overly forthcoming discussion of additives, a poisoner by proxy to boot).[68]

Strikingly, Thomson's agenda did not attract any public criticism from brewers. James Hinton Baverstock (1785–1837), son of the early hydrometrist, not only endorsed the *Britannica* piece but proclaimed the 1806 'Scotch Report', on which it was largely based, the 'best work on brewing and malting that has appeared'.[69] The contrast with the vociferous brewery denunciations of Humphrey Jackson in the 1770s, and Friedrich Accum in the 1820s, is easily explained. First, though Thomson asserted chemical authority to overrule internal brewery wisdom, he was usually careful not to be seen to be *selling* that authority: he did not promote secret manuscripts to brewers, or nine-shilling duodecimos full of second-hand process accounts to the public at large. It helped, of course, that he had beyond doubt overseen operations in real brewhouses, albeit in uncommon circumstances. It helped, too, that he did not foster stories of toxic adulteration at the expense of any class of brewers, large or small; indeed, the *Britannica* article suggested such tales were unfounded or outdated.[70] There remained the fact of Thomson's public attack on Richardson and other brewer-authors, but the *Britannica* piece appeared soon after Richardson's death: with the possible exception of the shadowy Shannon, none of Thomson's named targets were living at the time of publication, and the rising generation was free to view itself as untainted by the problems Thomson raised.

Indeed, Thomson had some common ground with the 'scientific' brewer-authors against whom he defined himself. In opposing the idea that the brewery was a sphere of separate knowledge, he also opposed the brewery-specific legislation they found most frustrating. Thomson had no time for malt-and-hops purity (chapter 5), arguing that any non-toxic practice should be encouraged if it led to efficiencies. Cheaper forms of colouring should be legalized, he argued, as should quassia, and since all sugar was equivalent, there was no requirement to prefer malt: unmalted barley was a perfectly valid material for making small beer, as had been demonstrated by several Edinburgh brewers until prosecutions interfered. John Richardson would have agreed entirely with Thomson's judgment that a 'wise government would have permitted the improvement, and would have levied the malt-tax in a different manner'.[71]

Thomson's *Britannia* article, with its interspersed accounts of theories of matter and process narratives, laboratory analyses and brewhouse records, served as an important prototype for the writing of brewhouse chemistry by brewers and non-brewers alike in the years that followed. His contributions to the actual theoretical content of the field, however, were notably short-lived, as the next section demonstrates. In ruling the whole internal literature of the brewery out of consideration, Thomson overlooked or dismissed significant interpretations which others, in time, were able to incorporate into the chemical mainstream.

Fermenting the Unfermentable: Brewers' Debates in Print

Thomson considered it self-evident that the established form of brewing publication could not, unaided, extend human knowledge. A scientific professional, he decreed, must keep abreast of the work of his peers to inform the direction of his novel research, and must publish that research promptly to guide others in turn. The brewers, by contrast, not only had obvious commercial reasons *not* to publish all they knew, but also lacked suitable channels for routine communication. In Thomson's eyes, their literature consisted of a sparse handful of treatise-length books, each with its own system or secret to sell, compiled and promoted in isolation. Yet Thomson's judgment overlooked the brewers' ingenuity in discussing their research in venues he would have considered unorthodox: three brewers managed such a debate through the letters columns of the *Political Register*, William Cobbett's Radical weekly, in 1808. Brewer-authors likewise used pamphlets, and sometimes longer works, as forums to discuss, refine and systematize theoretical concepts. The ideas of public chemists fed frequently into their discussions, and on at least one occasion, as we shall see, Thomson's own *Annals of Philosophy* provided a vector for brewhouse insight.

In 1808, John Reynoldson, a maltster and former brewer who was campaigning to overturn a ban on customary malting practices imposed as part of the 1802 restrictions, published a treatise entitled *Practical and Philosophical Principles of Making Malt*. As his title suggests, Reynoldson was inspired by John

Richardson, taking up his picture of a progressive, scientifically informed trade struggling against hidebound authorities. Reynoldson nonetheless dissented respectfully from Richardson on several points, most notably in asserting that the saccharometer did not do what its name implied. Where Richardson had assumed that the increased gravity of worts compared to water was due entirely (for practical purposes) to dissolved sugar, Reynoldson pointed out that malt worts also contained other matter. In particular, there was a gluey principle or 'mucilage', found in varying quantities, which raised the saccharometer reading without necessarily improving the value of the wort.

Reynoldson was here drawing on categories and terminology which chemical authors had introduced in attempting to define vegetable products more generally. Thomson, for instance, had earlier floated the idea of a continuum of vegetable principles, 'running on the one hand into *starch*, and on the other into *sugar*', with mucilage and gluten as stages in between.[72] The 1806 Report affirmed that malt extract contained mucilage, gluten and starch in detectable quantities, and that the proportion of mucilage obtained rose as the mashing process was repeated, but Thomson's experiments had convinced him that the saccharine matter 'bore so great a proportion to the other ingredients, that they may be overlooked' (thus safeguarding the validity of saccharometry, in which, like Richardson, he now had a commercial interest).[73] Reynoldson, by contrast, reported how a fellow malting reform campaigner, William Bent, had isolated the mucilage from solution, probably by treatment with alcohol, and verified that it significantly affected the saccharometer reading. The extracted mucilage was obviously distinct from true 'saccharum', since it did not taste sweet; it seemed more inclined to ferment to vinegar than alcoholic spirit, and was surely not a desirable ingredient.[74]

The mucilage question provoked lively debate among brewers. James Baverstock, who had read the 1806 Report soon after it appeared, supported Thomson's view that the measurable extract indicated fermentable sugar. On this basis, he mounted a public defence against the ever-present rumours of adulteration: no reasonable brewer would want to use cane sugar or treacle, he said, because hydrometric extract determinations showed that genuine malt was actually cheaper. A Radical in politics, Baverstock broadcast this argument in a letter to the *Political Register*, and there found himself challenged by another correspondent citing Reynoldson's objections.[75] 'Candidus', as he signed himself, was evidently a fellow commercial brewer, for his own argument appealed to trade considerations. How, he asked, would Baverstock characterize a third wort – the product, that is, of mashing a grist that had already been mashed and drained twice? By the saccharometer, it might show one-third the extract of the first wort; but no serious brewer would rate it anywhere near one-third as valuable, given its large portion of non-fermentable 'mucilage or viscous matter'.[76]

Thomson, as a non-brewer, had not addressed such issues: the brewers under his supervision at Edinburgh had considered two mashes sufficient, and it seems the first and second worts were generally mixed before the analysis began.[77] Baverstock, born to the trade, was obliged to acknowledge the seriousness of Candidus's objection. Performing distillation trials, however, reconfirmed his belief that wort gravity was reliably proportional to final spirit yield. To resolve the apparent contradiction, he speculated that the mucilage rested in a kind of suspension which did not buoy up the instrument.[78] His findings, he stressed, were in line with Thomson's 'very valuable' conclusions, and also with evidence given to the malt-duty committees by the 'scientific and practical' John Martineau, a partner in Whitbread's.[79] Replying, Candidus professed himself a fellow admirer of Thomson's efforts, but disappointed that Baverstock had 'shrouded himself under the protection' of the chemist's authority, rather than looking to his practical context. To appeal to distillation data was to ignore the obvious point that brewing and distilling were different trades: a practical brewer never tried to maximize the yield of alcohol, since this prevented a full-bodied taste and encouraged cloudiness.[80]

An attempt to reconcile the two combatants came from a third correspondent, signing himself 'I. H.': this was almost certainly John Ham, a brewer and distiller on the Channel Island of Guernsey.[81] Candidus was correct, he said, in believing that the hydrometer was affected by mucilage, but had gone astray in assuming that mucilage was non-fermentable. All the materials registered by the device – 'saccharine, mucilage, and perhaps a little oil' – could be fermented under laboratory conditions, which explained Baverstock's confidence in his distillation trials. Some, however, fermented more easily than others: under the less forceful processes of the real-life brewhouse, only the saccharine was converted in full, leaving most of the mucilage unfermented in the finished beer.[82] Thus, Baverstock's hydrometric comparison was unreliable, although his ultimate claim about substitutes was correct for another reason: the mucilage of malt was needed to give the beer its body, so cane sugar, being pure saccharine, was inappropriate to brewing. Ham's account involved a novel revision of the concept of 'mucilage': where both Baverstock and Candidus saw it as an unwholesome inconvenience, Ham made it a defining component of well-brewed beer.

Thomson, as we have seen, did not reciprocate the brewers' attention to his published work, and the whole debate seems to have passed him by. In 1814, he advised a correspondent to the *Annals* that the saccharometer was a reliable indicator of fermentable value, on the grounds that the proportion of fermentable sugar in the extract was constant for practical purposes. Just as in the *Political Register* debate, a response to this position came under the initials 'I. H.'. Thomson's assurances, said the challenger, were all very well 'for *similar* processes in the *same* brewhouse', but different materials or methods would lead to the constituent parts of malt being drawn out in very different proportions.[83] Thomson made no reply to this objection (which told against the reputation of all saccharometers, including his own), and did not address it in the *Britannica* essay, prepared a couple of years later.

Unfermented matter, however, remained of lively concern to brewer-authors. James Hinton Baverstock, who edited his father's collected works for posthumous publication in the 1820s, did not know the identity of 'I. H.' but accepted his correction: mucilage, he concluded, not only affected the hydrometer but was, in limited quantities, necessary in beer.[84] David Booth, in 1829, stated that the beer's characteristic flavour was determined by the relative proportion of sugar, mucilage, oils and 'earthy particles' in the wort: different temperatures were best suited to extracting these different principles, and so careful control of the mashing heats could determine the flavour.[85] This was just the kind of approach which Thomson had rejected as needlessly obscure: so far as he was concerned, mashing was little more than a dissolving process, and brewers should simply use the hottest water that would not risk setting the goods.[86] According to Booth, however, every skilled brewer a decade later knew otherwise.

Some brewer-authors, while accepting the concept of a non-fermentable principle in beer, saw confusion in classifying it as 'mucilage'. George Adolphus Wigney, who prided himself on observing the conventions of chemical literature, restricted the term to the gummy matter which could be physically isolated, and which was largely removed in the hop-boil.[87] The substance remaining in the beer, on the other hand, could be dried to a sticky sweet mass like wort: it was surely best treated as unfermented sugar. Here, though, the old riddle remained. If the 'saccharine matter' of the malt was a uniform, homogeneous substance, why did part of it ferment, and part remain? One response might have been that the increase of alcohol in the fermenting beer itself affected the progress of the fermentation, halting it at a certain level; but this raised the question of why the level varied widely from one brewing practice to another. Brewers who focused on this puzzle were well placed to engage with a reconceptualization which began to take hold, in chemical literature, not long after Thomson's *Britannica* piece.

Opening the Enclosures

The source of this reconceptualization was a series of investigations into vegetable starch, undertaken largely in France. Starch, it was generally accepted, was the dominant component of barley and other unmalted grains: the key change in malting was the transformation of some starch into sugars and mucilage. From 1825 a string of papers in French-language scientific journals explored the observation that starch was not a homogeneous material, but appeared to consist of two substances. Microscopic membranes or envelopes of the first substance, which was insoluble in water, enclosed and protected the second, which was soluble and resembled gum.[88] In 1833, Jean-François Persoz (1805–68) and Jean-Baptiste Biot (1774–1862) announced that they had isolated and analysed the gummy principle; observing that it rotated polarized light strongly to the right, they called it *dextrine*.[89] The same year, Persoz collaborated with Anselme Payen (1795–1871), a chemical industrialist turned lecturer, in isolating from barley malt a substance they named *diastase*. A tiny quantity of diastase, they reported, produced two significant effects when applied to starch in water at around 70°C. Instantaneously, it tore open the envelopes of insoluble matter, releasing all the dextrine; and more gradually, it converted the dextrine into fermentable sugar.[90] Both activities, the collaborators noted, were highly dependent on the temperature being close to optimum.

This research drew some of its inspiration from growing brewery interest in the use of unmalted grain. It was well known that if raw barley was substituted for a portion of the malt in the mash, it would still produce a fermentable extract.[91] The practice, indeed, was commonplace among distillers, but had historically been avoided as a source of sourness in the more sensitive beer process.[92] The large malt-duty increases of 1802 led to renewed interest in the approach, but the purity regulations imposed under the same legislation (chapter 5) formally banned it, a point confirmed by prosecutions around 1811.[93] Nothing, however, could stop the brewers from experimenting privately. John Ham, the brewer and miscellaneous industrialist who had challenged Thomson in the *Annals of Philosophy*, was one of the most enthusiastic. Where Thomson had believed that it ought to be possible to brew safely with a grist made entirely of raw grain, Ham's trials confirmed that a proportion of sweet malt was needed for success: in his words, '[b]arley that has been malted seems to possess the power of *malting* a further quantity in the mash-tub', making it fermentable.[94]

By 1822, Ham believed he had arrived at a combination of timings and procedures which would allow an experienced brewer to manage with one part of malt to two of raw grain. With no legal market in the trade, he sought to profit by retailing a small pamphlet of instructions for private brewers – apparently successfully, as the work ran quickly through four editions.[95] In 1829, he recast this pamphlet into a Richardsonian polemic against the Excise prohibition on non-malt brewing, strikingly underlining his point by proposing a nationwide shift to making beer from potatoes. Potato-farming, he considered, gave a markedly better yield than grain relative to the land required for cultivation; an enthusiastic deviser of poor-relief schemes, he added that potato-processing would provide useful labour for children in areas of high agricultural unemployment. If unmalted barley could be brewed, then so, chemical analysis showed, could the 'fecula' (starch) of potatoes. Either, added to a due portion of malt, would render fermentable sugars.[96]

Such brewery observations formed part of the inspiration for Payen and Persoz's search for an active principle in malt. One of their sources was a French translation of Friedrich Accum's brewing treatise, which reported eighteenth-century work on unmalted barley by the chemist William Irvine. They apparently also had more direct access to trade interpretations, judging from their opening acknowledgment to 'the scholarly research and painstaking efforts of Leeuwenhoek, Saussure, Kirchhoff, Vauquelin, English brewers, Dubrunfaut, Raspail'.[97] They did not name the brewers in question: evidently, philosophy and art were not to be addressed equally in all respects.

In a further paper of 1834, Payen and Persoz elaborated on their scheme's implications for the practice of brewing, among other industrial processes. Because diastase converted dextrine into sugar only gradually, the wort from the mash would always contain a mixture of both substances. The dextrine, being non-fermentable, would survive into the finished beer: it was, they suggested, the component responsible for the retention of carbonic acid, the stability of the foam, and the distinctive flavour and consistency of the beer, properties they described as 'mucilagineuse'.[98] Payen and Persoz had, therefore, recapitulated most of the properties assigned to mucilage at the end of the Baverstock–Candidus debate.

The first known comment on the diastase discovery in any English-language publication came from a brewer.[99] William Black (fl. 1830–44), whom I shall discuss more fully in chapter 7, made it his business to heed Thomson's call for authors to keep up with the current chemical literature. Learning of the discovery just as his 1835 *Practical Treatise on Brewing* was about to go to press, he added an appendix endorsing the Payen–Persoz account as fully in line with practical brewery observations: the sudden rupturing of the starch accounted for an 'instantaneous change of colour' seen during a well-conducted mash, and the temperature-dependence explained the crucial importance of the first mashing heat. A similar theory, he added, had been devised 'long ago' by a fellow brewer, Robert Stein.[100]

Thomson himself showed no apparent interest in this new interplay of practical brewing and chemistry. His own first account of the diastase discovery, for an 1838 survey text, ignored its brewery dimension; in the same work, he included a section on beer-brewing which merely recycled his early, pre-diastase observations.[101] This was by no means the general attitude of chemical professionals: Andrew Ure, by contrast, used William Black's brewing manual as a source. Mashing, said Ure, not only dissolved the sugar and dextrine pre-existing in the malt, but converted its starch into what he termed a 'sweet mucilage' or 'saccharo-mucilaginous extract', the temperature critically determining the relative proportions of saccharine and dextrine.[102] Much the same account appeared in the 1842 *Elements of Chemistry* of Thomas Graham, who, as we shall see in chapter 7, became an advisor to Black and oversaw the chemical content of his publications.[103] The literature of professional chemistry, then, ultimately endorsed elements of the practically derived brewery framework which, according to Thomson, had no place in a properly chemical theory of beer.

Thomson's *Britannica* account nonetheless endured, ironically, through the kind of recycling he had striven to abolish. The seventh edition of the encyclopedia (issued 1830–42), again edited by Napier, reproduced it entirely unchanged, and in 1849 the *Britannica*'s publishers reissued the text as a stand-alone work, supplemented with 'practical instructions' by William Stewart, 'practical Brewer'. The suggestion of a definitive theory-and-practice compendium was deceptive: Stewart's essay in fact combined a very general (if updated) survey of commercial methods with a laudatory commentary on Thomson's work, reaffirming his assumptions on the nature of saccharine matter and mashing temperature.[104]

By the time of Thomson's 1852 death, the role of commercial brewers in chemical authorship was assured. This was underlined in the eighth (1854–62) edition of the *Britannica*, for which the brewing entry was revised by the Edinburgh medic, James Stark. While paying warm tribute to Thomson, enshrining the 1806 Report as 'the first great step to get science to bear on the operations of the brewer', Stark not only replaced Thomson's mashing chemistry with interpretations based on the diastase concept, but erased most of his historical assault on the brewery's internal literature, softening the thrusts against Combrune, bringing Baverstock into the picture and praising William Black.[105]

Thomas Thomson, then, was the first chemist to work publicly and systematically in a brewhouse, but the last to do so without closely engaging the priorities and interpretations of practical brewers. The Excise consultancy was a gift to his particular breed of professionalization: he could demonstrate that his hands had been caked with malt without being dirtied by trade. His scorched-earth approach to the literature of brewing affirmed the point, and undoubtedly had some influence as a template for later chemical analysis. The generation of chemists who followed him, however, found that innovation depended on a much closer integration with observations and explanations that originated inside the trade, as we shall see in the final two chapters.

7 TREATISES FOR THE TRADE

> [A] brewer, on going from one brewhouse to another, often finds it impossible to produce beer which is equally good in his new situation ... How is this to be accounted for? In days of yore it was attributed to witchcraft ... Chemistry has, however, superseded witchcraft, in every process dependent upon its own laws. The process of brewing being strictly chemical from beginning to end, must be subservient to the laws of chemistry, and until these laws are understood and applied, no uniformity can be expected.
>
> W. Black[1]

This chapter traces the changing character of published literature on brewing between the 1820s and 1850s. The backdrop to this change was the slowly emerging acceptance, in a variety of manufacturing industries, that chemically trained managers could make a decisive commercial difference.[2] The largest breweries first began to appoint chemists of their own from the 1830s, although it was not for another three decades that a group of 'brewing chemists' emerged large enough to function as a community, a development I will discuss in chapter 8. Across the period covered by the present chapter, only a handful of texts was published by men brewing on their own account.

In the eighteenth century, it had been possible for a brewer such as Michael Combrune to theorize about the trade, before a public audience, with no other goal than establishing himself as a credible philosopher. The increasingly restricted expert culture of nineteenth-century science closed off this pathway. It was not that a brewery background presented any crucial social barrier: the sons of even moderately wealthy proprietors were, unproblematically, educated gentlemen. The gatekeepers of elite science, however, retained a residual suspicion of trade knowledge, and increasing demands for laboratory precision ruled out investigations developed in workshop environments. Among manufacturers, of course, there remained the usual wariness about releasing production details. Most scientific devotees from industrial backgrounds, therefore, published chiefly on fundamental theory or side-interest applications, rather than the particular processes which paid their way. The iconic case is James Prescott Joule (1818–89), sprig of a successful Salford brewery. Joule kept up an active managing role whilst performing the thermodynamics research which made his international reputation, but published nothing on beer.[3]

Those brewers who did publish scientific and technical publications on brewing, then, usually did so with the aim of making a commercial profit. Most took fellow members of the trade as their primary audience. The case of John Ham, whose innovations would have been illegal in commercial brewing (p. 153), illustrates the difficulties of selling novel information to domestic brewers, who, though much more numerous, were unlikely to invest in consultancy, expensive equipment or large treatises. Ham's strategy of promoting explicit directions widely through a small, cheap pamphlet carried the obvious risk that it would swiftly be pirated and enter the common literature: he could only appeal to his reader to 'consider it a point of honour, and justice to the author, not to lend it to any one'. Although the pamphlet was initially successful, it appears that its market was soon saturated, and Ham focused his next publication on canvassing to legalize his methods in the trade.[4]

Equally problematic were the more detailed publications, aimed at both trade and general audiences, which were fostered as part of 'useful knowledge' campaigns to raise the literacy of working people and the nation's capacity for trades improvement. My first case study considers the remarkable David Booth, a former commercial brewer who had moved into a precarious career as a professional scholarly writer, who compiled a treatise on brewing for the Society for the Diffusion of Useful Knowledge (SDUK) in 1829. Booth evidently valued the security offered by an organization which paid its authors irrespective of sales; ironically, however, the Society's patrons came into conflict with Booth precisely because of his financial dependency and commercial mindset. More typical of the useful-knowledge agenda was a rival publication by the chemist Michael Donovan, a non-brewer who replicated the standard objections to brewers' authority.

Books for the trade, by contrast, were often constituent parts of long-term commercial ventures, running through multiple editions over a decade or more. Sales of the books themselves were seldom intended as the only source of profit, although a few authors, hoping to retain some of the appeal which had once clung to secret manuscripts, risked promoting printed works privately within the trade at premium prices: George Lloyd Worthington demanded five guineas for his *Brewer's Guide*, a short and unexceptional octavo of 1812.[5] Most authors, however, published mainly to promote private consultancy, equipment, or both, on the principle laid down by John Richardson. Some began to rely on patents as a respectable form of enclosing knowledge, leading to the emergence of the identity of the 'brewer's engineer' around 1850. Yet the most notable difference from the culture of the eighteenth-century instructors (chapter 3) was that all now affirmed a knowledge of chemistry as essential to proper brewery management, claiming close acquaintance with chemical authority through reading or, increasingly, in person.

One consequence of the ever-cheapening print culture of the nineteenth century was that literary productions of all kinds were increasingly visible. The various accusations familiar from past disputes – stubborn illiteracy in brewers, effete impracticality in chemists, mercenary quackery and nostrum-mongering all round – were played out with ever greater regularity, to the point where authors automatically offered pre-emptive defences as they wrote. This is particularly evident in the work of William Littell Tizard, the first self-styled 'brewers' engineer' and author of the largest and most influential mid-century treatise. Though unapologetically unconventional in his written style, Tizard made himself thoroughly familiar with the prior literature and used it masterfully in the service of his own agenda. No chemist himself, he not only accepted the consensus on the need to respect chemical authority, but was actually able to harness it to assert his own credibility over that of a leading public chemist.

Lastly, I address the prospects for alternatives to the treatise form. The emerging convention of the scientific journal article – short, specialized, novel and topical – was ill-suited to the ambitions of brewer-authors who relied on consulting income: the length and comprehensiveness of the treatise were valuable, not only as testimonies to the author's breadth of experience, but as a means to minimize casual borrowing and piracy. Journal publication was more attractive to the small minority of proprietors and employee brewers, not working as consultants, who wanted to explore tentatively the extent to which knowledge could safely be shared in the general interest of the trade. A possible forum was presented by the early publications of the Chemical Society, founded in 1841 with aims including the encouragement of industrial chemistry, but in the event, only a handful of papers emerged by this route. Only in the 1860s, when a new breed of periodical literature internal to the brewery emerged (chapter 8), did published writing on brewing cease to be shaped overwhelmingly by the conventions of the treatise.

'All *Science* but Nothing *Useful*': Brewers, Broughamites and Useful Knowledge

The SDUK was founded in 1826 at the instigation of Henry, Lord Brougham (1778–1868), the Whig statesman and author. Its central aim (not, in the event, notably realized) was to take advantage of the falling cost of print to fulfil Whiggish aspirations for gradual reform from above, giving working-class readers a means for respectable self-improvement through reading in the sciences, history, biography and other topics, as a safe alternative to the more socially militant radical press.[6] The SDUK's largest project, begun in 1827, was the 'Library of Useful Knowledge', published at two-weekly intervals in the form of sixpenny pamphlets, so densely printed that each achieved something like the coverage of a short treatise.

'The Art of Brewing' was the title of a two-part contribution to the Library, issued as volumes 57 and 60 in late 1829, and one of relatively few texts devoted to a particular practical art. The early focus of the Library had been on surveys of theoretical fields; accounts such as 'Brewing' were integral to the overall plan, but of explicitly secondary importance.[7] Brougham's paternalistic agenda was revealed in a piece boosting the project in the *Edinburgh Review*: the practical treatises, he said, were a strategic necessity rather than an end point. The operative 'whose whole life is devoted to Bleaching, and has no thought beyond it', he supposed, would respond only to print material appealing directly to his commercial self-interest: the writer's task was to lead him in stages from such material towards a true appreciation of theory.[8]

'The Art of Brewing' as published, however, thoroughly subverted Brougham's vision. Whereas the mathematical and scientific entries were mostly the work of university men or medical professionals, 'Brewing' was penned by the self-taught David Booth (1766–1847). An enthusiastic philologist, now living mainly by compiling dictionaries and other reference works, Booth had pursued a varied career including at least two stints in the beer trade; 1827 saw him angling (unsuccessfully) for the Professorship of English at the new University of London whilst proprietor of a brewery in King Street, off London's Snow Hill, on the fringes of the literary labyrinth around Paternoster Row.[9] The text he produced, though certainly reformist, was anything but paternalistic, and strongly favoured the authority of direct experience in the brewhouse.

Where most Library authors took on the conventional persona of a disembodied scientific authority, Booth adapted the posture of the informed colleague found in most brewery manuals, referring frequently to personal experience. Like John Ham, he was particularly inspired by John Richardson's fiery philosophical-practicalism and polemics against the Excise laws. The 'absurd prohibitions', he said, were incoherently framed, made 'no distinction between useful and poisonous ingredients', and operated mainly to stifle beneficial innovation.[10] Booth had also acquired the manuscript directions which Richardson had sold to private correspondents under conditions of secrecy, and considered himself free, fourteen years after Richardson's death, to print them in full. He was careful, however, to establish points of dissent from his chief authority: Richardson had been mistaken in rejecting sparging, or rinsing of grains after mashing, a familiar Scottish practice whose efficiency was now increasingly accepted in England.[11]

Booth also drew on Thomas Thomson's *Britannica* entry, but Thomson's attempt to re-found the theory of brewing on the basis of current chemical literature was very far from his purpose. The two had corresponded briefly during the preparation of Booth's manuscript, and when Thomson objected to some of his terminology, Booth indicated that he would happily stand correction: 'No man', he told the chemist, 'can be less a theorist than I wish to be'. Similarly, Booth conceded that Thomson's saccharometer scale was more philosophically accurate than Richardson's familiar pounds-per-barrel, but felt no advantage 'would arise to the *practical* Brewer' in making the change. The correspondence was seemingly amicable, notwithstanding that Booth saw chemical theories as 'merely certain doctrines of faith which string the facts into bundles'. Booth respected Thomson for his history of hands-on experiment, while Thomson could not consider Booth as any kind of a rival. Thomson concluded his reply on a practical note, emphasizing their common ground with a reference to the parlous state of the trade in Scotland: 'Unless some alteration is made in the laws it will be ultimately destroyed'.[12]

Booth's commercial context was the cause of persistent tensions, however, in his relationship with the SDUK. His correspondence with the Secretary, Thomas Coates, reveals him as a perfect type of the necessitous author, constantly seeking subventions to ward off creditors and starvation. This, and his self-taught background, set him apart from the Broughamite academic and medical professionals who more typically wrote for the Library of Useful Knowledge. Their preoccupations, he said, were bound to frustrate the Library's purpose: 'You are all *science*', he twitted Coates, 'but nothing *useful*'.[13] 'Brewing', in the event, was one of only a few accounts of particular arts to appear in the Library series. Booth himself tried to expand the scope, soliciting a text from a Fife millwright, but eventually became convinced that the Society's managers were merely paying lip service to such work.

Booth also, unsurprisingly, saw treatise-writing as a branch of trade in its own right. To a degree, the SDUK – or more probably its publishers, Baldwin and Cradock – accommodated this position. In another Richardsonian touch, the collected edition of the two parts advertised a 'Description and use of the brewer's saccharometer', written by Booth, to be sold with the instrument by Baldwin and Cradock at two guineas all in.[14] The Society's proprietors proved less amenable, however, to Booth's proposals to extend his material from two treatises into three, the author's fee being a flat £40 per treatise. Relations ultimately soured after the Society requested an additional treatise on wine- and cider-making, then vacillated at length about publishing it. The situation festered until 1834, when Booth was given leave to release the wine-making treatise, and two further instalments of 'Brewing', through another publisher.[15] Four years later, he was to be found advertising as a brewing consultant, apparently on a private basis.[16]

More typical of useful-knowledge authorship was a text compiled by Michael Donovan (1791–1876), Professor of Chemistry to the Apothecaries' Company in Ireland, for the Cabinet Cyclopædia, a series of 6*s*. treatises issued from 1829. The Cabinet's director, Dionysius Lardner, founding Professor of Natural

Philosophy and Astronomy at University College London, felt that 'useful applications' rested inescapably on expertise in the sciences and mathematics, and also that secrecy was inevitable and (he grudgingly conceded) rational among commercial producers.[17] Accordingly, he assigned the brewing entry not to a brewer but to Donovan, who had previously written on chemistry for the Cyclopædia, as part of a volume on 'Domestic Economy' which also encompassed distilling and baking. Donovan's text, which appeared the year after Booth's, was at pains to establish that a chemist was best placed to describe these processes. His opening affirmation of the paradox of the trade writer could almost have been written in the seventeenth century:

> To learn the manipulative parts in which concealments are attempted can scarcely be effected with certainty. Such knowledge can only be obtained from those whose necessities compel them to make sale of it; and who, after all, are seldom capable of communicating what they know in such a manner as will render their skill available, without great trouble to the enquirer.[18]

Yet Donovan was obviously familiar with the standard counter-arguments, and freely admitted that past accounts by 'scientific men' had been misleading or incoherent. His resolution imitated Thomas Thomson's (chapter 6): he had, he assured his reader, personally experimented 'on the scale on which the manufacturer himself works' over a twelve-year period.

In fact, Donovan's text was heavily modelled on Thomson's *Britannica* entry and those of other encyclopedias, to the extent that one reviewer called him out for hack work.[19] A fifty-page historical prelude was followed by a succession of theoretical and practical accounts in which Donovan's first-hand research jostled with long sections on chemical matters peripheral to the brewery, such as the nature of putrefactive fermentation and the preservation of bodies. Where Thomson had attempted to exclude the brewer-authors from chemical legitimacy, however, Donovan claimed to interweave the best of all the available sources.[20] Thus, he sometimes showed the influence of practical authors, accepting that approaches that might seem to be optimal in terms of yield must sometimes be rejected where they were difficult to engineer.[21] Yet his lack of first-hand engagement showed, for instance, in his belief that the malt-and-hops purity regime of 1802 still applied, and that 'preparations of fish' were no longer used.[22]

Donovan differed most sharply from Booth in downplaying the value of brewers' special expertise. Trade secrets, he said, were usually more laughable than sinister, offering no real improvements on common knowledge. Worthwhile innovations could only come from outside the trade, as the paradox of experiment made obvious:

> To make experiments in brewing on the large scale is a dangerous occupation of capital. No brewer ever thinks of making such an experiment if he apprehends a failure:

> when untoward circumstances occur in a brewing, his object is very naturally to correct and modify them, rather than to ascertain what would be their results if allowed to proceed as they threatened ... Throughout his whole career he is cautious, and never willingly varies from that *norma* which he has once found successful: hence the results of other modes of operating are known chiefly by conjecture; disputed points remain for ever matters of opinion, and, in fine, the art remains pretty much what it was a century since.[23]

Commercial brewers, Donovan alleged, remained blind to their greatest shortcomings, which included (in a distant echo of Jeffrey Boys) working on an uncontrollably large scale.[24] Booth, who took his cues from Richardson and John Tuck, showed much greater sensitivity: a call for change must not be open to reading as an attack on the trade.

From opposing sides, Booth's and Donovan's treatises recapitulated the gulf between the established personas of the practical brewer and the chemical authority. As we might expect, Booth, the self-proclaimed insider, was a strong influence on the later brewer-authors I discuss below. Most took up some version of an argument he added in 1834, justifying his recent decision to start charging his readers for personal advice. Recalling the misadventures of past instructors, he was careful to clarify:

> I hate quackery; and I should not have written at all on the subject, had I not meant to be explicit: but written directions alone, though they may assist, will never make a practical man ... many peculiar obstacles stand in the way of the young practitioner, which can only be removed and prevented, for the future, either by the oral instructions of some one who has trod the path before him, or by his own (often dear-bought) experience.[25]

What is more surprising is that the brewer-authors of the 1830s and 1840s also affirmed part of Donovan's platform: secretism was worthless, and valid improvements must come from work not customary to the brewer's practice. Key to this change was the increasing stability of professional identities which shared the brewhouse's knowledge culture, but not its precise commercial aims: consulting brewers, brewers' engineers and, in later years, brewers' analysts.

'Applying to Men of Science': William Black's Division of Authority

William Black's 1835 *Practical Treatise on Brewing*, noted in chapter 6 as the first text in English to discuss diastase, claimed forty years' experience for its author. Formerly a partner in the successful Devanha Brewery in Aberdeen, noted for its nationally distributed porter, Black had moved to London, taking on a small brewhouse of his own, but working increasingly as a consultant and tutor.[26] His treatise targeted an exclusively commercial audience, and he did not consider it necessary to buttress his credentials with a show of reformist zeal. Though he found the legal restrictions on malting and brewing absurd, he avoided the Excise-baiting of Richardson and Booth, preferring to emphasize instances of 'intelligent' Excisemen turning a blind eye to necessary infractions.[27] In other respects, his self-justification was conventionally philosophical-practical. He was moved to write, he said, by the lack of an up-to-date treatise charting a middle course between 'homely' inanities and abstruse 'scientific technicalities'. He faithfully reproduced Richardson's insistence on the paramount intelligence of the experienced brewer and Tuck's invocation of copper-side testimony, and asked that the reader should excuse any stylistic deficiencies in the writings of a practical man. Nevertheless, said Black, his language must reflect the fact that brewing was at root a chemical process, albeit one which remained in its 'infancy, as a science'.

This, he said, was because the connections between 'eminent chemists' and 'practical men' were not secure. Most brewers failed to apply themselves to the chemical literature, while those few 'scientific gentlemen' who had considered practical questions had not been well guided: like Tuck and Ham he considered that Friedrich Accum, in particular, had been led far up the garden path by his canny informants.[28] Black went beyond most earlier sources in proclaiming openly that 'drugs', while unnecessary in the normal course of business, were the appropriate remedy for beer which had fallen '*sickly*' – an occurrence which, despite ongoing research, remained common and unpredictable, and told nothing against the brewer's competence. Chemical expertise among brewers, therefore, was to be fostered rather than feared. Sugar, Black pointed out, was 'by a very simple chemical process' converted into poisonous oxalic acid; a chemically trained brewer would be on guard against such combinations.[29]

The chief novelty of Black's approach, however, was a very strong emphasis on the need to apply different practices to different brewing sites. He could testify to this at first hand, as a Scots brewer in England; so could his friend and collaborator Robert Stein, successful in Edinburgh, who had almost failed when he came south. The arrangement and management of the brewing plant, Black believed, was most responsible for determining success or failure, although the whole trade was 'still very much in the dark' as to the mechanism.[30] It was thus important to use practical experience gained in as many different sites as possible – the experience, that is, which a man in Black's own position could offer. Black later added a refinement: diagnosis depended crucially on 'aromas and appearances' which could not be described in print.[31] This self-justification was close to, and perhaps modelled on, David Booth's; it was also a clear negation of Michael Donovan's claim that experience led only to common knowledge.

Black's insistence on site-specific treatment was closely integrated with his main theoretical interest: the effect of electricity on beer. As a chemical agent, thought Black, electrical fluid had the ability to increase the rate of fermentation in worts. During thunderstorms, therefore, the uncontrolled charge of the atmosphere would often result in the beer turning sour, though this depended crucially on the layout of the brewhouse. Here, Black suggested, was an answer to the awkward question of why seventy years' worth of brewer-authors' exhortations to measure, systematize and record had not turned large-scale brewing into a predictable business. Foxing was a disease of progress: as brewers increasingly replaced their wooden fittings and vessels with cast iron, they unwittingly created ever-greater electrical interactions between the atmosphere and the beer. The solution was careful insulation, for which Black offered himself as consultant. The approach, though unknown to most of the trade, was fully in line with contemporary chemical theory: 'I have the authority of a scientific gentleman to whom I was lately introduced to say, that he can at any time immediately communicate to the soundest beer the acetous principle, by sending through it a shock of electricity'.[32]

Why did Black place a boundary between the 'scientific gentleman' and himself? All significant brewers styled themselves 'gentleman' by this period, and Black claimed explicitly the status of a 'scientific' practical writer, in the tradition of Richardson. By a 'scientific gentleman', however, Black meant a gentleman publicly devoted to science, covering a much broader range of enquiries than could be accomplished by a competent representative of any one trade. The archetype, still, was Humphry Davy, who had died in 1829: Black's first mention of the relationship between electricity and fermentation stressed Davy's endorsement.[33] Black carefully distinguished the theory behind his account, which was due to 'the most eminent men in Europe', from electrical management in 'the operative department of brewing', which was his own innovation.[34] He thus presented himself as both modest and eminent, a competent follower of theory but a leader of practice. The key difference was that whereas the flow of respectable chemical knowledge must always be unrestricted, chemically informed brewhouse knowledge was restricted by nature, requiring a degree of personal communication.

Though he did not place himself among the gentlemen of science, Black was at pains to establish that his connections with them were personal. He was in contact by 1838 with Thomas Graham (1805–69), a former Glasgow student of Thomas Thomson who had recently been appointed Professor of Chemistry at University College, London.[35] It was perhaps through Graham's influence that Black came to present a paper on electricity and fermentation to the 1837 Annual Meeting of the British Association for the Advancement of Science (BAAS), held in Liverpool. Black's motive in attending, he explained, was to solicit members of the Association 'minutely to examine the nature and extent of the electrical agency'. The 'well-devised experiments' of laboratory specialists would achieve a level of understanding which his existing audience of practical brewers could not;

more importantly, scientific authority was needed to convince brewers to reform their practice, raising the quality of beer in the national interest.

The BAAS, in fact, offered limited opportunities for this kind of intervention. Founded in 1831, its chief stated goals were to break down geographical barriers to the circulation of scientific ideas; to put scientific enquiry on a more systematic basis; and to encourage support for such enquiry among policymakers and the educated public. Affirming the mutual relevance of the sciences and manufacturing arts was part of this project – hence, in maritime Liverpool, a showpiece demonstration of 'the application of Lightning Conductors to Ships' – but papers on practical topics were relatively unusual. The star contributor to the Chemistry and Mineralogy Section was Justus Liebig (1803–73) of Giessen, then on his first British tour. Liebig's focus on organic chemistry was strongly endorsed by Thomson and Graham.[36] His whole approach was fundamentally concerned with industrial processes, and he was to visit the major London brewery of Truman's later in his trip;[37] yet Liebig's Liverpool presentation, on uric acid, was framed purely as an approach to settling theoretical questions of elemental composition.[38] In a section dominated by university chemists, physicians and civil engineers, the only contributions which focused on industrial processes came from two ironmasters and Black himself.

There was thus no particular climate of encouragement for Black's audience to translate his research, which focused on the practicalities of brewhouse insulation, into the laboratory. Black did gain something from his inclusion: his account, chemically orthodox for its time, drew praise from both Thomas Thomson and Michael Faraday, making a rare BAAS appearance as Section President. The encounter also helped to circulate Black's name in print, as the substance of his paper was reproduced in the *Literary Gazette*, which had favourably reviewed the 1835 *Practical Treatise*.[39]

Ultimately, though, Black found the *Treatise* itself more useful for building his chemical credibility. A second edition of 1840 was greatly revised and expanded: now said to be 'based on chemical and economical principles', it showcased the support of George Birkbeck (1776–1841), the physician and natural philosopher best known for his support for the extension of education via the London Mechanics' Institution, the early University of London and the SDUK. Personal authority mattered: Black stressed both that he had been 'materially assisted by his respected friend' in the volume's 'scientific arrangement', and that he was 'authorised by Dr. Birkbeck to state his general coincidence with the subsequent opinions'.[40] The chemical patron's role passed upon Birkbeck's death to Thomas Graham ('whose eminence in the science of chemistry is universally acknowledged'); the preface to a third edition of 1844 suggested that Graham had directly revised the text. All the *practical* content, Black repeated, derived purely from his own experience, 'in all parts of the United Kingdom'.[41]

Black also used the later editions to hone the exposition of his central point. What every brewer really wanted, he said, was a reliable output. To get it, he must reject the traditional 'witchcraft' of the trade, the 'pretended secrets', the 'adages, old saws and dogmas': the answers lay in chemical and electrical theory, rightly applied. Rather than 'throwing away money for the assistance promised by the secrets of empirics', brewers should 'adopt the superior plan of applying to men of science when any difficulty occurs, and we have little doubt they will be courteously received, and the required information be freely imparted'. Few brewers, of course, were in any position to do this directly; yet they could consult Birkbeck or Graham by proxy, Black implied, by reading his books.

Yet theory, to Black, did not give easy or universal solutions: it provided the ground rules for detailed investigation of the brewing site, resulting perhaps in its complete reorganization, based on long-term experience such as his own.[42] Experience, in this picture, always implied a degree of experiment. Where Michael Donovan had argued that brewing innovation must depend on experimental chemists, because tinkering with established practices was not credible or financially sustainable in the trade, Black saw no reason why a brewer should not follow the example of 'the great Sir Humphry Davy', who had even 'represented himself to be a bad manipulator', arguing that it was through interesting mistakes that productive new discoveries came to light.[43]

Black's approach depended not only on chemists' willingness to endorse brewhouse knowledge, but on the increasing accessibility of their work to brewers, a consequence of the general wider circulation of chemical texts through the useful-knowledge movement and cheap print more generally. Though he seldom resorted to humour, it was presumably with tongue in cheek that Black referred to production getting 'out of trim, (technically so called) or in plainer language, into irregular fermentations'.[44] Such apparently plain-spoken brewhouse argot as 'out of trim' was now significantly more esoteric than the language of chemistry, at least as far as literate audiences in general were concerned. Imitating chemical works for brewing audiences was still no straightforward business, however, as one of Black's contemporaries was meanwhile discovering.

George Adolphus Wigney and the Paradox of Experiment

At first glance, the agenda of George Adolphus Wigney (1789–1872), a Brighton-based brewer and consultant, closely resembles Black's. Wigney's *Theoretical and Practical Treatise on Malting and Brewing*, issued, like Black's *Practical Treatise*, in 1835, presented a particularly eloquent evocation of the potential for reform. Over five pairs of facing pages, Wigney contrasted the 'Old System', reliant on 'guess', with his 'New System', determined by 'rule'.[45] Mashing heats were to be set by calculation, using correction tables he had drawn up by experiment; mashing times and procedures were to be modified according to the evidence of systematic trials, based on chemical understanding, to achieve the greatest yield of sound beer with the least outlay of fuel and time.[46] Elsewhere, Wigney meditated on the folly of successful brewers' sons, raised as gentlemen, who thought themselves above the 'minutiæ' of production, and failed to keep up with incessant practical innovation.[47]

A major difference, however, was that where Black had focused on the practical consequences of the new chemistry, Wigney took it as part of the brewer-author's business to lead his reader through a general course of chemical instruction. His earliest book, *A Philosophical Treatise on Malting and Brewing* of 1823, opened with a long section on the nature of heat, proceeding through a didactic sequence of definitions of terms to build up an outline of the oxygen-caloric theory, touching only briefly on its relevance to brewing. He then addressed the elemental composition of malt and its constituents, before speculating at length on the nature of the transformations involved in malting and fermentation, at one point breaking into rhyme:

> Man dies! and putrefactive ferment gives
> In gaseous form, those elements that wrought
> His wond'rous frame; and transmigrated lives
> Fresh forms, with the same ærial substance fraught.[48]

This esoteric approach was probably a simple consequence of the balance of Wigney's chief sources: 'Messrs Brande, Murray, Nicholson, Parkes, Blair and Richardson'. The first four authors had produced synoptic works on chemistry at various dates since the 1790s; 'the Reverend David Blair' was an industrious but invented scholar, dreamt up by the publisher Sir Richard Phillips as a reassuring title-page garnish for his *Grammar of the Principles and Practice of Chemistry* and other works.[49] The *Grammar*, like Samuel Parkes's *Chemical Catechism*, was an elementary survey, intended mainly for juvenile audiences. Presuming his reader to be a chemical novice, Wigney aped the stylistic and structural forms of this beginners' literature: his excursions into verse were perhaps inspired by the poetical quotations interspersed through the *Chemical Catechism*'s footnotes. The ubiquitous John Richardson was Wigney's only model among the brewer-authors.

Nonetheless, Richardson's influence predominated, as it had for Booth and Black, when Wigney asserted the special advantages of a philosophically aware practical brewer. In particular, Wigney attacked the promoters of systems based on invariable rules and directions, such as fixed mashing heats, in terms which closely echo Richardson's and George Blake's attacks on chemical projectors (chapter 3). A rigid system, said Wigney, would always go astray: sound management was a matter of continually responding to variations in materials, atmospheric temperature and other circumstances, guided by careful record-keeping. Proprietary secret systems, therefore, were as useless and dangerous as the 'trash' peddled by nostrum-mongers. There was no short cut to success: the ideal training was a blend of formal education and wide practical experience.[50]

Above all, Wigney challenged the paradox of experiment head-on, recording with increasing frankness his most expensive production disasters.[51] To experiment with unorthodox materials and methods, he argued, was enlightening precisely because the results were typically far worse than traditional practice could achieve: the failure was often traceable to some factor which the terms of enquiry had overlooked, pointing the way for further research. Past innovators, for instance, had tried to abolish the customary boiling of hops in the wort, which was known to reduce the yield of fermentables, and instead boiled the hops separately in water and mixed the resulting extract into the wort; but the approach, in practice, always left 'something wanting' in the flavour of the beer. Wigney conceived that there must be an essential oil in the hop which was extracted only in the presence of the 'saccharum'.[52] Later, he explored an alternative solution to the problem, using a hydraulic press to recover wort retained by the hops.[53]

The risk, of course, was that practical brewers would write off such experimentalism as dilettante tinkering. Wigney's opportunities to forestall the charge by pointing to his personal experience were not ideal. Although his wealthy and influential family had long-standing brewing interests, he was not personally raised in the trade: it was his cousin, plain George Wigney (1789–1847), who inherited the large-scale brewery of Wigney and Sons. George Adolphus had been adopted by George's father as a young boy, but had seemingly been kept out of the family business altogether, instead setting up as a grocer in nearby Chichester; he was a total novice when he added a small brewhouse to this venture around 1815, meeting with frequent disasters which were presumably a factor in his 1818 bankruptcy.[54] Subsequently, he worked as an employee at various breweries including his cousin's, later claiming to have focused 'sedulously' on the improvement of malting and brewing equipment, some of which he promoted for sale. In 1830, however, when his affairs were in difficulties again, Wigney was identified as a stationer.[55]

Perhaps surprisingly in the circumstances, Wigney did not publish the 1835 treatise heralding his 'New System' at large, but promoted it by private subscription via postal circulars to the trade. He named his price at £100 per copy, of which half was to be paid in advance. Such fees had precedent – Richardson's manuscripts had been priced higher decades earlier – but the sum was extraordinary in the era of ever cheaper print: William Black's published treatise cost a guinea in 1835, falling to 12*s*. for the second edition and 10*s*.6*d*. for the third. Wigney freely accepted that his price might appear 'startling' and 'preposterous', and that it was an occupational hazard in the trade to be 'often waited upon by itinerant professing communicators of important information' which turned out to be 'trash'. His unusual strategy was to convince readers that no brewery insider would dare to set such a fee, and risk the annihilation of his reputation, unless his guidance could really be relied on to give satisfaction. Ingeniously, Wigney's prospectus also turned his employee status, and lack of a consulting practice, to advantage: since his 'engagements' at the Brighton brewery prevented any 'itinerant excursion', a printed book was the only possible recourse to foster the adoption of his scheme.[56]

Wigney clearly felt the need for some practicalist self-justification in the treatise itself. He repeatedly pressed the point that he was literally writing *in the brewhouse*, with the processes under discussion going on around him.[57] Yet his expertise, he emphasized, rested on twenty years' research among books and brewhouses alike. Simply giving a practical account of his processes would not be enough, for the reasons he had earlier outlined in attacking secret systems: a treatise on brewing should be a spur to the reader's own research and understanding, hence the continuing need for chemical generalities.[58]

Wigney's initiative succeeded to the extent that he met his initial subscription target.[59] In corresponding with his readers, however, he found that many reported that they could not make sense of the chemistry underlying his directions.[60] In their requests for pointers to suitable chemical primers, Wigney saw another publishing opportunity: the material he deemed useful was scattered across 'many portly volumes', and might be gathered and provided in 'concentrated form'.[61] The result, published in 1838, was a unique production which Wigney termed a *Cyclopædiæ* [*sic*]. Superficially resembling a reference encyclopedia, with entries under alphabetically arranged headings, its content ranged widely from the technical to the highly abstract: 'Acetous acid', for example, was preceded by entries on 'Accuracy' and 'Accustom', meditative essays which argued for the reform of brewing practice in the most general of terms.

Unsurprisingly, the *Cyclopædiæ* did not resolve Wigney's problems with his correspondents. He eventually concluded that he was beginning too late: brewery training must be completely revised, with formal technical education at the apprenticeship stage, including such typical useful-knowledge preoccupations as 'Pneumatics, Hydrostatics, Mechanics, &c'.[62] To proclaim the inadequacy of his past publications, of course, he issued another. *An Introductory Treatise on the Theory and Practice of Malting and Brewing*, published 1850, explicitly targeted the beginner brewer with elementary accounts of the process, set amid further exhortations to supplement the traditional brewers' 'mechanical' knowledge with the lessons of chemical insight.

Remarkably, Wigney's appeals for universal chemical instruction were coupled with an increasingly unabashed attitude to putting a price on knowledge. In the 1820s, he had been careful to mount a version of the justification later refined by Booth and Black: some kinds of private instruction were not secretive, because it was simply not possible to reduce them to print.[63] By 1836, however, he was promoting sections of his writing piecemeal, at £10 or £20 per chapter, to brewers unwilling to part with the full £100; the 1850 *Introductory Treatise* is a sustained exercise in withholding, repeatedly hinting at advantageous possibilities and referring the reader to the more expensive *Theoretical and Practical Treatise* or *Cyclopædia* for details.[64] The final forty pages of the *Introductory Treatise* are filled with specifications for equipment designed or distributed by Wigney, culminating in a complete list of his wares. Numerous private printings – brewhouse plans, wheat and barley values, a narrative of a water-softening process, an 'arithmetical formula for the computation of the proper first mashing heat' – appear amid the brewing vessels, refrigerators and thermometers: print, to Wigney, had become another form of proprietary apparatus.[65] Unfortunately, we have no direct evidence of brewers' response to Wigney's schemes, although this may in itself be telling.

The irony of Wigney's case is that for all his championing of experimentalism, the new chemical order, and the didactic form of the elementary chemical text, he showed far less engagement with current chemical research than any of his brewer-author contemporaries. The *Introductory Treatise* shows that he had ignored, or missed altogether, the diastase discovery which had been assimilated into Black's account fifteen years previously.[66] Others, however, were meanwhile beginning to seek opportunities to create, rather than interpret, fully fledged chemical literature.

'The Help of the Mechanist and Chemist': Authorship and Consulting in the Mid-Nineteenth Century

The largest and most prominent treatise of the mid-century was the work of William Littell Tizard, formerly an 'operative brewer' in Vauxhall. In 1841, Tizard took out a patent specifying several pieces of brewery apparatus including the 'mashing attemperator', a spinning arrangement of submerged pipes which, he claimed, could regulate temperature far more effectively than conventional static devices.[67] Thus equipped, he set up as an independent consulting engineer specializing in brewery work, operating from a workshop on Aldgate in central London. His chief business was the installation of his own patent equipment, but, like William Black, he consulted more generally on site layout and brewing methods.

Tizard published his *Theory and Practice of Brewing Illustrated* in 1843, having raised subscriptions among commercial brewers. Priced at 25*s*., the volume ran to more than five hundred pages, around twice as long as either Black's or Wigney's most recent productions.[68] Much of the text was devoted to his own patent technologies, including such wonders as the 'hystricon', a rotary device to agitate the grain whilst sparging; the 'Caloriphagon', a lamp enclosed in a cooling water jacket to avoid heating the fermenting room; and the 'Pneumatic Life-protector', a signal light to show excess build-up of carbonic acid gas in subterranean vaults, which he proposed should be purpose-built as the ideal fermentation environment.[69] Tizard went on to acquire further patents, which he worked into further editions of the *Theory and Practice* until 1857.[70]

In using a comprehensive-looking treatise as a vehicle for patent promotion, Tizard was almost certainly inspired by Richard Shannon's even larger volume of 1805. Yet where Shannon had parroted material from chemical texts, earning the hostility of Thomas Thomson (pp. 145–6), Tizard took a far shrewder line. Most importantly, he was the first brewer-author to see the rhetorical value of surveying the prior literature in detail. At what must have been considerable trouble and expense, he familiarized himself with all the major treatises of the nineteenth century and several from the eighteenth. He could thus provide frequent comparative synopses such as the following:

> In adjudicating the preference in brewing waters, writers have differed, the greater number being neutral, as Shannon, Donovan, Black, Combrune, and his disciple Hayman; of the soft-water advocates are Richardson, Wigney, and Roberts; and for employing hard are Levesque, Ham, and, under his own system of general management, the present author.[71]

Tizard claimed to have taken the best from his various sources, while improving on all of them in some respects. He described all the widely used saccharometers, proclaiming the methods behind each to be less accurate than his own (the Thomson–Allan device was particularly deficient).[72] Likewise, he used history to exclude ideas he did not approve of: a competitor's 'Calorific Infusor' could be dismissed because it was merely equivalent to part of John Long's system of 1790 (p. 67), which the whole trade had long ago realized offered no special advantage.[73]

The survey also gave Tizard a keen sense of the *faux pas* of earlier authors. To forestall charges of puffery, he acknowledged upfront that he wrote to promote his patent apparatus, and in apologizing for the length of his work, stated that the nature of patent protection made it reasonable that he should explicate his equipment as far as possible. He stressed that he did not presume to teach any competent practical brewer his business: when he presented common trade knowledge, it was for 'young brewers, and those called sleeping partners' who 'may lack practical information'.[74] Tizard was at pains to contrast himself with rivals such as Thomas Hitchcock, a self-proclaimed 'Professor of Brewing and Malting', who (said Tizard) had set up as an instructor using fake testimonials from men posing as brewers in country towns. Tizard's own claims were supported by testimonials from satisfied brewers, and by invitations to witness the machinery in operation.[75]

Tizard was committed to the usual picture of brewing as a chemical art, subject to continual improvement through scientific discovery: there was cause enough to publish a new treatise, he said, in the fact that many brewers were still unaware of the theory of diastase, on which he referred to standard sources such as Liebig and Edward Turner.[76] Where Thomson and Donovan had condemned the older brewer-authors as unscientific and mercenary, however, Tizard instead used his extensive literature survey to depict a cumulative tradition of credible brewhouse science dating back to the days of 'Dad Richardson', the 'veteran' banisher of old-womanly superstition, evidently his favourite authorial model.[77] Scientific attainment must, of course, be tempered by brewhouse experience. Here, Tizard followed Wigney and Black, citing the case of a 'clever young man' from a soft-water district who came to grief by importing his chemical knowledge rigidly to a hard-water locality.[78]

One attainment which Tizard certainly shared with leading chemists was his ability to conduct an entertaining public feud with Andrew Ure.[79] Although the quarrelsome chemist took the internal brewery literature more seriously than his great rival Thomas Thomson, drawing in particular on William Black, an 1845 supplement to his *Dictionary of Arts, Manufactures, and Mines* contained harsh words for the trade in general. British brewers, said Ure, were hidebound and unscientific in ignoring the superior low-temperature fermentation and storage methods of Bavaria, which, thanks chiefly to Liebig, were 'more philosophically studied and incomparably better understood'.[80] Compounding the felony, Ure annexed an adulteration narrative, alleging that London porter was routinely dosed with '*Calfini*, a preparation made with the oil distilled from the outer bark of the birch'. Ure's accusations were also excerpted in the *Mechanics' Magazine*, a threepenny weekly providing one of the few outlets for topical discussion by literate artisans, often of a disputatious nature.[81] Here they drew a sharp response from Tizard, 'on behalf of the trade'.

No brewer, Tizard said, had ever even heard of the alleged '*Calfini*', while in another passage, Ure had confused the old 'essentia bina' with the new and legitimate patent malt. Aware that he was addressing a readership for whom Ure's industrial authority carried some weight, Tizard offered the standard mitigation: the professor must have been 'woefully betrayed by some ignorant, malicious, and designing knave'. The ultimate problem could be summarized in six words: '*Dr. Ure is not a brewer*'.[82] In a revision to the *Theory and Practice*, where he could assume a sympathetic audience, Tizard was less restrained about the chemist's 'delusions'. It was not good enough to write as Ure did, concerning a process to maximize extraction, 'this I believe is seldom practiced by respectable brewers, as it impoverishes the grains which they dispose of for feeding cattle'. An authoritative chemist would deal in concrete evidence, not vague reports cloaked with an 'I believe'. Deliberately leaving saccharine value in the grains was actually a hallmark of old-fashioned, dilettantish and financially illiterate brewers: grains which sold for £250 for pigs might, if properly extracted instead, contribute £800 or £1,000 to the value of the beer. Only a naïve outsider, said Tizard, would mistake the laggards of the trade for its 'respectable' element.[83]

Tizard's ire was stoked by Ure's involvement in a new scheme for brewing from malt extract. The nineteenth century had already seen several such projects, usually patent-based: none had succeeded commercially, presumably because they left, as Wigney had put it, 'something wanting' in flavour or stability.[84] This did not prevent the consulting engineer Henry Dircks (1806–73), later known as the promoter of the 'Pepper's Ghost' illusion, from mounting a 'British National Brewing Company' to work two patents for hopped extracts around 1846.[85] The Company's published advertisements relied on a testimonial from Ure, who professed to have analysed the product, stressing its freedom from 'those noxious ingredients too frequently present in the beers, porters and ales of commerce'. Tizard concluded that Ure (whose postal address was prominently featured) was himself the director of the project.[86] In this, he was probably quite mistaken; the projectors were, however, relying on Ure's authority to emphasize suspicion of commercial brewers among private householders, who – since the process involved unmalted grain – constituted their only legal market.[87]

Again, the *Mechanics' Magazine* became the seat of public controversy. Dircks was challenged by an anonymous 'Brewer' of Oakham, Rutland, who was probably one of Tizard's faithful clients, John Crowson of the Patent Steam Brewery.[88] It should be obvious, protested the Brewer, that the process' 'needless and expensive concentration and dilution' made it hugely inefficient compared to conventional brewing, and the insinuation in the promoters' guarantee of 'a truly wholesome and undrugged beverage' was farcical when they themselves were selling an unfamiliar, sticky mass manufactured behind closed doors.[89] The Brewer, in turn, was attacked by a 'Chemist' of Kennington – a locality which would have suggested associations with the Chemical and Agricultural College directed by John Collis Nesbit (1818–62). Strikingly, the Chemist simply assumed that his opponent was Tizard, or a shill acting on his behalf, and that the only purpose of the objection was to puff Tizard's wares in the *Mechanics'*.

The Brewer had emphasized the thousands of pounds invested in specialist knowledge to secure the relative efficiency of the common brewery; the Chemist bracketed such behaviour with reliance on the itinerant beer doctors, to whom such a 'simple man' as the Brewer, lacking full understanding in his own right, must fall prey.[90] The Brewer retorted that the Chemist had failed to grasp the growing importance of consultants in nineteenth-century industry. Of course brewers employed men with specialist knowledge beyond their own, as would any right-minded commercial man: 'What would be the condition of our factories, our iron and glass works, our bleaching and dyeing establishments, &c.,

but for the help of the mechanist and chemist[?]' If his opponent did not understand this, wondered the Brewer, 'Can he really be 'A *Chemist*'?'[91]

This line of attack was unsurprisingly picked up by Tizard himself: working the controversy into the third edition of the *Theory and Practice*, he contrasted the Chemist's reliance on outmoded sources and sneering invective against the restrained and informed Brewer.[92] Something crucial had changed in the quarter-century since John Tuck, responding to Accum's publications, had seen chemists as a breed fundamentally opposed to the brewers' interests. Chemical literacy was now taken as a universal condition of authorship, to the extent that a professing brewer could lacerate a professing chemist for failing to be chemical enough. Such literacy did not require the author to be a chemical researcher in his own right; but it did necessitate active immersion in the current chemical literature, of the kind Thomas Thomson had argued was entirely lacking in the trade. No brewer was more aware of this than Tizard, who drew his readers' attention to the heavy reworking that went on between successive editions of the *Theory and Practice*.

Tizard was also careful, however, to avoid claiming the status of a chemical theorist or analyst. Indeed, he was one of few chemically minded authors to reject the argument for the permission of harmless additives. Malt and hops, he said, gave the soundest and most saleable product: substitution could only be justified in desperate circumstances, and there was neither point nor profit in 'illegitimate infringement', which was often the result of untutored country brewers paying too much attention to quacks.[93] Tizard's proprietary interest focused entirely on machinery, layout and process innovations: no mention of sugars or drugs crept into his 'Brewers' Repertory', even in later years when additives began to be deregulated.[94] Fear of being bracketed with the nostrum-mongers was probably likewise what led Tizard to alter his title-page styling, around 1850, from 'Professor of Brewing' (used also by his enemy Hitchcock) to a form which displayed him as a disinterested practical insider: 'Late operative brewer at the Nine Elms Brewery, Vauxhall, and now brewers' engineer'.

Nor did Tizard see any advantage in emulating the sober formalisms of chemical treatises and papers. His scrupulously careful self-positioning sat alongside a pronounced strain of whimsy, demonstrated both in the naming of his patent apparatus (*hystricon*, for instance, 'from ὕστριξ, a porcupine or hedgehog, so called from the structure of its agitating rake'), and in his conceit of treating authorship as a brewing process, piling on the analogies to the point of collapse:

> As his volume is of the magnitude by some considered bulky, though he trusts that he has not drawn a greater length than his grist required; and as he has imbibed the notion that a Book without an Index resembles a Vat-room without a Registry, he has added an appendage of that nature, as the concentrated essence of his work ... his readers will oblige him by brewing a little of that wonted patience and clemency

> which they are accustomed to extend to other authors under circumstances similarly unattenuable.[95]

This particular device did not survive past the first edition, and was there confined to the preface (the 'last drawn but least valuable' part of the book, Tizard could not resist pointing out, 'like the blue wort of a spent mash'). He was more attached, however, to his self-presentation as an innovator, featuring wildly overblown – and surely self-mocking – invocations of the martyred genius of Galileo and Socrates.[96] Although practical brewing treatises had shared rhetorical conventions for very nearly a century, as Tizard knew better than anyone, the genre remained fluid enough to tolerate such eccentricities. By the time of his first publication, however, several brewer-authors were already exploring the rather different possibilities of formal journal publication.

A Space for Brewing Science? Beer and the Early Chemical Society

A possible new channel for the work of scientific brewer-authors, more promising than the BAAS or SDUK, opened with the founding of the Chemical Society in 1841. The Society represented a coalition of chemically committed industrialists, commercial analysts and Thomas Thomson's disciples in academic chemistry, united in seeking alternatives to an elite London chemical culture which in their view perpetuated two hangovers from Davy's time: insular rejection of the theoretical insights of Liebig and other Continental chemists, and snobbish refusal to share a footing with the knowledge-makers of trade.[97] Davy, for all his earnings as a consultant to industry, had kept firmly aloof from efforts by operative chemists and mechanics ('chimney sweepers', as he privately termed them) to publish on their own account.[98] In the 1830s, trade authors such as William Black had accommodated such lofty disinterest through a sharp distinction between theoretical and 'operative' leadership, but the Chemical Society suggested the alternative prospect of an entirely shared research culture.

The Society's founding Secretary, Robert Warington (1807–67), had himself spent eight years in the brewery. Son of a wine merchant, Warington had served an apprenticeship with the London chemist John Thomas Cooper, whose interests spanned industry, consultancy and lecturing, before becoming assistant to Edward Turner, Professor of Chemistry at the new University of London, who recommended him when the major London brewery of Truman, Hanbury and Buxton sought 'an able young chemist' around 1831.[99] Warington left Truman's a couple of years before the Chemical Society's foundation, later becoming Chemical Operator to the Society of Apothecaries, but two serving brewery employees feature among the thirty-one active founding members identified by Robert Bud (as against only three 'industrial chemists' from other fields combined): Thomas Hetherington Henry, also of Truman's, and William Ferguson of Reid's, another of the great London firms. At least seven more brewers joined the Society before 1850.[100]

By appearances, then, the largest breweries were beginning to accept the value of bringing 'scientific gentlemen' – or, at least, men trained by such gentlemen – directly into the brewhouse. Unsurprisingly, given the growing influence of Thomas Thomson's students, such figures typically satisfied Thomson's key demand to keep up with current international developments. Those with formal chemical training could, by definition, negotiate the French- and German-language literature. Warington, for instance, at one point acknowledged Henry's help in pointing him to the *Ökonomische Chemie* of the Breslau apothecaries Adolf Duflos and Adolf Hirsch, a work little read in England, whose section on beer partly anticipated some of his results.[101]

Just how far such individuals were able to integrate their chemical and brewing careers is unfortunately unclear. Whatever chemical work Warington undertook for Truman's has not left its trace in the brewery records: hired as 'under Brewer', he was apparently treated little differently from his fellow brewhouse clerks. Things started awkwardly, as he was under threat of dismissal by 1832, and despite improved relations he was still judged to need periodic warnings in later years. When he resigned in 1839, 'having an Interest in a Patent for Making Soap', the firm was 'sorry to lose a person whose conduct had been so respectable', but with the implication that his chemical pursuits had ultimately proved more a distraction than a help. Henry, who was performing 'Experiments' on hops and on the 'real quality of malt' around 1840, had a similarly chequered career at Truman's, his master in 1841 (probably Robert Hanbury) noting that 'he does not seem to have that real interest in our prosperity which he ought'.[102] More successful years followed, but Henry eventually left Truman's around 1850 to resume his former profession as a consulting analyst.

Nevertheless, it is in the early publications of the Chemical Society that we find the first journal-format publications reflecting the context and concerns of the commercial brewhouse. The Society's initial goals, as conceived by Warington, included 'the reading of notes and papers on chemical science (in the most extensive meaning of the term)'.[103] Its earliest publications, titled *Memoirs* and then *Proceedings of the Chemical Society*, offered only brief summaries of presentations given at the Society's meetings, but from 1843 the format allowed relatively detailed presentation of experimental results, similar to those found in established philosophical journals.

The first brewing paper was by an author whose connections to the trade have left no trace: Septimus Piesse (1820–82), later a successful perfumer and a popular author on chemical topics.[104] His subject, however, was a classic question of the practical school: 'Is anything left in the grains which ought to be in the wort?' Piesse reported, on the basis of chemical analysis, that the spent grains from mashing routinely still contained some starch which had not been converted to sugar. The standard philosophical-practical explanation was to blame inappropriate mashing temperatures: Piesse's view was that there was a shortage of the diastase needed to perform the conversion, because it was run off in the first wort and unavailable for re-mashing. His simple solution was to supply more by adding fresh malt to the second wort – thus justifying, by the diastase theory, an old practice known as 'capping on', which some theorists had sought to abolish as inefficient. Piesse also proposed laying a wooden 'float' over the underback during filling, to protect the wort from the air.[105]

The first known paper by a brewery proprietor followed in 1843. Educated at Oxford, John Furze (1817–59) was the son of a bootmaker, but apparently had family brewing connections. By 1841 he was a partner in Manners and Co., a moderately sized brewery in Whitechapel.[106] Like Piesse, he began with an established debate in practical brewing: should fermenting vessels be kept open to the atmosphere, or enclosed? Some projectors had patented enclosure systems which promised savings by collecting alcohol said to be lost with the exhaust vapours, but Ham, Booth and Black all doubted their practicality.[107] Furze's experiment involved a fully airtight fermenter, with the vapours carried away along rubber pipes and bubbled through a water-bath. Some alcohol could, he found, be recovered, but separating it from waste products also carried into the water was more trouble than it was worth. A better approach, he said, was to apply a certain level of back pressure on the exhaust pipe, preventing most of the alcohol from leaving the fermenter to begin with.[108] Throughout, the focus was on how established chemical techniques could be used to optimize brewhouse operations.

The Society's role as a forum for the circulation of such manufacturing theory was, however, limited and short-lived. Piesse's and Furze's papers were isolated contributions which proved to be increasingly unrepresentative of the Society's journal as it fell into line with philosophical convention, minimizing or excluding from discussion the commercial contexts which had prompted the research. This reflected the governance of the Society, which was always, in the words of Robert Bud and Gerrylynn Roberts, a 'coalition under academic leadership'. Warington's initial plans had featured a highly inclusive definition of 'notes and papers', and a range of chemical facilities for members' use; it was apparently Thomas Graham, the Society's founding President, who shifted the vision for the Society firmly towards novel research, and towards the norms of philosophical publication. Active industrialists formed a minority of the membership, and published relatively little through the Society.[109]

Accordingly, Robert Warington's published chemical output bore little trace of his commercial background; a eulogy of 1868 tellingly observed that his 'connection with the brewery did not prevent the pursuit of chemistry in his leisure'.[110] Though he assisted in Furze's experiments, Warington never published directly on brewing: indeed, his closest approach demonstrated the polar opposite of the kind of experimentation that would pass in the practical brewing literature. In a paper of 1845, he described how a friend, noticing the growing popularity of pale ales, had suggested treating 'ordinary beer' with animal charcoal to remove its colouring material. Warington knew that such treatment would render the beer vapid and unfit for practical management; on trying it anyway, he found the result even worse than expected, since the hop bitterness was also completely removed. Curiosity led him to repeat the experiment on beer dosed with the most notorious of illegal hop substitutes, finding that quassia was likewise removed by the charcoal, whereas nux vomica was unaffected. The discovery, he suggested, might aid analysts in determining adulteration.[111] It would, of course, have been an act of lunacy for Warington's masters to allow him publicly to associate such research with the brewery, given the ever-present ambiguity which surrounded all discussion of substitutes in beer (chapter 5).

Overall, by 1860, the trade accepted that brewing was a chemical business, not only in theory, but in the connections it must make to achieve practical improvement; and yet the public identities of 'chemist' and 'brewer' remained largely separate. Most brewing consultants and authors found their interests were best served by appearing as loyal adherents, rather than contributors, to the culture of chemical theory. The knowledge-reform campaigners of the Mechanics' Institutes, SDUK and BAAS alike offered notably little that might change this picture; even the Chemical Society, founded with an eye to the growth of industrial applications in various fields, did little to bridge the divide at an institutional level. In its early publication culture, however, the Society suggested opportunities for change which were ultimately realized across the later nineteenth century. One key element in the realization of these changes, unsurprisingly, was the increasing presence of trained chemists in breweries, itself an artefact of the overall rise in the reputation of chemistry in the era of Liebig. Equally important, however, was the growth of more regular and more formalized brewery publications, as we shall see in the final chapter.

8 ANALYSIS AND SYNTHESIS

> It was in 1876 that I first published what must now seem a very elementary work on Brewing ... brewing was then a very simple matter, as compared with the more complex process of the present time. Increasing competition, comparative inferiority of material, and a growing public taste for weaker beers, have certainly combined to render the manipulation of necessity more and more difficult, and to bring it into more perfect connection with theoretical teaching.
>
> F. Faulkner, 1888[1]

Almost all of the nineteenth-century authors I have discussed so far were keen to be seen either as brewers informed by chemistry, or as chemists attuned to the brewery. If these standard self-definitions made it possible to collaborate across the divide, they also reinforced it. The separation was probably at its strongest around the mid-century, as the last of the philosophical amateurs gave way to devotees and professionals. A brewer-author such as William Black (chapter 7) might go as far as to co-produce published material with a university chemist such as Thomas Graham, but could not become a chemist in his own right: doing so would have meant the total reorientation of his life and work. The second half of the century, however, saw the growth of a brewhouse laboratory culture including researchers able to move in both spheres.

A key institution for this change was the Royal College of Chemistry in London, founded in 1845. The College's founders presented its chief agenda as being to transplant to England the laboratory research and teaching culture of Justus von Liebig (as he now was). Liebig was closely involved in planning the College, and nominated its first director, the Giessen-trained August Hofmann (1818–92). As Gerrylynn Roberts has pointed out, however, the College was also responding to a movement for technical education which had existed in England before Liebig's rise, emphasizing routine chemical teaching to support medicine and agriculture. There was sometimes tension here. For all that Liebig was closely associated with practical applications, he saw what was increasingly termed 'pure' research as integral to any effective laboratory; Hofmann, likewise, articulated his own status chiefly through his research achievements.[2] In the beer industry, the effect of the collision was to create a self-consciously new generation of hybrid brewer-chemists.

Around the time that Hofmann arrived at the Royal College, another Giessen-trained German, Heinrich or Henry Böttinger (1820–74), was appointed resident chemist to Allsopp, one of the major Burton breweries. Hofmann and Böttinger were good friends, and when Böttinger was promoted to head brewer in 1862, Hofmann recommended his own assistant, the Marburg-trained Peter Griess (1829–88), to take his place. From 1866, the emerging laboratories of Burton were firmly dominated by men who had passed through the College. In that year, Cornelius O'Sullivan (1841–1907), another of Hofmann's former assistants, was hired by Allsopp's competitor Bass, while another Burton firm, Worthington, appointed the very young Horace Tabberer Brown (1848–1925), fresh from a 'student's year' at the College. Hofmann's role had meanwhile passed to Edward Frankland (1825–99), who had worked at Marburg, Giessen and Manchester, and had consulted for a wide range of industries including brewing.[3] Brewers trained at the Royal College under Frankland include James O'Sullivan (1856–1938), Cornelius's brother and assistant at Bass; John Heron (1850–1913), like the O'Sullivans a native of Bandon in County Cork, who went to Worthington's in 1877; and Adrian Brown (1852–1919), half-brother to Horace, who went to a fourth Burton brewery, Salt's.[4]

The O'Sullivan–Brown generation contrasted sharply with earlier brewer-chemists such as Robert Warington (pp. 176–9), who had kept their publication careers firmly separate from their commercial research. Of the Burton chemists, only Griess had a similar attitude: he arrived at Allsopp's with a strong reputation as the discoverer of the diazo compounds, and it was on related substances, useful chiefly for dyestuffs, that he published and patented for the remainder of his life. For Griess, it seems, the chief advantages of a brewery career were the relatively short working hours of senior staff and the break in production over the summer, giving time for focused research.[5] The younger men, however, saw it as a professional obligation to circulate novel research on the chemistry of malt, beer and water composition. They had, of course, to negotiate with their employers on the boundaries of disclosure, but were in practice able to publish widely on 'theoretical' or 'fundamental' questions of material composition. The major outlet for both Horace Brown and Cornelius O'Sullivan was the Chemical Society, whose early *Memoirs* and *Proceedings* had been superseded in 1849 by a journal on the conventional academic plan, co-founded by Hofmann, who encouraged his industrial pupils to write for it.[6]

Conventional scientific publication was only one element, however, in a larger system of circulating knowledge which encompassed the brewing and related trades in general. The monthly *Brewers' Journal*, launched in 1865, and its fortnightly rival the *Brewers' Guardian* of 1871, were both founded to cover miscellaneous matters of trade concern, but increasingly featured a preponderance of scientific and technical material. By republishing chemical papers, they connected their readers to scientific discourse, and in professing to speak for the trade, they strongly shaped its scientific direction. Two particularly influential mediators were Frank Faulkner, responsible for much of the *Journal*'s scientific coverage, and E. R. Southby, who filled a similar role at the *Guardian*: both encouraged a militantly scientific manifesto for the late nineteenth-century trade.

The new agenda was partly a response to the growing threat of the temperance movement, as the traditional anti-spirits lobby was marginalized by the teetotal and prohibitionist United Kingdom Alliance from 1853.[7] Science was pressed into service in the brewers' defence – not merely as a source of intellectual validity, but as a guarantor of respectably moderate alcohol levels. The vision also introduced a new scientific hero, the French chemist and microbiologist Louis Pasteur; a revival of trades improvement rhetoric, now turned towards imperial exports and a vigorous endorsement of free trade; and a demand – inverting, respectfully but firmly, the key practicalist tenet of the past century – that the domains of the laboratory and the brewhouse must be brought together once and for all.

Liebigian Chemical Authority and the 'Great Bitter Beer Puff'

In the mid-nineteenth century, despite a growing public literature of chemistry as the domain of respectable university scholars and gentlemen devotees, the old associations of the drugshop and the poison bottle remained alive in the minds of many drinkers. The unforgettable sensationalism of Friedrich Accum (chapter 5) inspired several imitations. Chemical and medical professionals occasionally challenged such performances: reviewing one particularly histrionic tome put out by an 'Enemy to Fraud and Villainy' in 1830, the *Lancet* noted that the author's allegations were not backed up by 'names and dates and places', and stressed the difference between harmless drugshop additives and poisons.[8] Yet the campaigning *Lancet* was itself not above harnessing the power of public alarm: from 1851 its 'Analytical Sanitary Commission', consisting chiefly of the microscopic researches of Arthur Hill Hassall (1817–94), focused attention on the adulteration of various foodstuffs.[9]

Periodic adulteration panics clearly affected the scope for institutionalized brewing chemistry. Horace Brown, recalling the Burton of 1866 at a distance of fifty years, recalled that his masters at Worthington's shunned laboratories for fear of giving customers 'the horrible suspicion that the beer was being "doctored"'; a small room was eventually fitted out on condition that its windows were 'carefully obscured'.[10] Brown overstated the case, however, in painting the whole trade as sclerotically resistant to science in general: by the 1850s, leading brewers recognized that chemical authority, properly mobilized, could serve their cause in ways that had been impossible in the time of Accum. Their commercial context, however, still led to tensions with the dominant etiquette of chemical and medical discourse.

In 1852, a major controversy unfolded after Anselme Payen, the co-discoverer of diastase, stated that large quantities of strychnine were being manufactured in Paris for use by English brewers as a hop substitute. Payen's claim, apparently made quite casually in a public lecture, was reported by the French medical press, and thereby prominently covered by the *Medical Times and Gazette*, a weekly rival to the *Lancet*.[11] For the major brewers, this was a grave concern: the 'narcotic poisons' paragraph of 1807 (p. 116) had illustrated how easily such stories could spread. Teetotal campaigners, then on the verge of becoming a serious nuisance to brewery interests, were quick to build the allegation into a long-term history of toxic adulteration.[12]

The Burton brewers were particularly affected: the *Medical Times* write-up had contained references to unwarranted levels of bitterness, inferentially singling out their characteristic product. Bass, Allsopp and others vehemently protested their innocence, but the paper responded that their acknowledged respectability did not extend to the 'small dealers' and 'middlemen' who handled their product.[13] When Michael Bass, an MP and leading public spokesman for Burton interests, accused the *Medical Times* of scaremongering, its editor, J. Stevenson Bushnan, retorted that his duty as an '*ex officio* custodian of the public health' was to publish such information when it came from a respected authority. If there were concerns that Payen was somehow mistaken, this was a matter for official investigation; it could hardly be resolved on the say-so of an interested party.[14] Bass's rival, Henry Allsopp, responded more strategically, endorsing the idea of a public enquiry employing 'the most acute and scientific investigators'.[15] Böttinger, his house chemist, meanwhile rushed to London to consult Hofmann, who cautioned that the defence would require analytical evidence, and the support of a yet more authoritative chemist not linked to the firm. The only man suitable in England was Thomas Graham of University College London, whom we met in chapter 7 as one of the advisors to William Black.[16]

In fact, Graham and Hofmann, having previously worked together on water analyses, were already collaborating on a beer-related consultancy.[17] In 1847, some forms of sugar had been reauthorized for beer use, following pressure from the West Indian planters. Proposals for this measure had reopened the question of taxation based on the properties of the end product, rather than the raw materials: as a first step, the approach was applied experimentally to the system of 'drawback', the duty rebate received by brewers who exported their beer.[18] The assessment required a means of establishing the original gravity from a sample of finished beer, which could be closely estimated from the saccharometric final gravity and the spirit content. The Excise was increasingly able to manage such investigations in-house via the Laboratory of the Government Chemist established in 1842, which duly devised a process based on evaporation of the spirit fraction: when brewers disputed the valuations, however, the Board turned as usual to elite university chemists as a court of appeal. Assembled in 1851, the investigating committee consisted of Graham, Hofmann and Theophilus Redwood (1806–92), Professor of Pharmacy and Chemistry at the Pharmaceutical Society.[19]

Working on the drawback and strychnine enquiries simultaneously, Graham and Hofmann unsurprisingly discussed the drawback work with Böttinger. In their final report – which vindicated the Excise position – he is singled out as having lent particular assistance, drawing on his research at Burton.[20] Böttinger also accompanied the London chemists as they trawled Allsopp's stores, and the sites of various distributors (Bushnan's 'middlemen'), drawing half-barrel samples which they analysed at the Royal College and pronounced to be free of strychnine.[21] The report on the latter investigation, issued in April 1852, couched these findings among the common-sense arguments usually deployed against adulteration rumours: strychnine did not taste like hops, the quantity needed for the required bitterness was well above the fatal dose and applying it throughout the breweries of Burton would take many times the current world supply.[22] Independently, the *Lancet*'s Analytical Sanitary Commission pitched in with a report of its own, specifically exonerating both Allsopp and Bass.[23]

The ultimate chemical authority, of course, was Liebig himself. Hofmann directed Allsopp to send a consignment of pale ale out to Giessen, with a payment of £100 to secure a testimonial.[24] Trusting to his acolytes' judgment, however, Liebig did not perform any independent analysis: 'my own experience', he wrote elliptically, 'enables me to recommend it, in accordance with the opinion of the most eminent English physicians, as a very agreeable and efficient tonic, and as a general beverage both for the invalid and the robust'. Subverting the usual commentary on the relative state of English and German technical prowess, he added that 'continental brewers have much to learn' in approaching the sophistication of Burton methods.[25]

Henry Allsopp had realized that chemical and medical authority could be used not merely to deflect adulteration charges, but as a direct promotional device. He had thousands of copies of the Graham–Hofmann report printed up and distributed, and wrote a string of open letters to editors to ensure its insertion in the major newspapers, as well as in the *Medical Times* and other medical and chemical periodicals.[26] Allsopp went on to monitor the presses carefully for several months, chasing those medics who had publicly entertained the story through the letters columns and offering to throw open his brewhouse and stores for inspection.[27] Soon afterwards, his firm began to place prominent advertisements in any periodical with a large or influential readership, from the

Lancet to *Punch* to the ongoing serial publication of *Bleak House*, and on street posters besides. The headline endorsement of Liebig was followed by those of Hofmann, the Analytical Sanitary Commission and a procession of surgeons, medics, chemists and analysts, ultimately around thirty in number, from Sir Charles Clarke, 'Physician to her late Majesty the Queen Dowager', to John Henry Pepper, the populist showman-educator of the Royal Polytechnic.[28]

Was this a respectable use of scientific authority and professional communication networks? Similar questions had plagued English responses to Liebig's programme since the Royal College of Chemistry was first founded: two of its promoters had, with Liebig's connivance, quietly patented one of his discoveries in Britain and used the College as a platform to promote the product, Hofmann also taking a financial stake.[29] Liebig's relatively direct promotion of Allsopp led to some embarrassment, as rival firms drew from him a clarification that his remarks applied to Burton bitter beer in general, and that the impressive methods he referred to were witnessed at 'another brewery' (inferentially, Bass, which may also have paid for his support). There followed something of a backlash against the 'Great Bitter Beer Puff'. The *Monthly Journal of Medical Science* whimsically pictured a brewer from outside Burton pleading to be accused of poisoning so that he, too, could be impressively vindicated; a later price rise by Allsopp's saw *Punch* politely enquiring of Liebig as to the scientific validity of the move.[30]

To Bushnan's successor at the *Medical Times*, the matter was more serious. It was, in essence, an inflated version of the commercial temptation which captured unwary medics every day, causing immeasurable harm to the profession; and it had sadly degraded the reputation of the world's leading chemist.[31] Moreover, the exercise did not entirely quench the rumours. In December 1852, a letter to the *Times* described druggists' travellers continuing to leave cards boasting of mysterious powders to communicate a 'delightful flavour and pungency'. Their stock-in-trade also included 'humuline', a patent hop extract promoted as a solution to the wort retention problem; the correspondent suspected this to be a cover name for strychnine, despite all the chemists had said.[32]

Despite these difficulties, Allsopp on balance regarded the exercise as a major success. The strychnine scare, in Hofmann's later estimation, 'put chemistry on the map in Burton-on-Trent', forging long-term social links between the Midlands brewery and the London chemical establishment.[33] It was also from around this time that Burton practices began to dominate the encyclopedia and other public accounts which had formerly showcased the London porter giants. This literature provides illuminating material on the contemporary state of the trade writer's paradox.

The Case for an Experimental Culture

Among the successors to the useful-knowledge initiatives of the 1830s was a part-work encyclopedia issued from 1853 to 1860 under the editorship of Sheridan Muspratt (1821–71).[34] The son of a powerful soda manufacturer in Liverpool, Muspratt had struck out on his own as a consulting analyst, establishing a College of Chemistry in the city in 1848. He was a fellow Giessen graduate and close contemporary of Henry Böttinger, and it was to Böttinger that he turned in preparing an entry on 'Beer' for the encyclopedia. Horace Brown later recalled that Böttinger wrote the piece himself, which is probably an overstatement: it shows clear signs of editorial compilation, drawing from several published texts. Böttinger's personal experience was the single most important source, however, and the account accordingly favoured the practices of the major Burton breweries.[35]

In accounting for its sources, the text shows a move away from the hack-work borrowing still typical in the 1830s (chapter 6) towards the survey and citation conventions which were now prevalent in the chemical journals. The account of the composition of malt opens with a long section on analytical chemistry, introducing the interpretations of various chemical authorities; there follows a series of process descriptions drawn from William Tizard and John Levesque's books, supplemented with new guidance from Böttinger. Threaded throughout is a rhetoric of improvement through chemical awareness, taking its cues from Tizard and William Black. There was more of an echo of Michael Donovan, however, in Muspratt's dismissal of trade secrecy as little more than a 'foible'. Remarking that '[e]xperiments are readily performed now-a-days', he argued there was little point in hiding an improvement which occurred to one brewer, since it would occur to others soon enough. Rather, it would be in the interest of the whole trade for brewers to share and corroborate their implementations across different sites.[36]

This picture, which at first glance recalls contemporary campaigns to abolish the patent system,[37] sits oddly alongside Muspratt's enthusiasm for several patent innovations, chiefly those of the energetic Tizard. The contradiction disappears when we understand that Muspratt implicitly saw brewers' engineering as having an intellectual property regime quite different from that of brewing itself: Tizard's machines were in the tradition of those mechanical arts which had been shown to improve under a patent system, but the practical management of beer would flourish better in a knowledge culture which was free as well as open. Muspratt, of course, was not a brewer, and in no position to dictate to the trade, but Böttinger's involvement (presumably with Allsopp's blessing) shows that his position was to some extent shared.

Muspratt also put forward a new resolution to the paradox of experiment. Discussing humuline, Muspratt acknowledged that brewers had a right to be suspicious of untested 'modern innovations', which might very well profit nobody but the promoters. The brewer could best serve his own interests, however, not by rejecting all such innovations, but by making a thorough practical test of any 'plausible process' which was 'not inconsistent with the testimony of scientific men'.[38] Building on Black's distinction between the authority of brewers and of 'scientific gentlemen', this was a convenient piece of rhetoric which flattered all sides. If the brewer alone must finally decide what could constitute good practice, science could clear his way by telling him what could not: its universality made it a first court of authority to winnow out unworkable schemes.

Such routine investigation, however, would clearly be impossible without the open knowledge culture Muspratt had recommended. For this to be achieved, it was not sufficient for brewers to be willing: they would also need a forum for ideas to circulate. The useful-knowledge movement, as we saw in chapter 7, was not well suited to providing such a forum; neither, at the level of practical investigations, was the Chemical Society. Brewer-authors had had some success in transmitting research through their stand-alone publications and occasional contributions to general periodicals (chapter 6), and the *Mechanics' Magazine* had sometimes served as a forum for industrial discussions. Such sources, however, could not guarantee regular communication to, and among, the community of brewers in general. What the trade needed was a periodical of its own.

Trade Journals as a Vehicle for Scientific and Technical Communication

The *Brewers' Journal* was the brainchild of William Reed (1830–1920), a one-time sugar merchant who launched his first specialist trade journal, the *Grocer*, in 1862. He quickly added a supplement, the *Wine Trade Review*, which drew sufficient interest to become a separate publication. Thanks partly to the enduring independence of his firm, Reed has acquired a somewhat legendary status as a spotter of gaps in markets: it was supposedly a chance encounter with a brewing acquaintance which inspired him to repeat the formula in 1865.[39] If the plaint which supposedly inspired him ('I want to buy a mash tun, and for the life of me, I don't know where to get one') absurdly understates the existing scale of marketing and private correspondence networks, it is true that the industry lacked a regular, nationally distributed forum. The *Brewer's Journal* flourished as had its parent, from which it separated in 1867.

The initial mainstays of the monthly paper were much as for other trade journals: lists of prices in the London markets; reports from Parliament on issues affecting the trade, or from the general meetings of the major firms; details of tenders accepted, new firms established and bankruptcies; and short miscellaneous items of potential interest, from warnings of confidence tricksters targeting publicans to comic tales and antiquarian titbits. Much material came from France and the German lands, such as a regular column of 'Foreign Hop Intelligence'. Readers were encouraged to contribute via letters and short response pieces, and responded enthusiastically. The tone was often one of gentlemanly curiosity, recalling the popular periodical *Notes and Queries*: responses to the question 'What is a hop?', raised in September 1866, ranged widely over the plant's historical categorization and use as a foodstuff.

From the outset, the *Journal* made it a priority to report on new developments in brewing technology. This approach, besides helping to convince the trade community to accept the *Journal* as an essential regular read, ensured a reliable stream of advertising revenue. William Tizard (chapter 7), not content to rely on his instructional publications alone, was an early and frequent advertiser, as were rival brewers' engineers such as Henry Pontifex and Sons of King's Cross.[40] One of the first regular sections in the *Journal* was titled 'The Brewers' Engineer', offering a monthly round-up of innovations, patent and otherwise, often described at length with engravings. The listing of patents became increasingly systematic: 1869 saw an attempt to chronicle every relevant patent from 1635 to 1868 under categorized headings in monthly instalments, a project which was updated periodically in later years. There were also instructional accounts of practical brewing, often delivered as monthly series (a standard device, of course, for ensuring subscriptions). An unsigned account of 'Summer brewing by a Burton brewer' was under way by 1866, querying some of the conclusions of William Black. Since the audience, by definition, could be taken to consist entirely of fellow experienced operatives, the author was at pains to point out that he did not 'pretend to dictate to any brewer how he should manage', stressing only the reliability of his method.

Where did this leave the paradox of the trade writer? We must conclude, from the increasing frequency with which brewers wrote casually to pass on the benefits of their experience, that the perceived value of trade secrecy had sharply declined. The reasons had less to do with the wider public culture of science than with changes in the trade. One factor was that medium- and large-scale brewers (who, as ever, were enclosing more and more of the market) relied increasingly on hired consultants, engineers, architects and analysts, usually choosing among a limited variety of known approaches: the partisans of Tizard's and Pontifex's systems could enjoy chiding each other in print without risk. In this they were encouraged by the *Journal*, whose declared policy was to promote awareness of all potentially promising innovations without taking sides – inevitably, since it could not afford to alienate potential advertisers.[41]

More importantly, authors could increasingly invoke the common interests of the trade. Teetotalism had lately emerged as the great enemy; the *Journal* also fixed on what it saw as scaremongering claims of toxic adulteration by public analysts and self-appointed chemical experts, claiming (as had its parent journal the *Grocer*) that all honest traders should be mobilized to understand the business of analysis for their own protection.[42] It was equally possible to articulate common opportunities, through an export-minded vision of British imperialism; reports of the growing success of German brewing methods provided a lesson and a warning. When the *Journal* urged that open communication would be to all brewers' advantage, many heeded the call. Indeed, the *Journal*'s editorial line was significantly determined by active brewers, particularly where brewing science was concerned. After three years in which William Reed pulled the paper together himself, the editorship passed to F. W. Lyon, who also acted as publisher; both relied on a series of chemically and technically trained contributors who wrote both signed articles and, certainly in the Lyon era, much of the editorial content.[43]

One opportunity for scientific instruction immediately presented itself. Black's treatise – 'the best we know', according to the *Journal*'s answer to a reader's question – had not been updated since 1844, and serial publication in the *Journal* was an attractive vehicle for any possible replacement.[44] After an abortive first attempt in 1867,[45] a work entitled 'Practical and theoretical brewing' was completed over nine parts in 1872–3. Its proclaimed mission was to present to the *Journal*'s readers, 'who are all practical men', a version of the current state of chemical brewing knowledge 'devoid of unintelligible nomenclature'. The approach was somewhat derivative, drawing its stance from Muspratt and Black. The only significant novelty was an increased role for analysis: it was 'of primary importance to every brewer' to understand the salt profile of his water.[46] The overall presentation, however, was self-consciously modern, citing the names (if not always in the conventional spelling) of Berzelius, Liebig and other Continental authorities.

In the 1870s, as we will see, the theoretical and technical emphasis predominated to the extent that the *Brewers' Journal* became almost a fusion of trade paper and scientific review journal. The same may be said of its principal rival. The *Brewers' Guardian* was launched in 1871 as the fortnightly 'official organ' of the Country Brewers' Society, a hitherto relatively inactive body formed half a century previously to seek Excise reform.[47] The *Guardian* was specifically a response to the rise of the teetotalist United Kingdom Alliance, aiming to unify and co-ordinate the trade; its editorial preoccupation, in the early years, was licensing reform. The Society had, however, a significant connection to analytical chemistry in the shape of its house analyst, Dugald Campbell (1818–82), a former pupil of Thomas Graham, who held numerous consultancies to private and public bodies on topics such as water pollution and the adulteration of food. One of the principal advantages of membership was access to analyses for 'a nominal sum', paralleling the legal advice offered by the resident solicitor.[48]

Both Campbell and his successor, E. R. Southby, whom I discuss below, contributed scientific material to the *Guardian*, advising readers on how best to deal with public analysts who were empowered to inspect beer under recent anti-adulteration laws.[49] From around 1876, the paper regularly published or republished books and articles on scientific topics, in serial form, at a page or so per issue: readers were, presumably, expected to clip and file the pages over the year or more it might take a full treatise to emerge. These were supplemented by a regular section of 'Scientific notes', and increasing numbers of short papers and reviews. By February 1880, the *Guardian* was simultaneously carrying installments of three scientific publications, including an article by the prestigious Danish fermentation theorist, Emil Christian Hansen, '[t]ranslated specially' for the occasion.

'Twin Sisters' in the Revitalization of British Brewing

The issue of engagement with industry had meanwhile become something of a battleground at institutions such as University College London. Alexander Williamson (1824–1904), the Heidelberg- and Giessen-trained Professor of Chemistry, literally delivered a 'Plea for Pure Science' in 1870: industrial improvement clearly depended on scientific knowledge, he argued, but formulating new scientific principles was a distinct art from applying them, and the surest path to progress was to keep the traditional 'division of labour' between university scholars and factory managers.[50] Diametrically opposing Williamson was his former assistant, Charles Graham (1836–1909), who in 1877 obtained a professorship at UCL in the newly designated field of 'Chemical Technology'. A product of John Collis Nesbit's agricultural college, Graham had worked on agricultural and mining chemistry in the German lands, and believed university teaching should be industrially engaged.[51]

The disparity was well illustrated in the forum presented by the annual Cantor Lecture series of the Society of Arts. Though the Society had been somewhat eclipsed by other institutions since the eighteenth century, it had in the 1870s a vigorous Chemical Section, consciously inspired by the trades-improvement agenda of the founders. The Cantor Lectures, usually presented by academic chemists, gave the opportunity for discussion with an audience which included numerous industrialists.[52] What Williamson understood by this was illustrated by a series he presented on 'Fermentation' in 1870: his syllabus focused on fundamental properties, with little space for practical context. In stark contrast, Graham's lectures of 1873–4 concerned the 'Chemistry of Brewing', directly targeting an audience of commercial brewers.

Graham was in one sense traditional: he upheld the distinction between theorists and practical men. He was at pains, however, not to draw the usual hierarchical conclusion. The chemist did *not* have the innate right to redirect the knowledge culture of the brewer: authority had to be earned. Brewers need feel no shame at the trade's unscientific past, for science had been slow to demonstrate its practical advantages: beermaking, like agriculture and metalworking, was a crucial enterprise whose past experts had rightly sought to make progress by marshalling their own skills and resources. Now, however, science was well mobilized to assist, and the 'modern high-pressure conditions' of trade made it necessary. Continental advances showed that the model he proposed – '"Practice with Science", twin sisters'[53] – was effective; the chief threat came from Germany, 'the birth-place of modern beer', which was well on its way to dominating the global export trade.[54]

Correspondingly, if less had been achieved in Britain, the responsibility lay not only with laggard brewers, but with scientific men who valued 'the phenomena of the moon or equally distant objects' over the more immediate challenges of industry. Graham used his Cantor Lectures to showcase his alternative ethos, carefully pointing out instances where discussions with working brewers had influenced his approach. The structural division of the series – 'Malting', 'Mashing', 'Boiling', 'Fermentation' – showed a clear debt to brewer-authors' treatises; in a Richardsonian touch, the chemist even managed to find a maltster to bemoan for relying on his 'very horny' hand as a makeshift thermometer. Graham also stressed his belief – 'coming to the subject merely as a chemist' – that Excise restrictions on sugar and malting imposed unreasonable restraints on the trade.[55]

Graham's proposed solution to the national problem ran along the lines proposed by philosophically minded practitioners such as George Adolphus Wigney: formal chemical training for young brewers, and dedicated experimental facilities focused on the art. Brewers, he said, should take on some operational skills which were currently the preserve of chemists. For instance, readings from the saccharometer (which, it was now generally accepted, detected dissolved matter indiscriminately) could be elaborated by routine testing with Fehling's solution, a standard reagent which indicated the presence of actual sugar.[56] Similarly, Graham argued, brewers should familiarize themselves with the microscope. Graham was ultimately able to put various of his proposals into action, determining the curriculum of the City and Guilds of London Institute's formal brewing qualifications initiated in 1880.[57]

Graham's key influence was the French chemist, Louis Pasteur (1822–95). Though little known to brewers at this stage, Pasteur had attracted attention among chemists by arguing that the alcoholic fermentation was not a chemical decomposition, as Liebig believed, but due entirely to the action of living organisms: the difference between successful and disastrous fermentation was made by the preponderance of the appropriate organisms (*torulae*, or true yeast cells), and the relative absence of harmful *bacteria*. Pasteur experienced the German threat rather more tangibly than his English contemporaries: greatly affected by the debacle of the Franco-Prussian War, he fixed on beer as an emblem of France's technological failings, experimenting towards what he called a '*bière de la revanche nationale*', free of bacteria and possessing the stability crucial to international success. More important for Graham, however, was Pasteur's demonstration of the microscope's diagnostic potential. Accounts of Pasteur's 1871 visit to England (not widely discussed at the time, but later greatly mythologized) depict him touring the major London breweries and showing the bacteria present in yeast samples, correctly diagnosing failings in the beer.[58] Here was a self-evident and visually striking demonstration of how the tools of science could serve the brewer: Graham arranged for microscopes to be arrayed around the demonstrating table at one of the Cantor Lectures, so that his audience could view the phenomena.[59]

In essence, Graham's scheme retained for the non-brewing chemical expert a status which, if not dominant, was still privileged: chemists should tell brewers where to look for useful insights, and how to apply them. Graham explicitly addressed himself not merely to those present, but to 'the readers of the Society's *Journal*, and of the two important brewers' journals': both the *Brewers' Journal* and *Guardian* by now routinely covered such presentations.[60] Many brewers responded just as Graham had hoped, and his reward was an extensive consulting practice that lasted through his time at UCL and into his later years as a private consultant. Though he had never, in the literal sense, been a brewer's chemist, the trade had enough need for his services to allow him to specialize almost entirely in the chemistry of brewing. The very fact of such specialization suggests, however, that his model of '"Practice with Science", twin sisters' contained the seeds of its own demise. Increasingly, 'practice' would co-opt 'science' within its own domain, as the rhetorical divisions between the laboratory and the brewhouse finally broke down.

Publication, Publicity and Patenting

In 1872, Cornelius O'Sullivan of Bass compiled the first of his several papers for the *Journal of the Chemical Society*. He began with a review of the past decade of French and German literature on the effect of diastase in breaking down starch into sugars, showing that the field was suffering from a build-up of incompatible interpretations. The field had first emerged when chemists drew on practical brewers' understandings in evolving a distinction between the fermentable sugars and the unfermentable 'dextrine' which provided the body of beer (chapter 6), and O'Sullivan stressed his industrial context as the reason for reopening the investigation: 'every fact connected with the transformation of starch is of importance to the manufacture in which I am engaged'.[61] O'Sullivan's experimental approach, however, displayed nothing of the brewhouse, in the traditional sense. Writing in the impersonal style now conventional in chemical journals, he described purifying his samples by repeated precipitation, to a degree of nicety unknown in commercial production, and testing with mineral reagents unheard of by the previous generation's brewer-authors.

Nonetheless, O'Sullivan's findings ultimately invoked the specialized knowledge of his position. Retrieving a neglected 1847 paper by Augustin-Pierre Dubrunfaut, he confirmed that the predominant fermentable sugar found in malt was physically different from the 'glucose' of grapes and many other substances: lack of discrimination between different production cases, he said, had led later authors into confusion. What was needed was a specific focus on the properties of the substance he referred to as *maltose*.[62] Maltose and dextrin – O'Sullivan came to prefer the German spelling – were, he later affirmed, the *only* products of the breakdown of starch by malt extract; maltose always predominated, but the proportion varied with the temperature of the application. Here, O'Sullivan was undoubtedly thinking of the control of the mashing temperature to influence the product, which, according to David Booth, was generally known to brewers before 1830.[63]

The end point of O'Sullivan's investigation was a patent. Commercializing the outcomes of theoretical developments was quite usual by this stage: Pasteur had obtained a patent in 1871 for a revised fermentation process which involved charging the wort with carbonic acid gas to expel the dissolved air that would foster bacterial growth. Such patents could be mere safeguards or promotional measures: Pasteur's found no favour among practical brewers, and there is no clear evidence that it was ever worked.[64] O'Sullivan, however, clearly intended from the outset that his work in analysing sugars should lead to a profitable line in synthesizing them. The episode illustrates how an influential brewer-chemist might not be strictly bound to his employer, but free to pursue extensive side-interests involving the ever-growing class of 'allied trades'.

The deregulation of 1847 had removed any prospect of a return to malt-and-hops purity, and subsequent duty changes opened the brewers' eyes to a variety of saccharine materials. Some were preparations of sugar or molasses, but others were manufactured by industrial chemists, using mineral acids to break down rice or other cheap sources of starch.[65] Amid much debate over the merits of particular substances, brewers generally agreed that if some could provide valuable savings, none could really replace more than a fraction of the malt: consisting almost entirely of fermentable sugar, without dextrine, they contributed nothing to the body or flavour of the beer. O'Sullivan's approach, however, was to convert rice starch while carefully controlling the acidity level, halting the process before complete breakdown and leaving a proportion of dextrine. The principle was not new, but O'Sullivan successfully argued, on the basis of his maltose determinations, that he had found the means to reproduce synthetically the maltose-to-dextrine ratio characteristic of malt wort, thus allowing much more malt to be substituted. On this basis he obtained a patent, which was soon being worked by a London sugar refiner, for the discovery known in the trade as 'dextrine-maltose'.[66]

O'Sullivan had a collaborator, and co-patentee, in William George Valentin (1829–79), yet another mainstay of the Royal College of Chemistry. A respected analyst and teacher who published little, the German-born Valentin had been Hofmann's senior assistant, retaining the role under Edward Frankland and through the reorganization of the laboratories on a new site at South Kensington.[67] He also had connections with the Böttinger family, who had brewing interests in Würzburg, where his son Basil was later to train before going on to a prominent career as a brewery manager and consultant in England and Wales.[68] Valentin's role in the sugar research is unclear, but he provided O'Sullivan with a crucial metropolitan connection: not to elite theoretical chemistry, as in previous generations, but to the increasingly solid institutions of the chemistry–technology interface.

In March of 1876, Valentin addressed the Chemical Section of the Society of Arts on dextrine-maltose. O'Sullivan was present but, Valentin explained, pressure of work had prevented him from preparing a lecture himself. Valentin began (like so many others) by redrafting history to suit his purpose. 'It is only in the last thirty or forty years', he said, 'that the progress of chemical science left its mark upon this art, as upon so many others': the new opportunities had, however, been assimilated swiftly by the leaders of the trade, of whom he mentioned Bass and Allsopp by name. None of the earlier published analyses of malt were trustworthy, he affirmed, because all lacked the special authority conferred by O'Sullivan's brewhouse context. The dextrine-to-maltose ratio was the foundation of success and 'the real root of the evil' in failure, occuping for Valentin and O'Sullivan the central role that thermometry, saccharometric extract, yeast, water profile and electrical action had occupied for the researchers of the past.[69]

Valentin anticipated that some in his audience would consider trade advantage to be in tension with open publication. He pre-empted them with a line from Robert Warington, Jr (1838–1907), son of the one-time Truman's brewer. Warington had, with the proprietor's permission, published extensive details of analyses he had carried out as house chemist to an acid works, commenting as follows:

> A large amount of information is acquired in the laboratories of our great manufacturing concerns; most of this might be published without any injury to the individual manufacturer. Especially is this true of analytical methods, and the publication and discussion of these would do much to remove the disgrace to which science is often subjected from the wide discrepancies of commercial analyses.[70]

Valentin's co-option of this pronouncement was disingenuous, of course: O'Sullivan was the very opposite of injured by publicizing the research behind a process in which he, not his employer, held the patent. Towards the end of the lecture, Valentin's tone became openly promotional, namechecking the manufacturer, John Schwartz of Mile End, and deeming his product 'the cheapest and most suitable malt substitute ever brought under the notice of brewers'. In an unconscious echo of John Richardson's 1775 appeal to 'casks of beer of great eminence' as providing the ultimate testimonial, bottled specimens of O'Sullivan's handiwork were available on the demonstration table for 'inspection' after the lecture.

Like the 1852 'beer puff', this combination of scientific exposition and self-promotion attracted controversy. The *Journal of the Society of Arts* carried a letter from William Lyttle, a telegraphic engineer, who suspected a monopolist on the make. 'Dextrine-maltose', he protested, was nothing more than starch sugar, the preparation of which was detailed in standard works on organic chemistry. Two correspondents leapt to Valentin's defence. It was extraordinary, said one, that Lyttle, 'although a Fellow of the Chemical Society', had failed to grasp 'the marked differences that exist between dextrin, glucose, and maltose', a mistake which no competent brewer with 'a simple but complete theoretical knowledge of the chemistry of brewing' could ever make: the thin taste of beer made using earlier synthetic saccharines was enough to establish Valentin and O'Sullivan's achievement. Lyttle retorted that his point did not concern the nature of the materials, but the novelty of the process: the innovators had merely come up with a new control mechanism, and this was not enough for a patent. The real sin, however, was to insert a 'neatly concealed threat of violated title' into the pages of the *Journal of the Society of Arts.* The editor responded that, indeed, the Society's remit was 'scientific and technical merit' alone, and curtailed the correspondence.[71]

The *Brewers' Journal*, however, was wholly sympathetic to the patentees. Most unusually, a special supplementary number appeared three days after Valentin's lecture, reproducing its text in full. The next regular issue carried a long editorial lauding O'Sullivan's discovery as the latest achievement of modern scientific brewing, a phenomenon whose roots lay in the Great Exhibition of 1851 and the new export-mindedness of the trade. Key to the *Journal*'s position was Frank Faulkner, who had entered the Lyttle controversy on Valentin's side. Faulkner was then beginning a long career as a scientific and technical contributor to the *Journal*, which was closely mirrored, for a time, by that of his opposite number at the *Brewers' Guardian*, E. R. Southby. Though not themselves products of the Hofmann–Burton complex, Faulkner and Southby are of key interest as men whose activities spanned brewery employment, private consultancy, patenting, treatise-writing and journalism – in other words, nearly all the available means of public communication on brewing science.

'Modern theoretical brewing' and the Pastorian Synthesis

Son of a Birmingham brewer, Frank Faulkner (1843–1925) took various roles in the trade before settling at the large St Helens firm of Greenall Whitley around 1870. Having distinguished himself as a 'frequenter of technical libraries' in his youth, he gained a reputation as an advisor on practical problems, patenting equipment and consulting privately for other brewers, apparently without objections from Greenall's.[72] Edmund R. Southby (1833–86) originally trained in medicine, but gravitated to chemical analysis, and was engaged by Allsopp's of Burton around 1860. He then spent several years working on coal-oil manufacture, but returned to brewing in 1874, this time setting up on his own account as a private consulting analyst in London, and as the patentee and promoter of various processes for treating grain. In 1882, he succeeded Dugald Campbell as analyst in residence to the Country Brewers' Society.[73]

Faulkner wrote regularly for the *Brewers' Journal* from around 1875.[74] Among his contributions was a series of 'Practical and theoretical essays', presented anonymously as the work of a 'practical brewer and chemist'. The plan was broadly similar to that of the *Journal*'s series of three years previously, but the section titles – 'Chemicals used', 'Arrangement of brewery buildings', 'Electricity' – make clear Faulkner's orientation around the chief fields of private consultancy, chemical and otherwise. Faulkner made the best of the currency afforded by serial publication. In an early contribution, not having heard of the recent O'Sullivan–Valentin patent, he wondered why the sugar-makers didn't sell a saccharine-dextrin mixture; the *Journal* reported on Valentin's paper as the series was nearing completion, and Faulkner was able to incorporate an enthusiastic mention into his text.[75] The series also began to draw correspondence from brewers during its run, some of which was inserted into the *Journal* along with Faulkner's replies.

In 1876, partway through the run of Faulkner's essays, Southby began a series in the *Brewers' Guardian* on 'Brewing, practically and scientifically considered'.[76] The two men's narrative personae were subtly different in ways that broadly reflected their career histories: Faulkner tended to write as a fellow brewer, Southby as a friendly advisor from an allied profession. In fact, there was strong overlap in their coverage and claims, but the two were essentially competitors for the same market, and prone to friction on the relatively minor issues which separated them. Both openly used their writings to promote their own apparatus, methods and services.

In the summer of 1876, Faulkner's essays were compiled into a short treatise, *The Art of Brewing, Practically and Theoretically Considered*, priced 10*s*. and published by Lyon, with extensive advertising in the *Journal*. With a ready-made market, the work sold rapidly. (The opportunity was so evident that we must won-der why the 1872–3 series had not received the same treatment. Perhaps the author made unrealized plans to take it to another publisher; Faulkner's close relationship with the *Journal* made his path more obvious.) The approach went one better than mid-century authors such as Black and Tizard, who had relied on their private consulting networks to promote their books, and vice versa: dropping his anonymity on publication, Faulkner could bring his name to the whole trade through the *Journal*, and could shape public correspondence with his admirers and critics as he chose. Though he described his writing as 'the task of my leisure', it clearly laid a foundation for his later consulting career.[77] Southby, whose *Guardian* serial had been intended from the outset as a treatise, followed suit in 1877.[78]

Both authors – like every significant writer on brewing of the nineteenth century – justified the need for their work in terms of a novel negotiation of the theory–practice boundary. Faulkner offered 'a bridge over which my readers might pass from practical to theoretical knowledge': his *Art of Brewing* was not a comprehensive treatise, but a primer to affirm that 'theory and practice, instead of being rivals, are in reality of mutual aid'.[79] Like Charles Graham, Faulkner felt that references to the 'mere practical brewer' – the conventional lost soul, sighed over by theorists and philosophical-practical brewers alike over the preceding century – must be balanced by recognition that the 'mere theoretical chemist' also had his limitations.[80] Southby did roughly the same, but balanced his more analytical persona with a more vocally practicalist manifesto. There was no point, he stressed, in importing scientific attainment merely to wear it as a badge of rank: practical brewers need not memorize the atomic composition of materials, nor waste time and money commissioning detailed laboratory analyses at a level of precision useless in the brewhouse. The level of theory which would actually benefit any working brewer, he said, was well within that brewer's grasp. At the same time, the brewer keen to make progress must be careful not to lose sight of those 'facts dependent upon invariable natural causes, which scientific investigations have brought to light'.[81]

Both authors also followed Graham in promoting the importance of Louis Pasteur, whose status among brewers had hitherto remained marginal. Pasteur's chemistry had been noted intermittently in the *Brewers' Journal* since the 1860s; his fermentation system, as a patent process, had been written up in the 'Brewer's Engineer' column for 1872, but without solid enthusiasm. The serial essays of the same year concluded that while 'the balance of evidence' supported Pasteur's fermentation theory, the question remained undecided.[82] In his early essays, Faulkner hedged his bets and attempted to reconcile Liebig's and Pasteur's pictures of fermentation, having only indirect knowledge of the latter. Shortly after his first edition went to press, however, Faulkner read Pasteur's recently released book *Études sur la Bière*, presumably in the original French, and became an instant convert.

In August 1876, the *Études* was the subject of the *Journal*'s first long book review, almost certainly by Faulkner. Pasteur, the author stressed, was 'no mere laboratory theorist and experimentalist', but had ranged widely among the brewhouses of France. He possessed, it was true, 'supreme contempt' for the typical methods of English brewers: for the reform-minded, however, his dismissals were double-edged. There was no crediting his claim that British beer (the focus of a booming imperial trade) did not travel well, for instance; yet the process was undoubtedly unreliable, and the low-temperature reforms he promoted would improve it. Pasteur's unique contribution was as a theorist of genius who had furthermore, via microscopic diagnosis, attained the most self-evidently practical improvement imaginable. Exceptionally, the *Journal* included an illustrative plate reproducing three of Pasteur's figures, allowing the reader to *see* the diseases of beer.[83] Southby, likewise, affirmed himself a committed Pastorian; though less effusive than Faulkner, he was ahead of him in preparing an abridged English translation of the *Études*, with notes, which began printing in the *Brewers' Guardian* in 1877.[84] When readers demanded a full translation, however, it was Faulkner who obliged.[85]

Faulkner had by this point published a second edition of the *Art of Brewing*, in which he set out an aggressive programme to introduce routine microscopic testing, systematically weed out diseased strains, and replace all Liebigian interpretations with Pastorian ones. The *Études*, he enthused, contained 'the secret, the very essence of successful work'.[86] This new edition brought to prominence Horace Brown's research on the 'forcing tray', a diagnostic innovation based directly on the Pastorian insight. The tray consisted of a heated water-bath containing test tubes and flasks, carefully prepared to exclude atmospheric air, in which beer samples could be subjected to a 'forced' secondary fermentation, at temperatures well above those found in practical brewing. This 'severe test' definitively sorted good samples from bad, and could also be used for trials of the various preservatives on the market.[87] A century earlier, John Richardson had skewered Humphrey Jackson for supposing that experiments conducted 'in a snug room, with a few half-pint phials' could teach anything to the practical brewer. To Faulkner, although the disjunction retained some of its force, the analytical approach both allowed, and demanded, that it be broken down:

> I am quite aware that a laboratory ought to be entirely distinct from a brewery, that miniature operations must not tempt us too often to imitation on a large scale; that chemicals in any shape or form are best away from the brewery, and absent as constituents of our beer; but, at the same time, as brewing in itself is a strictly chemical process ... there is no doubt that a well-grounded analytical knowledge of the composition of materials, worts, and beers will prove of the greatest service to the operator.[88]

Faulkner's thinking on this point determined the remainder of his career. In 1889, he set up as a full-time consultant in his native Birmingham, erecting 'within a stone's throw of his father's brewery' an experimental brewhouse simply known as 'the Laboratory'.[89]

Commitment to scientific authority did not mean deference to all who claimed to speak for science. Well into the 1880s, Faulkner accommodated William Black's theory of electrical influence on fermentation alongside the Pastorian picture, arguing that a metallic attemperator assembly could act as a Leyden jar. This claim drew the ridicule of William Stevens Squire (1834–1906), a Giessen- and Royal College-trained chemist who consulted on yeast production for distillers. Writing in a paper to the Society of Chemical Industry which was reprinted in the *Brewers' Guardian*, Squire sighed at the trade's persistent 'mystery-mongering' and refusal to abandon ideas which had been comprehensively dismissed in the scientific mainstream. Faulkner retorted that his interpretation rested on the advice of 'a very eminent electrician' – actually the junior but well-placed Victor Herbert Veley (1856–1933) of the Oxford University Museum laboratory.[90]

Faulkner's more detailed response in the *Brewers' Journal*, provocatively headed 'Theorists subdued', in fact typified the appeal to collaboration which had begun with William Black. Only a 'small band of scientific writers' would presume to tell the brewer his business from first principles, having 'never suffered anxiety from irregular fermentations' themselves: a true man of science would tackle the question alongside a practical brewer, as the collaboration with Veley demonstrated. Even here, Faulkner stressed, Veley's 'miniature experiments' on models had proved less productive than practical trial-and-error investigation in the brewhouse itself.[91] Awkwardly, however, Veley then objected to Faulkner's interpretations of his work, and after performing laboratory experiments with Squire, swung round to endorse him. Faulkner protested in the *Journal*, but his influence there was not unlimited: the editorial voice (presumably Lyon's) avoided any hint of endorsement of the electrical theory, and curtailed the correspondence. It immediately resurfaced in the *Brewers' Guardian*, where it followed the usual pattern of such altercations in sliding from questions of theory and practice to the conduct and character of the antagonists, each accusing the other of prolonging and inflaming a pointless dispute.[92]

Pasteur and the microscope provided more reliable rallying points. Horace Brown was a far more zealous Pastorian than either Faulkner or Southby: years later, in a highly selective reminiscence, Brown was to date the birth of brewing science to Pasteur's research in the 1860s (tellingly, the beginning of his own career), dismissing the entire Liebigian movement as a barrier to progress and Black's electrical theory as 'fatuous', for all that it had been endorsed by the chemical authorities of his day.[93] Yet Brown was probably alone in his single-minded devotion. Faulkner was more typical in attributing scientific advance to a constellation of important newcomers making contributions in different fields. Pasteur went too far, he said, in reducing the management of beer to a question of temperature: other considerations were equally important, such as the balance of fermentable to other matter.[94] O'Sullivan, of course, was the guiding light here, and Faulkner foresaw a synthesis: Pastorian methods could probe the complexities opened up by O'Sullivan's research, as the use of substitutes modified the form of the yeast itself.[95] Accordingly, in October 1876, the *Brewers' Journal* ran a piece attacking 'the scepticism respecting modern theoretical brewing', in particular microscopy and dextrine-maltose, professing confidence that the new technologies would soon be as fully accepted as the saccharometer.

The concern here was more than economic: the 'theoretical' approach, above all, was crucial to the moral standing of the trade, a key concern for the *Journal*. Conservative brewers, ran the argument, relied on high alcohol levels to ensure the preservation of their product; Norwegian and German competitors, using low-temperature methods, could achieve stability in a refreshing product unlikely to cause drunkenness. Though it was often unworkable to transfer the Continental method to British equipment, the same effect could be achieved through more careful management and a reliance on dextrines for body: such an effort would decidedly 'take the wind out of the sails of the Alliance'.[96] That the United Kingdom Alliance based some of its rhetoric on the age-old smears against the chemical character of beer, said a follow-up piece, confirmed the teetotallers' ignorance: modern chemistry was a sound, economical source of pure materials and hygienic decontamination.[97] Such arguments, and the vision of a forthright technological culture they conjured up, soon determined the trade literature's response to the largest legislative shake-up in half a century.

Freeing the Mash-Tun

Since adulteration panics could flare up at any time, most authors who promoted a chemical underpinning for brewing continued to tread warily around the subject of additives. Muspratt, in the 1850s, had linked proper chemical management to a retreat from toxic substances, the sins of past generations being 'frightful to contemplate'.[98] Similarly, Charles Graham amplified William Black's position that mineral additives should be used only to correct an unsound product: otherwise, 'the less perhaps you have to do with them the better'.[99] Faulkner saw the traditional objection to 'chemical substances' as rank superstition – nobody, he pointed out, objected to salted meat – while accepting that the manufacturing chemist was not necessarily the brewer's friend. It was hardly unreasonable for brewers to mistrust 'secret powders' offered as substitutes for hops or other materials, yet the brewer must acknowledge the chemist's difficulty, since such preparations could not usually be patented.[100]

On the other hand, governmental reliance on the authority of analytical chemists was leading to the ever-greater toleration of substitutes. Most experts firmly rejected the purity position, seeing malt extract, molasses and synthetic sugars alike as consisting of an assortment of sugary molecules which could, in many cases, be shown to be equivalent. The Boards of Excise, which had been the chief defenders of malt-and-hops purity as necessary to prevent more serious frauds, were increasingly persuaded that new laboratory methods could obviate the problems, an important test being the abovementioned drawback procedure which accompanied the derestriction of sugar in 1847. In 1862 the hop duty was abolished, replaced by a small increment in the malt tax, and hop substitutes explicitly permitted. In 1874, the remaining sugar restrictions – which covered liquid preparations and material intended for colouring – were removed, allowing 'any description of sugar' into the brewhouse.[101]

Pressure for such deregulation was usually couched in free-trade terms. The 1847 measure, for instance, had been urged in a pamphlet, *Good and Cheap Beer for the Million*, which argued that the emancipation of slaves throughout the British Empire had left the planters under heavy regulation and with few outlets: permitting sugar and molasses in beer, it was proposed, would free up grain needed for the Irish market, lowering prices and raising consumption all round.[102] Such thinking resonated with Liberals in Parliament, who also took seriously the case, pressed by farming interests, for removing the malt duty altogether and permitting unmalted grain. Faulkner, in 1876, observed the field in transition. Sugars from maize and rice were now freely promoted, and brewers were experimenting openly to determine the yields from wheat and other materials. The effect of further liberalization, he felt, would not be particularly drastic.[103]

The *Brewers' Journal* had supported 'the advantages derivable from free trade' from the outset,[104] and it seems that Faulkner was, at least with regard to his own trade, a devotee of the invisible hand of capitalism. The spur of competition, he wrote, led brewers to seek an ever greater yield of extract from the grist. Although this inevitably led to more 'injurious material' in the wort, it ultimately benefited everybody by prompting further refinements to remove the unwholesome content, or to find alternative processes and materials:

> Thus it has come about that brewers are expected to have not only a deep theoretical knowledge of the process they have to arrange and conduct, but also of chemistry, electricity, and the allied sciences; and this is justly so, if we think what a complex subject brewing is at the present, and how much more complex it will become under a liberal alteration of the excise laws.[105]

Much had changed since the time of brewer-authors such as John Ham, who (though diligent in the cause of reform on economic principles) had attacked schemes to maximize yield as symptomatic of the naïve approaches cooked up by non-brewers.[106] Faulkner argued that while the dangers of the approach (impaired flavour and excess acidity) were real enough, they must nevertheless be confronted and managed. Southby agreed: although fortunes had been made by unscientific brewing in the past, the finest 'rule-of-thumb' producer had no systematic means of adapting to radical change, such as was inevitably 'becoming daily more imminent from increased competition'.[107] In Charles Graham's work, this technical case took on a moral dimension. Deregulation would mean not only cheaper materials, and thus a cheaper drink, but full control of the process, allowing a full flavour from increased dextrin with less alcohol. 'What we want', said Graham, 'is beer for quenching thirst – a beer that we may safely drink a great deal of'. He included himself in the 'we', but stressed that he was thinking chiefly of working men.[108]

The 'Free Mash-Tun Act', as it became known, was enacted soon after the Liberals' return to government under William Gladstone in 1880.[109] Henceforth, said Gladstone, the brewer 'will brew from what he pleases, and he will have a perfect choice both of his material and of his methods'.[110] Unmalted grain, cereals other than barley and sugars of all kinds were now permitted. At the same time, the malt and sugar duties were scrapped entirely, but – since no government could conceive of losing the tremendous revenues due to beer – replaced with a tax on brewers' wort. This was assessed on an entirely saccharometric basis: the duty assigned for one barrel of wort, at a nominated standard original gravity, was to be varied directly 'in proportion for any difference in quantity or gravity'.[111] Aside from the fact that values were to be expressed on the specific gravity scale of chemists and distillers, not in brewers' pounds, the new regime was remarkably similar to the system John Richardson had proposed in the 1780s.

Many brewers were nonetheless hesitant, perhaps suspecting that the new duty rate, supposedly chosen for a neutral effect on revenues, favoured the Excise (a suspicion which turned out to be well-founded). In September 1880, two weeks into the operation of the Act, the *Brewers' Journal* offered a positive blitz of material urging the trade to accept the change and rise to its challenges. The new regime, said an editorial, seemed to be working well enough despite teething problems, and the objectors would certainly 'have to improve their theoretical knowledge before entering the lists to contest statements made by Inland revenue authorities'. The *Journal* despaired in particular of 'one authority' – Daniel Watney, of the major Pimlico brewery – who had written to the *Times* arguing a flaw in the scheme: he had actually confused pounds-per-barrel value with mass of dry extract, an elementary distinction which had been definitively cleared up in 'the shilling work on brewing by Mr. Booth' of 1829.[112] If the brewers valued their industrial reputation, said the *Journal* in October, they must

> cease snarling at regulations that in the end will induce more scientific working ... There is far too much inferior beer in England, and the new regulations certainly tend to encourage the use of superior materials, and the practical closing of small and inferior plants.[113]

This case for more consistent production quality was one part of the *Journal*'s campaign to improve the overall reputation of the trade, foreshadowing the 'fewer and better' doctrine employed to defend the public house against teetotalism into the twentieth century.[114] Faulkner added technical notes on how to get to grips with the new scheme, stressing that the right to use raw grain brought both challenges and opportunities: brewers should know more about the activity of diastase, and would probably need to shift to more complex ('semi-decoction') mashing schedules, inspired by Continental practices.[115] Southby, whose work was seen less frequently in the *Journal*, followed up in December with notes on the practical working of the Act.[116]

Enthusiastic support for the free mash-tun was valuable in another sense: it helped to give the brewing lobby the ear of senior Liberals, at a time when the United Kingdom Alliance was making concerted efforts to co-opt the Liberal Party for the teetotal and prohibitionist cause.[117] To Faulkner and his allies, however, the issue went beyond the contingencies of the defence of the trade to its fundamental nature. As he wrote in 1884:

> Increasing competition, comparative inferiority of material, and a growing public taste for weaker beers, have certainly combined to render the manipulation of necessity more and more difficult, and to bring it into more perfect connection with theoretical teaching, while, if any proof were needed for the statement that greater skill and attention to detail has gradually been becoming more and more necessary, what better one could be urged than the rapidly increasing employment of chemical antiseptics and preservatives that were entirely unused, and, indeed, unthought of, a few years back?[118]

The possibility of such a retrospect by no means indicates that the institutional integration of science and brewing was complete. Later developments, beyond the scope of this survey, were to place it on a far more formal basis. In 1886 the most active brewing chemists formed a 'Laboratory Club', soon to be succeeded by a national federation of Institutes of Brewing: its journal provided a dedicated forum for the communication of brewery research, overlapping and complementing the existing periodicals. Meanwhile, the universities and technical colleges responded to the call for systematic instruction and research, with Horace Brown's brother Adrian becoming Professor of Brewing and Malting at Mason College, Birmingham, in 1899.[119]

Yet the freeing of the mash-tun – or, rather, the decision to embrace the measure by the trade's self-appointed defenders – illustrates that by 1880, the question of the credibility of brewing science was effectively closed. In the new world of free trade, indeterminate exports and the militant Alliance, it was still possible to distinguish a 'man of science' from a 'practical man'; but practicality itself, as traditionally defined by commercial cost-consciousness and a close regard to the materials and fabric of the brewhouse, now depended on chemical and related considerations. Commercial brewers with production roles must now either become, to some degree, scientists themselves, or else submit to the authority of science produced elsewhere. From this point forward, it was the non-scientific brewer whose credibility was open to question.

CONCLUSION

To 'BUNGS'[1]
of Independency: –
This Volume has a tendency
To place *you* in ascendancy
Above undue monopolies;
And hence to *you*, whate'er your station,
Your creed, your colour, or your nation,
Its author makes no hesitation,
To tender this, his
DEDICATION.

W. L. Tizard[2]

In 1854, *Punch* introduced its readers to Mr John Paterfamilias, 'a man of an inquiring, but by no means robust mind', addicted to reforming his suburban household along 'rational and sanitary principles'.[3] Undeterred by disastrous experiments in gas-fitting, and alarmed by adulteration reports in the *Lancet*, Paterfamilias decides to brew for himself, aided by his son Newton – a priggish 7-year-old, similarly addicted to scientific authority – and a 'very clear little practical treatise', priced 2*s*. The instructions dictate that mashing should begin at the preposterously low temperature of 78°F: having no thermometer, Paterfamilias accepts his servant's approximation of 'one pail o' bilin' to three o'cold'. Having mixed the water and malt, Paterfamilias runs off the wort immediately, and is greatly surprised by its watery thinness: 'I don't think all the gluten can have been converted into saccharine', remarks Newton sagely. They press on regardless to the fermentation stage. Decanted into a household water-cask, the brew exhales a sufficient blanket of carbon dioxide to strike the inquisitive Newton insensible, and in six weeks proves to be perfect vinegar. 'The fermentation', puzzles Paterfamilias, 'must have been acetous instead of vinous'.[4]

The incongruity of inserting high-flown theory into practical operations, and the disaster that might result, was a familiar topic to readers in general across the period of my study, and certainly needed no introduction among either scientists or brewers. Every brewer-author from Michael Combrune to Frank Faulkner assumed a reader who doubted the credibility of his work; almost every treatise opened with a defence of its own coherence, plausibility and novelty. By the mid-nineteenth century, such justifications could be selected from a range of standard rhetorical devices. Before the turn of the twentieth century, however, they had entirely disappeared from print literature. The emergence of brewing science as a stable phenomenon was marked by the fact that brewers stopped discussing it, except as a matter of historical enquiry and reminiscence.

This happened partly because brewing embraced the same division of labour between books and serial works which had taken hold across the applied sciences. New discoveries relevant to the chemistry and biology of brewing were reported in mainstream scientific journals, whereas book-length publications increasingly existed only to abridge, collate or reinterpret material that had circulated in other forms, usually for educational purposes or as a ready reference. The promotion of apparatus, materials and services, meanwhile, fell chiefly to the trade press, and to private promotional literature, which could be distributed ever more cheaply. The 'treatise', as a vehicle for an author's agenda or wares, no longer existed: compiling a useful survey text might help to consolidate a promising researcher's reputation, but the credibility of new books generally depended more on their authors' reputations than vice versa.

In the 1840s, William Tizard had used a seven-page preface to present his treatise as something quite new, a comprehensive survey of current innovations which would place his reader in the driving seat. After 1880, most authors positioned their books as reliably conventional specimens of genres whose purpose and value was obvious to a moderately informed reader. In 1904, for instance, Adrian Brown issued a 'student's laboratory guide' intended to accompany a formal course under a 'competent instructor', on a pattern conventional in technical education.[5] Egbert Grant Hooper (1855–1935), analyst to the Inland Revenue, stated simply the purpose of his *Manual of Brewing*, first published in 1882:

> 1st, to furnish a text book to the Technical Brewing Examination of the City and Guilds of London ... 2ndly, to afford all those – Revenue Officers and others – whose business brings them in contact with the Brewing trade a guide to its technicalities ... and, 3rdly, to offer to Brewers, and especially those who have not had the advantage of a scientific training, a Manual containing an account of the discoveries of science which bear upon their profession.[6]

One factor which had not changed was a tendency for enterprise to run a little ahead of what could evidently be accomplished: Hooper also hoped that the *Manual* would appeal to 'the large class of private Brewers' spawned by Gladstone's abolition of the malt tax. The boom did not, in the event, live up to expectations; even if it had, it is hard to imagine even the most scientifically enthusiastic home brewers finding much to their purpose amid the arsenal of albuminoid chemistry, iodine tests, optical polarimeters, microscopes and decimal measures that Hooper's core audience required.

Indeed, 'Brewing', as a subject for publication, was now distinctly esoteric: even texts for general audiences were more likely to be simplified introductions to the science than guides explaining how to brew. In 1912, Alfred Chaston Chapman (1869–1932), a disciple of Charles Graham in brewery consultancy who was then President of the Institute of Brewing, contributed a text to Cambridge University Press's shilling popular science series. Where the useful-knowledge texts of the previous century had shown up profound tensions between instructional and commercial agendas, Chapman had no difficulty in meeting the series conventions, showcasing brewery investigation as a contributor to 'chemical and biological science' in general.[7]

My introduction spelled out the two chief paradoxes challenging the creation of brewing science; it is worth concluding with a survey of how they were dispelled. Responses to the paradox of the trade writer were crucially directed, at an early stage, by Peter Shaw's modification of the Bacon–Boyle heritage. In tying his philosophical agenda to the economic state of the nation, Shaw defined a mode of enquiry that could be both wealth-seeking and selfless. The assumption that his disciple, Michael Combrune, must be jeopardizing his own livelihood if he published information that could help his rivals (and that his sincerity was therefore in doubt) was countered with a vision of general improvement to lower prices, expand markets and substitute for imports. Under Shaw's patronage, Combrune could present himself as a disinterested philosopher. Such self-positioning was common before the academic professionalization of science in the nineteenth century, although it was never universally convincing either to philosophical gentlemen or to brewers.

Most of Combrune's successors, by contrast, professed openly that they had something to sell, be it instruction, consultancy, instrumentation or literature itself. The crucial task for such authors was to affirm a common interest with the trade community which defined the reader's hopes and expectations. This is what Humphrey Jackson signally failed to do: his scheme of restricted instruction, whilst working evidently to his own benefit, set the trade against each other. It also defied the usual paths of informal publicity. It is significant that, despite the evidence of pamphleteering and the Thrale case, we are still in the dark as to exactly *how* Jackson came to grief: we must assume he was convicted through the word-of-mouth proceedings of the court of personal inspection, like William Ellis before him. Friedrich Accum, too, was rejected as an unwelcome intruder, but the aloof Thomas Thomson was not, for the straightforward reason that he neither attacked the trade's probity nor attempted to involve it in the sale of improbable secrets: his Excise consultancies, though performed for money, did not belong in the category of commercial endeavour. Thomson, moreover, had spent time in the brewhouse, at least in the literal sense; the fact of his doing so mattered more than the reason why.

The master of mobilizing the common interest, of course, was John Richardson, whose rhetorical tricks shaped the brewhouse literature of the earlier nineteenth century and reverberated into even the more self-consciously scientific productions of later decades. Richardson recognized that a common interest demands a common enemy: his image of the benighted Exciseman worked precisely because it told his readers what they already knew, guiding them gently towards the flattering counterpart vision of the practical philosopher – a vision most appealing, perhaps, to the highly literate and well-connected gentleman brewers who could best serve Richardson's own interests.

What trade audiences demanded above all was evident care in the management of disclosure as a commercial activity: openness, in other words, about the limits of openness. To his core audience of late eighteenth-century brewers, it was quite reasonable for Richardson to declare himself to be withholding knowledge that might be reduced to print, a strategy which in Thomson's view showed precisely why the trade could not be trusted with its own self-definition. The brewer-authors of the following generation – most influentially, William Black – needed the support of Thomson's direct successors in an increasingly formalized and exclusive chemical establishment, and so were careful to describe the limits of print in different terms. Their distinction was between a print culture which was as open as it could possibly be, and a mode of practical instruction and consultation that could not, from the site-specific and tactile nature of the task, be anything other than private.

Key to this change was the increasing specialization of roles made possible by the growing concentration and interconnection of the trade. Mid-century brewers-turned-consultants such as Black and Tizard combined a reassuring brewhouse context with a sound commercial basis for exporting knowledge to as many clients as possible; this brought, in turn, the even more valuable credibility claim of knowing, from the inside, a *plurality* of brewhouses. The same advantages accrued to the dedicated brewers' analysts and scientific brewing consultants of the next generation, whether their training lay in the brewhouse (as in the case of Frank Faulkner), outside it (Charles Graham), or between the two (E. R. Southby); the laboratory content of their work rendered the distinction increasingly irrelevant. Patents, too, were important, especially for brewers' engineers. In referring in his dedication to 'undue monopolies', Tizard implicitly distinguished the patent as a wholesome, self-limiting alternative to selling improvements in secrecy to the highest bidder.

The paradox of experiment, meanwhile, was partially neutered by industrialization, market and site concentration, and the decline of private brewing. Experiments stuck out as obvious deviations from sound practice only so long as the process (or, rather, the preferred local variant on the process) was a matter of common knowledge. With the search for distant markets, most commercial brewers encountered products very different from their own and began to dwell on questions of efficiency and emulation, inviting attention from nostrum-mongers and quantifiers alike. Among the brewer-authors who addressed such questions, George Adolphus Wigney stands out for declaring that even the most farcically disastrous experimental sallies must be celebrated as steps on the road to improvement; yet all the brewers' consultants, whether committed to heterodox methods or not, were essentially experimentalists, as they travelled the country negotiating ways to adapt their standard guidance to local peculiarities and constraints.

The message of such consultants was initially that a brewer *might* profitably experiment. The line that emerged in anticipation of the Free Mash-Tun was harder: brewers *must* experiment, or face oblivion. The modern brewhouse (or so the *Brewers' Journal* preached) would achieve a new rationality. On the one hand, Pastorian insights would banish the curse of skilled and diligent brewers suffering random and disastrous failures; on the other, the man who failed to achieve maximum efficiency, who did not take specialist advice or scan the trade press for the latest machinery, sugars and disinfectants, would face certain bankruptcy. In 1876 John Tyndall, the leading advocate of professionalized science, published an article in which – so like many before him – he hailed brewing as a newly scientific discipline, freed by Pasteur from its 'hidden enemies'.[8] As a good Darwinian, Tyndall presumably embraced the wider picture of a dynamic order in which the brewers best adapted to their scientific surroundings were the best placed to survive.

Finally, both of the putative paradoxes of brewing science assumed that the interests of the researching brewer were aligned with those of the brewing firm which accommodated the research. While this was always to some extent the case, the house chemists employed after 1860 – most conspicuously, those in the first Burton generation – developed a disciplinary loyalty which in some ways transcended their commitments to their proprietors. As trained analysts with valued transferable skills, moreover, they could often negotiate considerable flexibility in their employment terms, running side-projects and developing the patterns of research communication which suited them. There were limits: Horace Brown, taken on by the giant Irish firm of Guinness in 1901, fought running battles over the company's preference for secrecy and left on bad terms five years later, though not before securing the right to publish his work on the soluble nitrogenous constituents of malt in a voluminous set of 'Transactions'.[9]

In 1907, Horace Brown's friend and fellow 'Hofmannic' graduate, Henry Edward Armstrong (1848–1937) of the City and Guilds Institute, recalled chiding Brown that it was 'time he did something useful – that he gave up brewing and turned to science'.[10] The comment was tongue-in-cheek: the 'science' Brown had delivered, namely the nitrogen research, was integral to his work as a brewer. The union of laboratory and brewhouse no longer required bridges, patrons or translators: Brown was straightforwardly a member of the scientific elite, having obtained a Fellowship of the Royal Society in the 1880s and its Royal Medal in 1903; in later years, he was a leading light in the Chemical Society, of which he would, but for ill health, have been President.[11]

The securing of brewing science, we should be clear, had not abolished the distinction between 'theoretical' and 'practical' outlooks, nor the commonplace friction between them. (Indeed, tensions over the relative benefits of theoretical learning and practical experience in training junior brewers remained disruptive after the Second World War.)[12] In 1884 the chemical consultant William Squire, who was soon to clash with Frank Faulkner over electrical action (p. 200), offered essentially the same jaundiced picture as Thomas Thomson or Michael Donovan half a century earlier. Even as he cited Brown, Cornelius O'Sullivan and other brewery employees as authorities, he protested that the brewing trade had long been 'a close corporation of rule-of-thumbists ... and to a certain extent still is'.[13]

What had crucially changed, however, was that practical men now moved freely in the world of men like Squire. As the *Brewers' Journal* pointed out, when Squire spoke at a meeting of the Society of Chemical Industry, 'the majority of the gentlemen entrusted with the management of the more important establishments' in beer-brewing London turned out to hear him. Most already agreed that the electrical theory could be dismissed: they were more interested in discussing Squire's endorsement of the botanist Carl Nägeli, whose microbiology of fermentation departed from Pasteur's. Besides, said the *Journal*, experience showed that beer really *did* tend to sour in thundery weather, a point non-brewers were overly willing to ignore. Rather than scoring points off an already discredited theory, the scientist should put himself at the service of industry by seeking the true explanation.[14] Squire, for his part, had to acknowledge the new common context to shore up his own credibility. His consulting work for distillers, he told the *Journal*, meant that his 'own operations in brewing and fermentation have been, and are, on a considerably larger scale than Mr Faulkner's'.[15]

Though brewer-authors such as Faulkner could no longer expect to hold out against established points of scientific orthodoxy, it was Faulkner and his contemporaries who ultimately defined a coherent basis for 'brewing science', unthreatening to settled interests and firmly tied to commercial priorities. The rhetoric of the Free Mash-Tun (which, like all such platforms, tweaked history to its own ends) actually legitimated the heritage of the 'rule-of-thumbists', even as it abolished it. The rule of thumb had served very well to turn out sound and saleable beer, in the days when materials were cheap and good, markets local, and theorists institutionally unengaged with production; but times had changed.[16] In 1762, Combrune's *Theory and Practice of Brewing*, its title and structure modelled on Boerhaave, had implied the bringing together of two separable bodies of work. Faulkner's *Theory and Practice of Modern Brewing* was distinguished not only by the adjective: to be credible now, individuals and institutions must deal with 'theory and practice' as a coherent enterprise. Brewing science had ceased to be a problem.

GLOSSARY

ale in early modern times, usually referred to unhopped or lightly hopped drink, often treated as an entirely distinct product from the hopped 'beer'. In the eighteenth century, often identified a pale drink, contrasted with the darker 'porter' (particularly around London, where porter was the staple product). Meanings became more fluid in the nineteenth century.

attenuation the thinning-out of a wort during the fermentation process, as the sugars are converted to alcohol. Quantified as the difference between the original and final gravities.

barrel in brewing parlance, specifically a cask of 36 gallons' capacity (on the old ale measure, 32 gallons).

bigg a crop similar to barley, grown in the north of Scotland.

blue wort in brewers' argot, the final, very weak wort, drawn from a grist which had already been mashed several times.

brewhouse any building dedicated to brewing, usually containing a mash-tun, copper and other vessels.

brewing victualler publican ('licensed victualler') who brews her or his own beer, on a small scale, for consumption on the premises.

bright clear and transparent; used of beer following the initial fermentation, when yeast and other materials have dropped out of suspension, usually indicating that it is ready to drink.

butt cask of three barrels' (108 gallons, beer measure) capacity, normally used for maturation.

cleansing period of fermentation in which the yeast is worked out of the beer prior to casking; also sometimes, by extension, transfer of finished beer from fermentation vessels into casks.

cocculus indicus the fruit of an Indian climbing plant, rich in the poisonous narcotic picrotoxin. After use of cocculus as a beer additive was documented in some prosecutions around 1800, the substance gained a reputation (probably disproportionate) as the chief weapon of the 'brewers' druggists'.

common brewer brewer producing beer, usually on a large scale, for public houses which do not brew on their own account, and sometimes for private customers.

copper boiling vessel (usually made of metal, not necessarily copper), used for the hop boil and for general heating of liquor in the brewhouse.

dextrin, dextrine term introduced in the 1830s to describe a gummy component of starch, later interpreted as an unfermentable sugar and the principle source of body in beer.

dextrine-maltose proprietary synthetic sugar preparation, patented 1874 and promoted to brewers as a partial malt substitute, providing body as well as fermentable content.

diastase substance identified in malt extract in the 1830s, capable of breaking down starch into dextrine, and thence into fermentable sugars.

Excise the principal agency for enforcing and administering the system of taxes on goods inland.

extract fermentable materials in a wort, extracted from malt during mashing.

finings preparation of isinglass, or sometimes other materials, used for clearing ('fining') beer rendered cloudy by fermentation.

finished beer beer in a drinkable state after fermentation.

fox taint affecting beer.

goods the grist once wetted in the mash-tun, or the soluble fermentables produced from it.

grains the spent grist left behind after the worts have been drawn off. Commonly used as animal feed.

gravity the density of a liquid, usually as determined by a hydrometer or saccharometer.

grist the ground malt, and sometimes other materials, mashed to produce wort.

humuline proprietary name for a concentrated hop extract, patented 1841.

hydrometer flotation device used for gravity measurement.

isinglass gelatinous substance prepared from the swim bladders of sturgeon (and various other materials, with varying degrees of success), used to prepare finings.

length the volume of wort drawn from a given mash.

liquor in brewers' argot, plain water used for mashing ('water' being what was used for cleaning of utensils and other tasks). Apparently a brewers' shibboleth: eighteenth-century sources suggest that London apprentices were fined for calling liquor 'water'.

malt malted grain, generally barley unless otherwise stated. The principal raw material of beer production.

malting the process of partial germination and drying, applied to barley and other grains to make malt.

maltose from the 1870s, the preferred term in English for the chief fermentable sugar found in malt worts.

maltster, malster, malt-factor manufacturer of malt.

mashing the infusion of malt, and sometimes other materials, in hot water to produce wort.

mash-tun any large vessel used for mashing.

mild beer in eighteenth-century parlance, new and immature beer (opposing 'stale').

mucilage term in occasional use around 1800 to describe the gummy, apparently non-fermentable component in wort.

original gravity the gravity of the cooled wort following the boiling stage, immediately prior to the start of the fermentation process; usually considered by saccharometrists to reflect the quantity of fermentable sugars present in the wort.

porter	the staple beer of London from the early eighteenth century, later widely produced around Britain and Ireland. Usually dark in colour; originally prepared from brown malt and aged for several months; later produced more rapidly from paler malts with colouring. Usually highly attenuated compared to the fuller and sweeter 'ales'.
quassia	bark or wood of the South American shrub *quassia amara*, used in medicine as an antipyretic, and in brewing as a substitute for hops, usually contrary to law.
saccharine, saccharine matter	in the eighteenth and early nineteenth century, the sweet material dissolved in wort; increasingly, its fermentable component only (as opposed to 'mucilage' or 'dextrine').
saccharometer	name coined by John Richardson in 1784 to describe a hydrometer adapted to measuring the gravity of brewer's wort.
set	(of taps, stopcocks, etc.) turn on, set running.
set mash	brewhouse calamity resulting when the mashing liquor is too hot. The malt clots, assumes the consistency of paste and retains most of the liquor, leading to an inadequate (and usually cloudy) wort.
small beer	weak beer for everyday drinking, originally made from the final mash in a multiple mashing process.
sparging	rinsing the grains after mashing to retrieve further sugars, as an alternative to repeated mashes. Initially a Scottish practice, becoming widespread in England in the second half of the nineteenth century.
stale beer	beer matured for a time, producing a characteristic acidic tang (opposing 'mild').
tun	brewery vessel, usually large and coopered, such as a mash-tun.
victualler	keeper of a public house for drinking; see also 'brewing victualler'.
wash	in the distillery, wort which has been fermented, prior to distillation.
wort	unfermented beer; the solution of fermentables produced by mashing.

NOTES

Introduction

1. 'Brewing', *Encyclopaedia Britannica*, 2nd edn (1778).
2. The line – supposedly Behan's proposal for a slogan encapsulating the merits of Guinness Stout – is flagged as 'perhaps apocryphal' in the *Oxford Dictionary of Humorous Quotations*.
3. R. Unger, *Beer in the Middle Ages and the Renaissance* (Philadelphia, PA: University of Pennsylvania Press, 2004), p. [xiii]; P. Mirowski, *The Effortless Economy of Science?* (Durham, NC: Duke University Press, 2004), p. 195.
4. P. Sambrook, *Country House Brewing in England, 1500–1900* (London: Hambledon, 1996) is the definitive study of this mode of brewing in England.
5. A. Chapman, 'From Alchemy to Airpumps: The Foundations of Oxford Chemistry to 1700', in R. Williams, A. Chapman and J. Rowlinson (eds), *Chemistry at Oxford: A History from 1600 to 2005* (London: Royal Society of Chemistry, 2009), pp. 17–51, on p. 36.
6. H. B. Carter, *Sir Joseph Banks, 1743–1820* (London: British Museum (Natural History), 1988), p. 339; Sambrook, *Country House Brewing*, p. 231.
7. J. Beckett, 'Land Tax or Excise: The Levying of Taxation in Seventeenth- and Eighteenth-Century England', *English Historical Review*, 100:2 (1985), pp. 285–308.
8. W. Ashworth, 'The Intersection of Industry and the State in Eighteenth-Century Britain' in L. Roberts, S. Schaffer and P. Dear (eds), *The Mindful Hand: Inquiry and Invention from the Late Renaissance to Early Industrialisation* (Amsterdam: KNAvW, 2007), pp. 349–77, on pp. 361–6; P. Mathias, *The Brewing Industry in England, 1700–1830* (Cambridge: Cambridge University Press, 1959), pp. 339–50.
9. L. Roberts and S. Schaffer, 'Preface', in Roberts, Schaffer and Dear (eds), *Mindful Hand*, pp. xiii–xxvii, on p. xiv.
10. J. Tuck, *The Private Brewer's Guide to the Art of Brewing Ale and Porter*, 2nd edn (London, 1822), p. [viii].
11. U. Klein, 'Experiments at the Intersection of Experimental History, Technological Inquiry, and Conceptually Driven Analysis: A Case Study from Early Nineteenth-Century France', *Perspectives on Science*, 13:1 (2005), pp. 1–48.
12. J. Secord, 'Knowledge in Transit', *Isis*, 95 (2004), pp. 654–72, on p. 659.
13. Mathias, *Brewing Industry* and T. Gourvish and R. Wilson (eds), *The British Brewing Industry, 1830–1980* (Cambridge: Cambridge University Press, 1994) are classic feats of the marshalling of corporate records in the service of, respectively, economic history and business history. Sambrook, *Country House Brewing*, makes ingenious use of the

limited physical survivals from private brewing, while L. Pearson, *British Breweries: An Architectural History* (London: Hambledon, 1999) surveys the architecture of the commercial brewhouse.

14. A. Johns, *The Nature of the Book: Print and Knowledge in the Making* (Chicago, IL: University of Chicago Press, 1998), p. 30.
15. W. Yworth, *The Compleat Distiller: or the Whole Art of Distillation Practically Stated*, 2nd edn (London, 1705), 'Epistle to the Reader'.
16. The seventeenth-century model for this position is laid down in, for instance, Gilbert Burnet's characterization of the Christian virtue in Robert Boyle's chemistry: S. Shapin, *A Social History of Truth: Civility and Science in Seventeenth-Century England* (Chicago, IL: University of Chicago Press, 1994), pp. 185–6.
17. G. A. Wigney, *An Elementary Dictionary, or Cyclopædiæ, for the Use of Maltsters, Brewers, Distillers, Rectifiers, Vinegar Manufacturers, and Others* (Brighton: printed for the author, 1838), p. iv.
18. Secord, 'Knowledge in Transit', influentially explored the lessons of this insight for historians of science.
19. For instance, the second part of Yworth, *Compleat Distiller* ('Pharmacopœia Spagyrica Nova') adopts an alchemical tone entirely absent from the distillery process account in the same work. The brewing account in W. Yworth, *Cerevisiarii Comes, or, the New and True Art of Brewing* (London, 1692), pp. 52–60, is similar enough to that in J. White, *Art's Treasury of Rareties, and Curious Inventions*, 4th edn (London, [*c.* 1690]), pp. 98–103, to suggest common ancestry. For Yworth's background, see K. Figala and U. Petzold, 'Alchemy in the Newtonian Circle: Personal Acquaintances and the Problem of the Late Phase of Isaac Newton's Alchemy', in J. V. Field and F. James (eds), *Renaissance and Revolution: Humanists, Scholars, Craftsmen and Natural Philosophers in Early Modern Europe* (Cambridge: Cambridge University Press, 1993), pp. 173–92.
20. Johns, *Nature of the Book*, pp. 444–542; R. Yeo, *Encyclopaedic Visions* (Cambridge: Cambridge University Press, 2001), pp. 195–221.
21. *Museum Rusticum et Commerciale*, 1 (1764), pp. 201–4; *Farmer's Magazine and Useful Family Companion*, 2 (1777), pp. 372–4.
22. *The Complete Family-Piece, and Country Gentleman and Farmer's Best Guide*, 2nd edn (London, 1737), pp. 276–85; *The Complete Family Brewer* (London, 1789); T. Threale, *The Complete Family-Brewer: or the Best Method of Brewing* (London, 1802).
23. A. Morrice, *A Treatise on Brewing; Wherein is Exhibited the Whole Process of the Art and Mystery of Brewing*, 2nd edn (London, 1802), pp. 22–38; [W. Ellis], *The London and Country Brewer* [1st edn of the first part] (London, [*c.* 1735]), pp. 8–21.
24. W. Brande, *The Town and Country Brewery Book* (London: Dean and Munday, [*c.* 1830]), the work of a self-described 'malster and brewer', reproduces without update or acknowledgment material first found in [Ellis], *London and Country Brewer*.
25. The definitive study in this mode is J. Secord, *Victorian Sensation: The Extraordinary Publication, Reception, and Secret Authorship of Vestiges of the Natural History of Creation* (Chicago, IL: University of Chicago Press, 2000).
26. Unger, *Beer in the Middle Ages*, p. [1]; H. L. Platt, *Shock Cities: The Environmental Transformation and Reform of Manchester and Chicago* (Chicago, IL: University of Chicago Press, 2005), pp. 234–43.
27. For example: Peter Shaw in M. Combrune, *An Essay on Brewing, with a View of Establishing the Principles of the Art* (London, 1758), p. [vii]; J. Richardson, *Statical Estimates of the Materials for Brewing, or a Treatise on ... the Saccharometer* (London, 1784), p. iv; *Monthly*

Magazine, 24 (1807), p. 346; G. A. Wigney, *A Theoretical and Practical Treatise on Malting and Brewing* (Brighton: printed for the author, 1835); W. L. Tizard, *The Theory and Practice of Brewing Illustrated*, 3rd edn (London: printed for the author, 1850), pp. 4–5; J. Tyndall, 'Fermentation and its Bearings on the Phenomena of Disease', *Popular Science Monthly*, 10:9 (1876), pp. 129–54, on p. 131; 'The Annual Banquet', *Transactions of the Institute of Brewing*, 5 (1892), pp. 14–27; J. H. Reynolds's comments to the Northern section, *Journal of the Federated Institutes of Brewing*, 8 (1902), pp. 219–25.

28. A useful corrective, and the only account to date which pays due attention to brewing chemistry between the 1830s and 1850s, is R. G. Anderson, 'The Pattern of Brewing Research: A Personal View of the History of Brewing Chemistry in the British Isles', *Journal of the Institute of Brewing*, 98:2 (1992), pp. 85–109.

1 The Curious Brewer

1. [Ellis], *London and Country Brewer*, 1st edn, part 1, pp. 34–5.
2. T. Birch, *The History of the Royal Society of London for Improving of Natural Knowledge*, 4 vols (London, 1757), vol. 1, pp. 152–63, quotation on p. 162.
3. K. Digby, *The Closet of the Eminently Learned Sir Kenelme Digbie Kt. Opened* (London, 1669), pp. 114–16.
4. J. M. Bennett, *Ale, Beer and Brewsters in England: Women's Work in a Changing World, 1300–1600* (New York: Oxford University Press, 1996), p. 148. For the long-term endurance of the brewing victuallers in many areas see Gourvish and Wilson, *British Brewing Industry*, pp. 64–75.
5. Pepys's diary entries for 17 August, 24 August, 21 September and 17 October 1667. Online at http://www.pepysdiary.com/ [accessed 12 January 2013].
6. W. B. Houghton, Jr, 'The History of Trades: Its Relation to Seventeenth-Century Thought', *Journal of the History of Ideas*, 2:1 (1941), pp. 33–60, presents the project as continuously developing (and continuously frustrated) from Bacon to Robert Boyle; W. Eamon, *Science and the Secrets of Nature: Books of Secrets in Medieval and Early Modern Culture* (Princeton, NJ: Princeton University Press, 1994), pp. 342–5, sees it as largely doomed to failure, as not only tradesmen but the virtuosi themselves sought to protect valuable knowledge. K. Ochs, 'The Royal Society of London's History of Trades Programme: An Early Episode in Applied Science', *Notes and Records of the Royal Society of London*, 39:2 (1985), pp. 129–58, is somewhat ahistorical but adds useful detail.
7. Shapin, *Social History of Truth*.
8. R. Hooke, 'General Scheme of Natural Philosophy', in *The Posthumous Works of Robert Hooke* (London: Royal Society, 1705), p. 62; see also Shapin, *Social History of Truth*, p. 235.
9. The point is made in Boyle's *Sceptical Chymist*, reproduced in *The Works of the Honourable Robert Boyle*, ed. T. Birch, 6 vols (London, 1772), vol. 1, p. 463.
10. T. Sprat, *The History of the Royal-Society of London for the Improving of Natural Knowledge* (London, 1667), p. 391.
11. *ODNB*.
12. M. R. Best, 'The Mystery of Vintners', *Agricultural History*, 50:2 (1976), pp. 363–76, on pp. 364–8.
13. *ODNB*.
14. R. Moray, 'An Account of the Manner of Making Malt in Scotland', *Philosophical Transactions*, 12 (1678), pp. 1069–71; Birch, *History of the Royal Society*, vol. 1, p. 162.

15. Birch, *History of the Royal Society*, vol. 1, p. 371.
16. [J. Beale], *Herefordshire Orchards, a Pattern for all England* (London, 1657); M. Stubbs, 'John Beale, Philosophical Gardener of Herefordshire. Part 1: Prelude to the Royal Society (1608–1663)', *Annals of Science*, 39:5 (1982), pp. 463–89; M. Stubbs, 'John Beale, Philosophical Gardener of Herefordshire. Part 2: The Improvement of Agriculture and Trade in the Royal Society (1663–1683)', *Annals of Science*, 46:4 (1989), pp. 323–63.
17. A. Rusnock, 'Correspondence Networks and the Royal Society, 1700–1750', *British Journal for the History of Science*, 32:2 (1999), pp. 155–69; R. Sorrenson, 'George Graham, Visible Technician', *British Journal for the History of Science*, 32:2 (1999), pp. 203–21; D. P. Miller, 'The Usefulness of Natural Philosophy: The Royal Society and the Culture of Practical Utility in the Later Eighteenth Century', *British Journal for the History of Science*, 32:2 (1999), pp. 185–201.
18. L. Stewart, *The Rise of Public Science: Rhetoric, Technology and Natural Philosophy in Newtonian Britain, 1660–1750* (Cambridge: Cambridge University Press, 1992), p. 111.
19. For instance, among historians, Mathias, *Brewing Industry*, pp. xvii–xviii; H. S. Corran, *A History of Brewing* (Newton Abbot: David and Charles, 1975), pp. 11–14; Sambrook, *Country House Brewing*, pp. 18–19; Unger, *Beer in the Middle Ages*, pp. 3–7.
20. The error goes back at least to the early twentieth century: H. C. Barnard, *The British Isles in Pictures* (London: Black, 1910), p. 36. A plausible explanation for its survival is the unchanging stress placed on the word 'hops' in mass-market advertising campaigns: G. Wheeler, *Home Brewing* (St Albans: CAMRA Books, 1997), p. 31.
21. Early examples are S. Child, *Every Man his Own Brewer* (London, [*c.* 1790]) – which was reprinted at least ten times – and W. Ploughman, *Oeconomy in Brewing* (Romsey, Hants, 1797).
22. [P. Leigh], 'The Chemistry of a Pint of Beer', *Household Words*, 2 (15 February 1851), pp. 498–502, on p. 499.
23. The manuscript holdings of the Wellcome Library, London (hereafter Wellcome), are strong in this respect. Recipes likely to date from before 1700 may be found at MS.1325/236; MS.1340/78; MS.2990/124; MS.2990/141; MS.3009/148; MS.3724/101; MS.4054/111; MS.7721/56.
24. Digby, *Closet*, pp. 114–16.
25. W. Harrison, 'The Description of England', in R. Holinshed, et al., *The First and Second Volumes of Chronicles* ([London], 1587), pp. 131–250, on pp. 169–70; C. C. E. Steuens and J. Liebault, *Maison Rustique, or the Covntrie Farme*, trans. R. Surflet (London, 1600), pp. 724–6.
26. Sambrook, *Country House Brewing*, pp. 22–88, presents a detailed survey of the construction of apparatus based on material and documentary evidence.
27. Harrison, 'Description of England', pp. 169–70.
28. Best, 'Mystery of Vintners', pp. xx–xxi.
29. G. M[arkham], *The English Hus-wife, Contayning, The Inward and Outward Vertues which Ought to be in a Compleat Woman* (London, 1615), pp. 120–4.
30. [T. Tryon], *A New Art of Brewing Beer, Ale and Other Sorts of Liquors* (London, 1690), pp. 40–'40' [*sic* = 42].
31. *London Chronicle*, 1 November 1760, p. 436; *Every Man his Own Brewer: or, a Compendium of the English Brewery* (London, 1768), pp. 107–8.
32. Yworth, *Cerevisiarii Comes*, pp. 30–5.

33. The copy held at the Clark Memorial Library, University of California, Los Angeles bears a Latin inscription from the author, 'Galfridis Boys', to his nephew 'Tho. Brett'. Jeffrey Boys of Betteshanger was uncle to Thomas Brett, later a nonjuring bishop: E. Hasted, *The History and Topographical Survey of the County of Kent*, 12 vols (Canterbury, 1797–1801), vol. 10: pp. 44–5.
34. M[arkham], *English Hus-wife*, p. 123; [J. Boys], *Directions for Brewing Malt Liquors* (London, 1700), p. 8.
35. K. Whitaker, 'The Culture of Curiosity', in N. Jardine, J. Secord and E. Spary (eds), *Cultures of Natural History* (Cambridge: Cambridge University Press, 1996), pp. 75–90; Eamon, *Science and the Secrets*, pp. 301–18.
36. [Boys], *Directions*, p. 15.
37. Ibid., pp. 3–5.
38. Ibid., pp. 22–23.
39. Sambrook, *Country House Brewing*, p. 198.
40. The British Library copy bears the 'Museum Britannicum' stamp in black, which typically (though not infallibly) indicates Sloane's ownership. See Sloane Printed Books Project, 'Identifying Sloane's Books', http://www.bl.uk/catalogues/sloane/Identifiers.aspx [accessed 12 January 2013].
41. N. Chomel, *Dictionaire Oeconomique: or, the Family Dictionary*, trans. R. Bradley (London, 1727), s. v. 'Brewing'; N. Bailey, *Dictionarium Domesticum, Being a New and Compleat Houshold Dictionary* (London, 1736), s. v. 'Brew-house'.
42. The title page of the first (1697) edition, which lacks the brewing material, names its author as 'S. J. More'. See F. Willmoth, *Sir Jonas Moore: Practical Mathematics and Restoration Science* (Woodbridge: Boydell, 1993), p. 62.
43. J. M[ortimer], *The Whole Art of Husbandry, or, the Way of Managing and Improving of Land*, 1st edn (London, 1707), introduction and pp. 565–74; quotation on p. 565. Italics original.
44. T. Curteis, *Essays on the Preservation and Recovery of Health* (London, 1704), pp. 61–9; T. Fuller, *Pharmacopoia Extemporanea* (London, 1710), pp. 1–4; N. Robinson, *A Compleat Treatise of the Gravel and Stone, with all their Causes* (London, 1721), p. 68.
45. J. Quincy, *Pharmacopœia Officinalis & Extemporanea: or, a Compleat English Dispensatory* (London, 1718), pp. 223–5.
46. J. Sedgwick, *A New Treatise on Liquors* (London, 1725); *Vinum Britannicum: or, an Essay on the Properties and Effects of Malt Liquors* (Lo[n]don, 1727).
47. Yworth, *Cerevisiarii Comes*, p. 54.
48. G. Smith, *The Nature of Fermentation Explain'd* (London, 1729), pp. 10–13.
49. *A Discourse on the Preparation, Preservation, and Restoration of Malt-Liquors* (London, 1733), p. 35. The author identifies himself as 'P. C. J. P.', but refers (p. 48) to a passage in 'my former Discourse', with page references consistent with Smith's 1729 text.
50. *Discourse on the Preparation*, pp. [i]–ii, 40–1; compare M[arkham], *English Hus-wife*, pp. 123–4.
51. The *Dictionarium Rusticum & Urbanicum* (London, 1704), for example, draws its entry for 'Malt' from Tryon's second edition of 1691, and for 'Brewing' from Yworth's *Cerevisiarii Comes*.
52. D. de Coetlogon, *An Universal History of Arts and Sciences*, 2 vols (London, 1745), vol. 1, pp. 385–9.
53. [W. Lewis], *The New Dispensatory* (London, 1753).
54. Unger, *Beer in the Middle Ages*, p. 211.

55. [Ellis], *London and Country Brewer*, 1st edn, part 1, p. 35; *A General Description of all Trades* (London, 1747), pp. 34–6.
56. Such practices are better evidenced for later periods: J. Reinarz, 'Fit for Management: Apprenticeship and the English Brewing Industry, 1870–1914', *Business History*, 43:3 (2001), pp. 33–53.
57. Bennett, *Ale, Beer and Brewsters*, pp. 9, 84–92; Unger, *Beer in the Middle Ages*, pp. 53–7, 97–103; L. B. Luu, *Immigrants and the Industries of London, 1500–1700* (Aldershot: Ashgate, 2005), pp. 259–99.
58. A. Burnaby, *An Essay upon the Excising of Malt* (London, 1696), pp. 34–5; *An Essay upon Excisi[n]g Several Branches that have Hitherto Escaped the Duty of the Brewing Trade* (n.p., [1699]), p. 10.
59. J. Sumner, 'Status, Scale and Secret Ingredients: The Retrospective Invention of London Porter', *History and Technology*, 24:3 (2008), pp. 289–306. My account modifies, but is framed by, Mathias, *Brewing Industry*, pp. 3–27. The name *porter*, probably a contraction of 'porter's beer', was initially a nickname indicating the drink's popularity with the labourers licensed for carrying and loading in London: M. Cornell, *Beer: the Story of the Pint* (London: Headline, 2003), pp. 101–2.
60. [Boys], *Directions*, p. 22.
61. Ibid., p. 23.
62. *Essay upon Excisi[n]g*, pp. 10, 21.
63. *An Impartial Enquiry into the Present State of the British Distillery* (London, 1736), p. 26.
64. Sumner, 'Status, Scale', p. 291 (Calverts); J. A. G. Harley, 'Samuel Whitbread's First Enterprise: A Survey of the Organisation and Methods of an Eighteenth Century London Brewery, 1742–50', *Guildhall Miscellany*, 1:9 (1958), pp. 3–26; Mathias, *Brewing Industry*, pp. 230–1; D. Rapp, 'Social Mobility in the Eighteenth Century: The Whitbreads of Bedfordshire, 1720–1815', *Economic History Review*, 27:3 (1974), pp. 380–94.
65. Cotton family of Madingley and Landwade collection, Cambridgeshire Archives, Cambridge (hereafter Cotton MSS): 588/DR/E78–E92.
66. James Best collection, U480/B874 (not itemized), Medway Archives, Strood, Kent (hereafter Best MSS): Henry Goodwyn's brewing book for 1758, MS annotation by James Best.
67. Mathias, *Brewing Industry*, pp. 31–5.
68. Ibid., pp. 259–60.
69. *London Evening Post*, 2 January 1734/5. Robert Pycroft, who brewed at the same site, was bankrupt in the 1750s: *London Evening Post*, 24 January 1756; *London Chronicle*, 18 September 1764.
70. *British Journal, or, the Traveller*, 28 November 1730; *Daily Gazetteer*, 18 August 1735; *Norwich Gazette*, 5 December 1741.
71. *Free-mason's Magazine, or General and Complete Library*, 6 (1796), p. 451.
72. Cotton MSS: 588/DR/E79.
73. Mathias, *Brewing Industry*, p. 61.
74. Ellis's birthdate has been suggested as around 1700. It was presumably nearer 1690, given the biographical details I present in the following paragraph.
75. [Ellis], *London and Country Brewer*, 1st edn, part 1, title page.
76. W. Ellis, *The Family Companion* (London, 1750), pp. 376–7; will of Richard Ellis, prob. 8 June 1713, The National Archives, Kew (hereafter TNA): PROB 11/533.

77. [Ellis], *London and Country Brewer*, 1st edn, part 1, pp. 37–43, quotation on p. 38, emphasis original; [W. Ellis], *The London and Country Brewer ... third and last Part*, 1st edn (London, 1738), pp. 60–1.
78. Sumner, 'Status, Scale', pp. 292–3.
79. W. Ellis, *Ellis's Husbandry, Abridged and Methodized*, 2 vols (London, 1772), vol. 1, pp. [iii]–ix.
80. [Ellis], *London and Country Brewer*, 1st edn, part 1, pp. 128–9.
81. P. Kalm, *Kalm's Account of his Visit to England on his Way to America in 1748*, trans. J. Lucas (London: Macmillan, 1892 [1748]), pp. 187–93.
82. John Martyn to the Royal Society, 17 May 1731, quoted in H. le Rougetel, *The Chelsea Gardener: Philip Miller, 1691–1771* (London: Natural History Museum, 1990), p. 90.
83. U. Klein and W. Lefèvre, *Materials in Eighteenth-Century Science: A Historical Ontology* (Cambridge, MA: MIT Press, 2007), p. 28.
84. F. W. Gibbs, 'Peter Shaw and the Revival of Chemistry', *Annals of Science*, 7:3 (1951), pp. 211–37; J. Golinski, 'Peter Shaw: Chemistry and Communication in Augustan England', *Ambix*, 30:1 (1983), pp. 19–29; J. R. R. Christie, 'Historiography of Chemistry in the Eighteenth Century: Hermann Boerhaave and William Cullen', *Ambix*, 41:1 (1994), pp. 4–19. For Shaw and Quincy, see N. Howard-Jones, 'John Quincy, M. D. [d. 1722], Apothecary and Iatrophysical Writer', *Journal of the History of Medicine*, 6:2 (1951), pp. 149–75, on pp. 150–1.
85. Golinski, 'Peter Shaw', pp. 23–4; P. Shaw, *Philosophical Principles of Universal Chemistry ... Drawn from the Collegium Jenense of Dr George Ernest Stahl* (London, 1730), preface, emphasis original.
86. *The Philosophical Works of Francis Bacon*, ed., trans. and rev. P. Shaw, 3 vols (London, 1733), vol. 3, p. 50, note u.
87. P. Shaw, *Chemical Lectures, Publickly Read at London, in the Years 1731, and 1732*, 1st edn (London, [1734]), p. 424.
88. Gibbs, 'Peter Shaw', pp. 223, 229.
89. Sprat, *History of the Royal-Society*, pp. 404–9; Stubbs, 'John Beale, Part 2', pp. 347–9.
90. Shaw, *Chemical Lectures*, 1st edn. Shaw presented the commercial as one of four branches of chemistry, the others being the philosophical, technical and œconomical (i.e. domestic).
91. P. Shaw, *Three Essays in Artificial Philosophy, or Universal Chemistry* (London, 1731), p. 69.
92. Ibid., pp. 79–80.
93. [W. Ellis], *The London and Country Brewer* [first part] (Dublin, 1735).
94. *The Dublin Society's Weekly Observations*, vol. 1 [*sic*: 1 vol. issued] (1739), pp. 298–341, on p. 299.
95. Itemized lists of premiums held in the Archives of the Royal Dublin Society; *The Application of Eight Thousand Pounds Granted by Parliament to the Dublin Society* (Dublin, 1764), p. 15, emphasis original.
96. *ODNB*.
97. D. Hudson and K. W. Luckhurst, *The Royal Society of Arts, 1754–1954* (London: John Murray, 1954), p. 8.
98. Gibbs, 'Peter Shaw', p. 235.
99. Miller, 'Usefulness of Natural Philosophy', pp. 200–1.
100. W. Reddington, *A Practical Treatise on Brewing* (London, 1760), preface; will of William Reddington, prob. 23 June 1755, TNA: PROB 11/816.

2 The Theorist and the Thermometer

1. From the index to J. Mills, *A New System of Practical Husbandry*, 5 vols (London, 1767), vol. 5, p. xvii.
2. Mathias, *Brewing Industry*, pp. 66, 218–22.
3. See in particular Stewart, *Rise of Public Science*; M. N. Wise (ed.), *The Values of Precision* (Princeton, NJ: Princeton University Press, 1995); J. Grabiner, '"Some Disputes of Consequence": Maclaurin among the Molasses Barrels', *Social Studies of Science*, 28:1 (1998), pp. 139–68; W. Clark, J. Golinski and S. Schaffer (eds), *The Sciences in Enlightened Europe* (Chicago, IL: University of Chicago Press, 1999); Sorrenson, 'George Graham'. Similar points are raised, with regard to an earlier period, by many of the authors in Roberts, Schaffer and Dear, *Mindful Hand*.
4. Harrison, 'Description of England', p. 169; Steuens and Liebault, *Maison Rustique*, p. 725; and compare M[arkham], *English Hus-wife*, p. 121.
5. A receipt book of the Boyle family, possibly the property of Robert Boyle's sister, Katherine, Lady Ranelagh (1614–91), contains two adjacent recipes, one specifying boiling water, the other requiring the water to cool a quarter of an hour: Wellcome: MS.1340/78. The hourglass method is recommended in J. Lightbody, *Every Man His Own Gauger ... To Which is Added, the True Art of Brewing Beer, Ale, Mum ...* (London, 1695), p. 64 and attacked in [Ellis], *London and Country Brewer*, 1st edn, part 1, preface.
6. Face visibility is recommended in White, *Art's Treasury*, pp. 98–9; Yworth, *Cerevisiarii Comes*, p. 52; [Boys], *Directions*, p. 12; M[ortimer], *Whole Art of Husbandry*, pp. 565–7; *Discourse on the Preparation*, p. 37; [Ellis], *London and Country Brewer*, 1st edn, part 1, p. 44, and many further sources into the nineteenth century.
7. T. Hale, *A Compleat Body of Husbandry*, 4 vols (London, 1756), vol. 1, p. 602.
8. H. O. Sibum, 'Reworking the Mechanical Value of Heat: Instruments of Precision and Gestures of Accuracy in Early Victorian England', *Studies in History and Philosophy of Science*, 26:1 (1995), pp. 73–106, on pp. 83–91; H. O. Sibum, 'Les gestes de la mesure: Joule, les pratiques de la brasserie et la science', *Annales: Histoire, Sciences Sociales*, 53:4–5 (1998), pp. 745–74, on pp. 747–62; H. O. Sibum, 'An Old Hand in a New System', in J.-P. Gaudillière and I. Löwy (eds), *The Invisible Industrialist: Manufactures and the Production of Scientific Knowledge* (Basingstoke: Macmillan, 1998), pp. 23–57.
9. L. Roberts, 'The Death of the Sensuous Chemist: The "New" Chemistry and the Transformation of Sensuous Technology', *Studies in the History and Philosophy of Science Part A*, 26:4 (1995), pp. 503–29.
10. Mathias, *Brewing Industry*, pp. 339–83; W. Ashworth, *Customs and Excise: Trade, Production and Consumption in England 1640–1845* (Oxford: Oxford University Press, 2003), pp. 210–14, 280–98.
11. [W. Ellis], *The London and Country Brewer*, 2nd edn (London: Thomas Astley, 1743), p. 221; Reddington, *Practical Treatise*, pp. 11–12. The approach was seemingly general, at one stage, in the beer-brewing town of Ringwood: G. Blake, *Theoretical and Practical Remarks on G. Blake's System of Malting and Brewing* (London, 1817), pp. 29–30.
12. J. Golinski, *British Weather and the Climate of Enlightenment* (Chicago, IL: University of Chicago Press, 2007), pp. 108–36.
13. *Tatler*, no. 220, 2–5 September 1710. For thermometric analogies see T. Castle, *The Female Thermometer: Eighteenth-Century Culture and the Invention of the Uncanny* (New York: Oxford University Press, 1995), pp. 21–43; J. Golinski, 'Barometers of

Change: Meterological Instruments as Machines of Enlightenment', in Clark, Golinski and Schaffer, *Sciences in Enlightened Europe*, pp. 69–93.

14. *General Evening Post* (London), 17 January 1740.
15. Reddington, *Practical Treatise*, pp. 51–9.
16. J. Golinski, '"Fit Instruments": Thermometers in Eighteenth-Century Chemistry', in F. L. Holmes and T. H. Levere (eds), *Instruments and Experimentation in the History of Chemistry* (Cambridge, MA: MIT Press, 2000), pp. 185–210, on p. 191.
17. For the rival 'authorized' and 'unauthorized' translations, and for Shaw's tendency to subvert Boerhaave in commentary, see Gibbs, 'Peter Shaw', p. 216; Golinski, 'Peter Shaw'; Christie, 'Historiography of Chemistry'.
18. [W. Ellis], *The London and Country Brewer*, 5th edn (London, 1744), pp. 262–3.
19. W. E. K. Middleton, *A History of the Thermometer and its Use in Meteorology* (Baltimore, MD: Johns Hopkins University Press, 1966), pp. 71, 74.
20. Golinski, '"Fit Instruments"', pp. 188–90.
21. Shaw, *Three Essays*, p. 71.
22. W. Black, *A Practical Treatise on Brewing, and on Storing of Beer*, 1st edn (London: Smith, Elder, 1835), p. 5.
23. le Rougetel, *Chelsea Gardener*, p. 48; P. Miller, *The Gardeners Dictionary*, 2nd edn (London, 1733), s.v. 'Fire'; compare H. Boerhaave, *A New Method of Chemistry*, trans. P. Shaw and E. Chambers (London, 1727), p. 222.
24. Dublin Society, *Instructions for Planting and Managing Hops* (Dublin, 1733), p. 53; *ODNB*.
25. R. Crawford, *Poetry, Enclosure, and the Vernacular Landscape, 1700–1830* (Cambridge: Cambridge University Press, 2002), pp. 207–10.
26. C. Mortimer, 'A Discourse Concerning the Usefulness of Thermometers in Chemical Experiments', *Philosophical Transactions*, 44 (1746–7), pp. 672–95, on pp. 672–4.
27. R. Clarke, *Notice is Hereby Given, to all Dealers ... that the Hydrometer, or Brandy-Prover ... is now Brought to its Utmost Perfection* (London, [1746]), pp. 26–37, on p. 37.
28. Mathias, *Brewing Industry*, p. 67.
29. M. Combrune, 'An Historical Accompt of the English Brewery' (unpublished manuscript, 1767: Beinecke Library, Yale University, Osborn c603), fols 18–19.
30. Best MSS: two brewing books donated by Goodwyn to Best, 1764. The Goodwyns seem to have displayed a strong technical bent throughout the century. The eldest Henry Goodwyn (1713–1805), originally of King's Lynn, was succeeded by his son, also Henry (1740–1824), who was a noted amateur mathematician, installer of the first brewery steam engine (1784) and patentee of an 'improved' mash-tub (1797).
31. For Combrune's background, see *ODNB*; J. Sumner, 'Michael Combrune, Peter Shaw and Commercial Chemistry: The Chemical Origins of Brewing Thermometry', *Ambix*, 54:1 (2007), pp. 5–29, on p. 11.
32. Best MSS: Combrune to Best, 17 August 1762.
33. [M. Combrune], *Some Account of the Rise, Progress and Present State of the Brewery ... in a Letter to a Friend* (London, 1757), pp. 5–6.
34. [J. Macky], *A Journey through England, in Familiar Letters from a Gentleman*, 2 vols (London, 1722), vol. 2, p. 168; White, *Art's Treasury*, p. 101.
35. Best MSS: Combrune to Best, 20 August 1762.
36. E. A. Underwood, *Boerhaave's Men, at Leyden and After* (Edinburgh: Edinburgh University Press, 1977), pp. 160–1.

37. For Maty's position in English philosophical, medical and literary culture, see U. Janssens, *Matthieu Maty and the Journal Britannique, 1750–1755* (Amsterdam: Holland University Press, 1975); A. Günther, 'Matthew Maty MD, FRS (1718–76) and Science at the Foundation of the British Museum, 1753–80', *Bulletin of the British Museum (Natural History), Historical Series*, 15:1 (1987), pp. 1–58.
38. J. Anderson, *The Constitutions of the Antient and Honourable Fraternity of Free and Accepted Masons* (London, 1756), pp. 228–31. For the Huguenot character of Newtonian natural philosophy in England, see J.-F. Baillon, 'Early Eighteenth-Century Newtonianism: The Huguenot Contribution', *Studies in the History and Philosophy of Science*, 35:3 (2004), pp. 533–48.
39. H. Boerhaave, *Elements of Chemistry: Being the Annual Lectures of Herman Boerhaave, M. D.*, trans. T. Dallowe, 2 vols (London, 1735), vol. 2, p. 245, emphasis original. Combrune's use of Dallowe's edition is confirmed by quotation: M. Combrune, *The Theory and Practice of Brewing* (London, 1762), p. 169; compare Boerhaave, *Elements of Chemistry*, vol. 1, p. 195.
40. Combrune, *Theory and Practice*, p. ii.
41. Combrune's glossary of technical terms, for instance, parallels the *Chemical Lectures*' directly, with a couple of entries borrowed verbatim. Elsewhere, substantial portions of text are based on Shaw, often with substantial revision. Combrune, *Essay*, pp. 25–8, for instance, is loosely updated from P. Shaw, *Chemical Lectures, Publickly Read at London*, 2nd edn (London, 1755), pp. 36–7.
42. Mathias, *Brewing Industry*, pp. 109–14.
43. Among these were the historical account offered in the first 'Obadiah Poundage' letter (*London Chronicle*, 1 November 1760), widely disseminated through reproduction in the *Gentleman's Magazine* and *Annual Register* (for which see Sumner, 'Status, Scale', p. 297); and the anonymous *The Present State of the London Brewery: Recommended to the Perusal of Those Concerned in, or with the Trade* (London, 1760).
44. The account – signed 'A. B.', a common subscription of the period – is generally assigned to Combrune on the authority of J. R. McCulloch, *The Literature of Political Economy: A Classified Catalogue* (London: Longman, 1845). Some of the text reappears verbatim in Combrune's later manuscripts.
45. [Combrune], *Some Account*, p. 8.
46. Janssens, *Matthieu Maty*, p. 190, n. 106.
47. Best MSS: Combrune to Best, 20 August 1762; Shaw to Combrune, 20 July 1758, quoted in Combrune, *Essay*, pp. [vii]–[viii], emphasis original.
48. Combrune, *Essay*, p. 26. I discuss this image in more detail in Sumner, 'Michael Combrune', p. 5. Combrune's advisors perhaps found it unduly eccentric: his enlarged treatise of 1762, in all other respects a far more lavish production, omits it.
49. Combrune, *Essay*, p. 69.
50. Ibid., pp. 76–8.
51. Ibid., p. 137.
52. Ibid., pp. 145, 151.
53. Ibid., pp. 172–8, 183.
54. Ibid., p. 188.
55. Klein and Lefèvre, *Materials*, p. 218.
56. *The Philosophical Works of the Honourable Robert Boyle*, ed. P. Shaw, 3 vols (London, 1725), vol. 1, pp. v–vi.

57. Combrune, *Theory and Practice*, p. 135. The composite quantity, though distinctly defined, has no clear name in the 1762 publication: 'governing medium heat' was the term Combrune nominated in a later revision which was published posthumously in 1804.
58. Combrune, *Theory and Practice*, pp. 138–40. The value of 151° in fact represents the mean of a graduated progression of heats for multiple mashes. Combrune developed more elaborate arithmetical schemes to determine these individual heats: for more details, see Sumner, 'Michael Combrune'.
59. [Ellis], *London and Country Brewer*, 1st edn, part 1, pp. 14, 71; *The Dublin Society's Weekly Observations*, 1 (1739), p. 301; Combrune, *Theory and Practice*, pp. 69, 79–80.
60. For the technical character of London porter see O. Macdonagh, 'The Origins of Porter', *Economic History Review*, 16:3 (1964), pp. 530–5; Sumner, 'Status, Scale'.
61. *Gentleman's Magazine*, 29 (1759), pp. 59–61, on p. 59.
62. Mathias, *Brewing Industry*, p. 66.
63. *Monthly Review*, 20 (1759), pp. 277–8. For Noorthouck's authorship of unsigned reviews, see B. Nangle, *The Monthly Review, First Series, 1749–1789: Indexes of Contributors and Articles* (Oxford: Clarendon, 1934), pp. 30–1, 187; B. Nangle, *The Monthly Review, Second Series, 1790–1815: Indexes of Contributors and Articles* (Oxford: Clarendon, 1955), pp. 48–9. For his background, see *ODNB*.
64. *Gentleman's Magazine*, 33 (1763), p. 592.
65. Combrune, *Essay*, pp. 44–5.
66. Sibum, 'Reworking the Mechanical Value of Heat', p. 85.
67. Best MSS: Combrune to Best, 17 August 1762; Combrune, *Essay*, pp. [v]–[vi].
68. Best MSS: Combrune to Best, 17August 1762; Combrune, *Theory and Practice*, pp. 178–9.
69. G. A. Lindeboom, *Herman Boerhaave: The Man and his Work* (London: Methuen, 1968), p. 296.
70. For Haberkorn see G. Jefcoate, 'German Immigrants and the London Book Trade, 1700–1770', in R. Vigne and C. Littleton (eds), *From Strangers to Citizens: The Integration of Immigrant Communities in Britain, Ireland, and Colonial America, 1550–1750* (Brighton: Sussex Academic Press, 2001), pp. 503–10, on pp. 505–6.
71. My thanks to Jon Topham for guidance on this point.
72. Journal Book for 1762, entry for 28 January, Archives of the Royal Society, London (hereafter RS): JBO/25; D. P. Miller, 'The "Hardwicke Circle": The Whig Supremacy and its Demise in the 18th-Century Royal Society', *Notes and Records of the Royal Society*, 52:1 (1998), pp. 73–92, on pp. 80–1.
73. Election certificate for John Glen King, RS: EC/1772/23. Combrune was not among the guests Maty brought to the Society's meetings around the time it received his books: Journal Book copies, 1758–62, RS: JBC/23 and JBC/24.
74. For Combrune's epistolary style see Combrune, *Theory and Practice*, pp. [i]–[ii], and Combrune to Best, 20 August 1762, Best MSS. For Martin's misadventure, see J. R. Millburn, *Benjamin Martin: Author, Instrument-Maker, and 'Country Showman'* (Leyden: Noordhoff, 1976), pp. 28, 35–8, 55–8.
75. Klein and Lefèvre, *Materials*, p. 218.
76. Mills, *Practical Husbandry*, vol. 5, pp. 3, 111, 122–3. Mills was equally scathing (vol. 5, pp. 43–5) about Philip Miller's habit of introducing material '*servilely* copied' into the 'unwieldy' *Gardeners Dictionary.*

77. Golinski, '"Fit Instruments"', pp. 193–200; D. P. Miller, 'Seeing the Chemical Steam through the Historical Fog: Watt's Steam Engine as Chemistry', *Annals of Science*, 65:1 (2007), pp. 47–72.
78. J. Long, *A Treatise on Malting* (Dublin, 1800), pp. 35–6.
79. 'Papers ... Relating to Experiments Made by Order of the Commissioners of Excise for Scotland, to Ascertain the Relative Qualities of Malt Made from Barley and Scotch Bigg, &c', session 1806, paper 202 (hereafter 1806 Hope Report), p. 63.
80. *Monthly Review*, 26 (1762), pp. 122–3, emphasis original.
81. D. Loch, *Letters Concerning the Trade and Manufactures of Scotland*, 2nd edn (Edinburgh, 1774), pp. 13, 24–6; D. Loch, *Essays on the Trade, Commerce, Manufactures, and Fisheries of Scotland*, 3 vols (Edinburgh, 1778), vol. 1, pp. 198–219.
82. Miller, 'Usefulness of Natural Philosophy', p. 200.
83. [M. Combrune], *An Enquiry into the Prices of Wheat, Malt, and Occasionally of Other Provisions* (London, 1768). For Combrune's authorship of this work, see H. Higgs, *Bibliography of Economics, 1751–1775* (Cambridge: Cambridge University Press, 1935), p. 426.
84. Combrune 'turned over' his apprentice, his nephew Gideon Combrune, to another master in 1767, suggesting he had left off trade: C. Webb, *London Livery Company Apprenticeship Registers, vol. 1: Brewers' Company 1685–1800* (London: Society of Genealogists, 1996), p. 4. In the months before his death, he gave evidence before a Commons enquiry into the consumption of barley and malt: see for instance *London Magazine*, 43 (1773), pp. 116–7.
85. Klein and Lefèvre, *Materials*, p. 29.
86. H. Jackson, *An Essay on British Isinglass* (London, 1765), p. 63; Mills, *Practical Husbandry*, vol. 5, p. 3; [T. Thomson], 'Brewing', *Encyclopaedia Britannica*, supplement to the 4th, 5th and 6th editions (1824), pp. 460–92, on p. 467.
87. This simple mode of use was also the basis for introducing the thermometer (presumed unfamiliar to the reader) in the 1733 hop-drying account: Dublin Society, *Instructions for Planting*, p. 53.
88. *Encyclopaedia Britannica*, 2nd edn (1778–83), vol. 2, pp. 1378–85. The entry is a scissors-and-paste production culled from, unacknowledged, the brewery material from Chambers's *Cyclopaedia*, and with acknowledgment, Combrune's *Essay* and John Richardson's *Theoretic Hints* of 1777. For the consolidation of the *Britannica*'s authority see Yeo, *Encyclopaedic Visions*, pp. 170–87.
89. I. Christie, 'Samuel Bentham's Library in Russia', *Slavonic and East European Review*, 65:1 (1987), pp. 26–37, on p. 32; S. Baron, *Brewed in America: A History of Beer and Ale in the United States* (Boston, MA: Little, Brown, 1962), p. 139.
90. F. W. Heun, *Versuch der Kunst, alle Arten Biere nach englischen Grundsätzen zu brauen* (Leipzig 1777); M. Combrune, *Theorie und Praxis des Bierbrauens*, trans. C. H. Reichel (Leipzig, 1796); M. Combrune, *L'art de brasser, traduit de l'anglais*, translator unidentified (Paris: le Normant, 1802), pp. v–vj.
91. *Monthly Magazine*, 5 (1797), pp. 25–6. The 1768 treatise *Every Man his Own Brewer* had likewise nominated a mashing heat ('140 or [1]45 degrees' depending on malt colour and season, p. 68). By 1797, however, the only print account likely to have been widely available was Samuel Child's short pamphlet, also titled *Every Man his Own Brewer* but unrelated. Child proposes 180°F for the first mash and 190°F for the second – extremely high figures which the least experience would have rendered doubtful.

92. M. Combrune, *The Theory and Practice of Brewing*, new edn (London: Vernor and Hood, 1804). The origins of this work are not certain, but the pattern of extensive but superficial revision matches that which Combrune applied to his unpublished 'Historical Accompt' of 1767, as compared to an earlier draft: M. Combrune, 'An Historical Account of the English Brewery' (unpublished manuscript, 1762: Beinecke Library, Yale University, Osborn c602).
93. E. N. Hayman, *A Practical Treatise to Render the Art of Brewing More Easy ... Inculcating the Use of the Saccharometer* (London: Longman, 1819); J. S. Forsyth, *The Farmer, Maltster, Distiller, & Brewer's Practical Memorandum Book* (London: D Cox, [*c.* 1823]); J. Levesque, *The Art of Brewing and Fermenting, in the Summer, and all other Seasons*, 2nd edn (London: Thomas Hurst, 1836), with similar content in subsequent editions to 1853; Blake, *Theoretical and Practical Remarks*, pp. 50–70.
94. See in particular Richardson, *Statical Estimates*; Morrice, *Treatise on Brewing*, 2nd edn; Tuck, *Private Brewer's Guide*. Thermometric accounts aimed specifically at the small-scale brewer are A. Shore, *A Practical Treatise on Brewing, Founded on Experiments Made with the Thermometer and Hydrometer*, 2nd edn (Nantwich, 1809) and W. Chadwick, *A Practical Treatise on Brewing, with the Mode of Using the Thermometer and Saccharometer* (London: Whittaker, 1835). I briefly survey contemporary claims for the spread of thermometry in J. Sumner, 'Early Heat Determination in the Brewery', *Brewery History*, 121 (2005), pp. 66–80, on pp. 72–3.
95. *Every Man his Own Brewer*, pp. 39–40; B. Martin, *A Description of the Nature, Construction, and Use of the Torricellian, or Simple Barometer* (London, 1766), pp. 19–20.
96. Richardson, *Statical Estimates*, pp. viii–ix; Ploughman, *Oeconomy in Brewing*, p. 20.
97. Morrice, *Treatise on Brewing*, contains a collection of recipes including Fahrenheit values which were widely borrowed by later authors. Independent values appear in F. Accum, *A Treatise on the Art of Brewing*, 1st edn (London: Longman, 1820); Black, *Practical Treatise*, 1st edn; and many texts after 1840.
98. *The Theory and Practice of Malting and Brewing* (Edinburgh, 1793), pp. 34–6.
99. Tuck, *Private Brewer's Guide*, p. 72; Levesque, *Art of Brewing*, 2nd edn, p. 40.
100. G. Adams, *A Catalogue of Optical, Philosophical, and Mathematical Instruments* ([London], [*c.* 1765]), p. 9.
101. [D. Booth], *The Art of Brewing*, Society for the Diffusion of Useful Knowledge partwork (London: Baldwin and Cradock, 1829), p. 11.
102. Chadwick, *Practical Treatise*, p. 6; Levesque, *Art of Brewing*, 2nd edn, pp. 38–41, [130]. [W. Loftus], *Loftus's Almanack for Brewers, Distillers, and the Wine and Spirit Trades* (London: Loftus, 1869), insert, pp. 7–10, advertises some thirty different thermometer patterns, all adapted to different brewery applications.
103. Morrice, *Treatise on Brewing*, appendix; Tuck, *Private Brewer's Guide*, pp. 32, 60.
104. [W. Loftus], *The Brewer: A Familiar Treatise on the Art of Brewing*, new stereotyped edn (London: Loftus, 1863), p. [143]; H. Janes, *The Red Barrel: A History of Watney Mann* (London: John Murray, 1963), p. 140.
105. Ellis's authorship is revealed in advertisements (e.g. *Public Advertiser*, 4 July 1761) for a final, posthumous miscellany: W. Ellis, *The New Art of Brewing and Improving Malt Liquors* (London, 1761). [G. Watkins], *The Compleat Brewer; or, the Art and Mystery of Brewing Explained* (London, 1760); G. Watkins, *The Compleat English Brewer, or, the Whole Art and Mystery of Brewing* (London, 1767).
106. Of the most active authors of the 1820s to 1850s, John Tuck presented himself as an ex-operative brewer turned consultant and valuer; William Black, as an operative with

long and varied consulting experience; George Adolphus Wigney, as an employee in the service of his brewery-owning relations; John Levesque, as an ex-proprietor turned consultant; William Tizard, as an ex-operative turned brewer's engineer; and William Loftus, as an instrument-maker specially concerned with brewing.

107. Thomas Jefferson to James Barbour, 11 May 1821, quoted in Baron, *Brewed in America*, pp. 147–8.
108. Best MSS: Combrune to Best, 17 and 20 August 1762.
109. Best MSS: notebook, 'Mr. Combrune's Calculations for Brewing a Guile of Porter. March 6th. 1766'. Other items in the collection evidence Best's continuing dealings with Combrune in 1768 and 1772.
110. Minute books of the proceedings of Council, Royal Dublin Society Archives (hereafter RDS Minutes): entry for 16 July 1772.
111. RDS Minutes: entries for 25 June and 16 July 1772. Combrune's manuscript (sent in care of 'a Gentleman') is untraced.
112. P. Lynch and J. Vaizey, *Guinness's Brewery in the Irish Economy, 1759–1876* (Cambridge: Cambridge University Press, 1960); Sumner, 'Status, Scale', pp. 300–1.

3 Brewery Instructions in Public and Private

1. [J. Richardson], *Observations on the Art of Brewing Malt Liquors, in a Series of Strictures on a Secret System Inculcated in a Private course of Lectures ... by a Practical Brewer* (1775), pp. 3–4, emphasis original.
2. Shapin, *Social History of Truth*, pp. 175–7.
3. For Shaw's early programmatic opposition to 'Nostrums' in medical practice, see [P. Shaw], *The Juice of the Grape: or, Wine preferable to Water* (London, 1724).
4. *Universal Chronicle*, 21–8 April 1759. For the rhetorical force of quackery allegations, see R. Porter, *Quacks: Fakers and Charlatans in English Medicine* (Stroud: Tempus, 2000). For chemical applications of the terminology, see for instance Boerhaave, *Elements of Chemistry* (1735 Dallowe edn), vol. 2, p. 181; [R. Dossie], *The Elaboratory Laid Open, or, the Secrets of Chemistry and Pharmacy Revealed* (London, 1758), p. 261; D. MacBride, *Experimental Essays on the Following Subjects: I. On the Fermentation of Alimentary Mixtures ...* 1st edn (London, 1764), p. 256, n.
5. J. H. Appleby, 'Humphrey Jackson, FRS, 1717–1801: A Pioneering Chemist', *Notes and Records of the Royal Society of London*, 40:2 (1986), pp. 147–68, on p. 147.
6. Combrune, *Essay*, pp. 196–9; *London Chronicle*, 1 November 1760, p. 436.
7. J. R. Millburn, *Benjamin Martin: Supplement* (London: Vade-Mecum, 1986), p. 161; Jackson, *Essay on British Isinglass*, p. 22.
8. H. Jackson, 'An Account of the Discovery of the Manner of Making Isinglass in Russia; with a Particular Description of its Manufacture in England', *Philosophical Transactions*, 63 (1774), pp. 1–15.
9. R. Dossie, *Memoirs of Agriculture*, 3 vols (London, 1768), vol. 1, pp. 275–7. See also Mathias, *Brewing Industry*, pp. 51–3.
10. H. Jackson, *Reflexions Concerning the Virtues of Tar Water* (London, 1744), pp. 3–5; see also Appleby, 'Humphrey Jackson', p. 148.
11. *An Answer to a Late Scurrilous Pamphlet, Published by one Baker and his Accomplices Respecting Dr. James's Powder* (London, [*c.* 1754]). See also J. K. Crellin, 'A Note on Dr James's Fever Powder', *Transactions of the British Society for the History of Pharmacy*, 1:3 (1974), pp. 136–43; J. H. Appleby, 'Erasmus King: Eighteenth-Century Experimental

Philosopher', *Annals of Science*, 47:4 (1990), pp. 375–92, on p. 391; Appleby, 'Humphrey Jackson', pp. 149–50.

12. Appleby, 'Erasmus King', p. 389.
13. Best MSS: 'Proposals' (June 1762).
14. *The Distiller of London, Compiled and Set Forth by the Speciall Licence and Command of the Kings most Excellent Majesty* (London, 1639); *The Distiller of London, with the Clavis to Unlock the Deepest Secrets of that Mysterious Art* (London, 1652).
15. Mathias, *Brewing Industry*, pp. 259–60; J. Entick, *A New and Accurate History and Survey of London*, 4 vols (London, 1766), vol. 3, p. 302.
16. H. L. Piozzi, *Thraliana: The Diary of Mrs. Hester Lynch Thrale (later Mrs. Piozzi), 1776–1809*, ed. K. C. Balderston, 1st edn, 2 vols (Oxford: Clarendon, 1951), vol. 1, p. 333; see also Mathias, *Brewing Industry*, pp. 265–70.
17. Piozzi, *Thraliana*, vol. 1, p. 307; J. R. Townsend, *Trade and Plumb-Cake Forever, Huzza! The Life and Work of John Newbery, 1713–1767* (Cambridge: Colt, 1994); Millburn, *Benjamin Martin: Supplement*, p. 40; J. Wiltshire, *Samuel Johnson in the Medical World* (Cambridge: Cambridge University Press, 1991), pp. 73–5 and 96–7. Society fellowship dates are recorded in the RSA card index of early members.
18. Jackson, *Essay on British Isinglass*, pp. 46–7.
19. Ibid., pp. 73–4.
20. Ibid., pp. 33–4; *Monthly Review*, 33 (1765), p. 427, emphasis original. For the tentative attribution see Nangle, *Monthly Review, First Series*, pp. 13–14.
21. Best MSS: Jackson to Best, 21 July 1770. Best had assisted on this point in the past, perhaps by referring Jackson to Montague Grover, organizer of a petition of country brewers to Parliament concerning London's preferential excise regime.
22. The full text of Jackson's 1770 syllabus has not come to light. The description presented here is surmised from [Richardson], *Observations*, pp. 64–5 (discussed below), and partially confirmed by *London Review*, 3 (1776), pp. 166–7. Caution is, of course, required in attempting to read Jackson through selective reproduction by an opponent. However, Richardson quotes Jackson extensively and, judging from several passages which coincide with the 1762 *Proposals*, verbatim.
23. Jackson, *Essay on British Isinglass*, p. 59, emphasis original. On the preceding page, Jackson vents his anger at a short work which raised such a question: compare *The Complete Maltster and Brewer: Being a Brief Dissertation in Defence of Long Grown Malts* (London, 1765), pp. v–vi.
24. Jackson, *Essay on British Isinglass*, p. 50.
25. J. Golinski, 'Chemistry', in R. Porter (ed.) *The Cambridge History of Science, vol. 4: Eighteenth Century Science* (Cambridge: Cambridge University Press, 2003), pp. 375–96, on pp. 378–80; J. Golinski, *Science as Public Culture* (Cambridge: Cambridge University Press, 1992), pp. 59–60; Gibbs, 'Peter Shaw'.
26. Jackson, *Essay on British Isinglass*, p. 52.
27. Ibid., pp. 54–7.
28. Ibid., pp. 60–1.
29. Ibid., pp. x–xi, 47, 51–2, 57, emphasis original.
30. Quoted in [Richardson], *Observations*, pp. 29–31. In quoting Jackson via Richardson I have erased emphasis, since it was probably added by Richardson.
31. Quoted in [Richardson], *Observations*, p. 38.
32. [Richardson], *Observations*, pp. 24–8; compare Shaw, *Chemical Lectures*, 2nd edn, pp. 195–6.

33. [Richardson], *Observations*, pp. 33–49; compare Jackson, *Essay on British Isinglass*, pp. 64–6.
34. *Monthly Magazine*, 24 (1807), p. 413.
35. [J. Beresford], *Observations on the Brewing Trade of Ireland, Submitted to the Publick* ([Dublin?], [*c.* 1774]); *Parliamentary Register, or History of the Proceedings and Debates of the House of Commons of Ireland*, 12 (1793), p. 267.
36. *Votes of the* [*Irish*] *House of Commons* ([Dublin], [1792]), pp. 365–6.
37. Patent 1754, 4 June 1790; Patent 1769 (with Thomas Harris), 28 July 1790.
38. Quoted in G. Blake, *Strictures on a New Mode of Brewing, &c.* (London, [1791]), pp. 37, 46–8.
39. Best MSS: Clarke and Cooke (Newport, Isle of Wight) to Best, 15 December 1770; Nath. Stevens (Cambridge) to Best, 15 December 1770.
40. D. G. C. Allan, 'The Society for the Encouragement of Arts, Manufactures and Commerce: Organisation, Membership and Objectives in the First Three Decades (1755–84)' (PhD dissertation, University of London (external), 1979), p. 201.
41. Jackson, *Essay on British Isinglass*, p. 83; RSA: Minute Book 5, entries for 27–8 May 1760.
42. N. Chambers, 'The Society of Arts and Joseph Banks: A First Step in London Learned Society', *Notes and Records of the Royal Society*, 61:3 (2007), pp. 313–25.
43. *Premiums Offered by the Society Instituted at London for the Encouragement of Arts, Manufactures and Commerce* (London, 1766), p. 58.
44. Dossie, *Memoirs of Agriculture*, vol. 1, pp. 203–5; RSA: Jackson to Templeman, 30 December 1760, Guard Book 10, fol. 79.
45. Appleby, 'Humphrey Jackson', pp. 152–6.
46. Best MSS: 'Proposals', 2; Jackson, *Essay on British Isinglass*, p. 65.
47. Piozzi, *Thraliana*, vol. 1, pp. 53, 312; H. L. Piozzi, *Autobiography, Letters and Literary Remains*, ed. A. Hayward, 1st edn, 2 vols (London: Longman, 1861), vol. 1, pp. 257–8.
48. For the shift to storage, see Sumner, 'Status, Scale', and Mathias, *Brewing Industry*, pp. 53–62. Thrale's Rest Book for 1767 records three vats of unknown size, valued at £1,010 in total, as against almost 12,000 butts: London Metropolitan Archives, Clerkenwell, London (hereafter LMA), ACC/2305/1/158. Production figures around 1770 are not extant, but Thrale brewed around 33,000 barrels in 1760 and 75,000 in 1776 (Mathias, *Brewing Industry*, p. 551).
49. Victualling to Admiralty, 24 January 1774, TNA: ADM 110/26, fols 101–15; Mathias, *Brewing Industry*, p. 551. At least four of the thousand-hogshead vats were in use at Thrale's in 1775: J. L. Clifford (ed.), *Dr. Campbell's Diary of a Visit to England in 1775* (Cambridge: Cambridge University Press, 1947), p. 51.
50. Patent 910, 9 December 1768; H. Jackson, *New Art of Hardening and Preserving Wood* (London, 1770), pp. 4–7; Best MSS: Jackson to Best, 21 July 1770.
51. MacBride, *Experimental Essays*, pp. 171–6; C. Lawrence, 'Disciplining Disease: Scurvy, the Navy, and Imperial Expansion, 1750–1825', in D. P. Miller and P. H. Reill (eds), *Visions of Empire: Voyages, Botany, and Representations of Nature* (Cambridge: Cambridge University Press, 1996), pp. 80–106, on pp. 86–9. My account is based largely on a review of the sources discussed in Mathias, *Brewing Industry*, pp. 204–9.
52. The general principle was perhaps first publicized in English in *The Works of the Highly Experienced and Famous Chymist, John Rudolph Glauber*, trans. C. Packe (London, 1689), pp. 302–5; and see also Shaw, *Three Essays*, pp. 65–7.

53. Correspondence between Admiralty and Victualling Office, January and May 1772, and Jackson to Victualling, 29 April 1772, TNA: ADM 110/25, fols 166–9, 228–32.
54. J. Cook, 'The Method Taken for Preserving the Health of the Crew of His Majesty's Ship the *Resolution* during her Late Voyage round the World', *Philosophical Transactions*, 66 (1776), pp. 402–6; J. Pringle, *A Discourse upon some Late Improvements of the Means for Preserving the Health of Mariners* (London, 1776), pp. 14–15.
55. Election certificate for Humphry [*sic*] Jackson, RS: EC/1772/23.
56. Hunter was 'chairman' of a convivial coffee-house society which included Banks, Solander and (when not voyaging) Cook: R. L. Edgeworth and M. Edgeworth, *Memoirs of Richard Lovell Edgeworth*, 3rd edn (London: Bentley, 1844), p. 118. The remaining signatories were Peter Woulfe, the Irish commercial chemist, and a Swiss Linnean, Rodolph Valltravers.
57. Piozzi, *Thraliana*, vol. 1, pp. 53, 312; Piozzi, *Autobiography*, vol. 1, pp. 257–8. Millburn, *Benjamin Martin: Supplement*, pp. 39–43 pronounces Piozzi's recollections 'devious', noting the lack of corroboration and time lag. Against the latter, it must be noted that she recorded the same picture of universal hostility to Jackson in 1773: MS book ('Lamentation & Mourning & Woe'), University of Manchester Special Collections, John Rylands Library, Manchester, GB 133 Eng. MS 616, item 3.
58. Johnson to Hester Thrale, 20 March 1773, in H. L. Piozzi (ed.), *Letters To and From the Late Samuel Johnson*, 2 vols (London, 1788), vol. 1, pp. 78–9. Piozzi states that she prevailed on her husband to 'get rid of' Jackson in 1772 (*Thraliana*, vol. 1, p. 312), but he returned at least once the following year ('Lamentation', note 57). Thereafter his influence was presumably curbed, given that threats by John Perkins to leave the firm did not materialize.
59. Mathias, *Brewing Industry*, pp. 271–5.
60. R. B. Schwartz, *Samuel Johnson and the New Science* (Madison, WI: University of Wisconsin Press, 1971), pp. 38–9.
61. *Daily Advertiser*, 23 October 1775; TNA: ADM 110/27, fols 415–6; ADM 110/29, fols 473–4.
62. Appleby, 'Humphrey Jackson', p. 164.
63. 'Practical Remarks' (MS book), Brakspear of Henley collection, Oxfordshire History Centre, Oxford: Brakspear II/i/1, fol. 2.
64. [Richardson], *Observations*, title page and p. 4, emphasis original.
65. Ibid., p. 5, emphasis original.
66. Richardson never came forward as author, but the *Observations* is identical in style to a 1785 pamphlet Richardson published against his later rival, James Baverstock (pp. 100–3). Two of the *Observations*' three booksellers also distributed Richardson's *Theoretic Hints*.
67. J. Richardson, *Theoretic Hints on an Improved Practice of Brewing Malt-Liquors* (London, 1777), p. 3, n.
68. There is no reason to doubt this, although our only source on Richardson's career prior to his 1783 removal to Hull is the unreliable C. Frost, *An Address, Delivered to the Literary and Philosophical Society at Kingston upon Hull* (Hull: I. Wilson, 1831), pp. 48–9.
69. [Richardson], *Observations*, p. 3, emphasis original.
70. Ibid., pp. 21–3.
71. Jackson, *Essay on British Isinglass*, p. 74.
72. [Richardson], *Observations*, pp. 11–12, 29–30.
73. Ibid., pp. 64–5, emphasis original.

74. *Monthly Review*, 55 (1776), pp. 322–3; *Critical Review*, 42 (1776), p. 400.
75. *London Review*, 3 (1776), pp. 166–70.
76. Blake, *Theoretical and Practical Remarks*, pp. 29, 39.
77. Victualling to Admiralty, 6 November 1780, TNA: ADM 110/29, fols 129–32; Blake, *Strictures*, p. iv.
78. Blake, *Strictures*, pp. 13–34.
79. Ibid., pp. 56–7, for instance, strongly recalls [Richardson], *Observations*, p. 3.
80. Blake, *Strictures*, pp. 66, 119.
81. Ibid., pp. 20–1.
82. S. Schaffer, '"The Charter'd Thames": Naval Architecture and Experimental Spaces in Georgian Britain', in Roberts, Schaffer and Dear, *Mindful Hand*, pp. 279–305, on pp. 280, 290.
83. [Watkins], *Compleat Brewer*, p. 3.
84. Ibid., p. 32.
85. The periods of experience mentioned are asserted on the title pages to, respectively, E. Hughes, *A Saving of Twenty Per Cent. A Treatise on the Brewing of Beer* (Uxbridge, 1796); *The Farmers' Real Friend, Shewing How to Brew with Three-Fourths Barley to One-Fourth Malt* (London: Groombridge, 1847); *The Young Brewer's Monitor, Comprising a Luminous and Scientific Summary of that Very Ancient and Important Art* (London: Baldwin, 1824); W. Moir, *Brewing Made Easy, Being a Compendium of all the Directions that have Hitherto been Published* (London: Scott, 1802); Levesque, *Art of Brewing*, 2nd edn. George Blake ultimately claimed 'near fifty years' study' as a consultant and instructor: *Theoretical and Practical Remarks*, p. [iii].
86. Tuck, *Private Brewer's Guide*, p. 28.
87. Blake, *Strictures*, p. 69.
88. [Richardson], *Observations*, p. 29.
89. Blake, *Strictures*, pp. 9–10; Tuck, *Private Brewer's Guide*, p. vi; Levesque, *Art of Brewing*, 2nd edn, p. [iii].
90. [Richardson], *Observations*, pp. 38, 66, emphasis original.
91. Blake, *Strictures*, pp. 4–8.
92. Ibid., p. 52.
93. Ibid., pp. 78, 99.
94. Watkins, *Compleat English Brewer*, pp. 123–4; *Monthly Review*, 38 (1768), p. 65.
95. Richardson, *Theoretic Hints*, pp. 1–2.
96. Ibid., p. 14; *London Chronicle*, 6 December 1777. For Joe's, see B. Lillywhite, *London Coffee Houses: A Reference Book of Coffee Houses of the Seventeenth, Eighteenth and Nineteenth Centuries* (London: G. Allen and Unwin, 1963), pp. 294–5.
97. [Booth], *Art of Brewing*, pp. [33], 40–9.
98. [Richardson], *Observations*, p. 65.
99. *London Review*, 6 (1777), pp. 235–6; *Monthly Review*, 57 (1777), pp. 322–3.
100. Tuck, *Private Brewer's Guide*, pp. [v]–vi.
101. Blake, *Strictures*, p. 2.
102. Tuck, *Private Brewer's Guide*, pp. 52–5.
103. Blake, *Theoretical and Practical Remarks*.
104. Tuck, *Private Brewer's Guide*, pp. [iv], vi, 40–1.
105. Levesque, *Art of Brewing*, 2nd edn.
106. Long, *Treatise on Malting*, pp. [1]–2, 7; R. Shannon, *A Practical Treatise on Brewing, Distilling, and Rectification* (London, 1805), p. 4.

107. [Richardson], *Observations*, p. 58.
108. Ibid., p. 24.
109. Richardson, *Theoretic Hints*, p. 12, emphasis original.
110. Aphorism 95 of the *Novum Organum*, reproduced in, for instance, *Philosophical Works of Francis Bacon*, vol. 2, p. 392.
111. Blake, *Strictures*, pp. 5–7.
112. Ibid., pp. 96–7, 110–13.
113. Richardson, *Theoretic Hints*, pp. 49–50, emphasis original.
114. Richardson, *Statical Estimates*, pp. vi–vii, n., emphasis original.
115. P. Aldabella and R. Barnard, *Hull and East Riding Breweries: From the Eighteenth Century to the Present* ([Beverley]: East Yorkshire Local History Society, 1997), pp. 80–1.
116. *Star*, 6 October 1796 (advertisement by 'R. B.' of Christchurch, Hampshire); *Sun*, 24 July 1798 (advertisement by D. Morton of the Strand, London).
117. *Star*, 6 September 1792; Ploughman, *Oeconomy in Brewing*, pp. 6–7.
118. J. Richardson, *The Philosophical Principles of the Science of Brewing*, 1st edn (York, 1788).
119. Tuck, *Private Brewer's Guide*, pp. [viii], 126, 182.
120. E. Rigby, *Chemical Observations on Sugar* (London, 1788), pp. 94–100; J. Richardson, *The Philosophical Principles of the Science of Brewing*, 2nd edn (Hull, 1798), pp. 46–54.
121. [Richardson], *Observations*, p. 19; Richardson, *Theoretic Hints*, pp. 3–4.
122. [Richardson], *Observations*, p. 38; Richardson, *Theoretic Hints*, p. 58.

4 The Value of Beer

1. [J. Noorthouck], review of John Richardsons's *Statical Estimates*, *Monthly Review*, 73 (1785), pp. 48–52. For Noorthouck's authorship of brewing reviews in the *Monthly Review*, p. 229, note 63.
2. Although Richardson mentions Hales only briefly and indirectly in his work: Richardson, *Theoretic Hints*, p. 203.
3. The pairing emerged through successive revisions of the *Encyclopaedia Britannica*, spreading to technical sources such as J. Yeats, *The Technical History of Commerce* (London: Cassell, Petter and Galpin, 1871), p. 234, and recurs in the *DNB*. From A. Clow and N. L. Clow, *The Chemical Revolution: A Contribution to Social Technology* (London: Batchworth, 1952), pp. 544–5, it influenced the coverage of technical topics in Mathias, *Brewing Industry*, and remains common in historical sections in brewery literature.
4. I discuss elements of this case in more detail in J. Sumner, 'John Richardson, Saccharometry and the Pounds-Per-Barrel Extract: The Construction of a Quantity', *British Journal for the History of Science*, 34:3 (2001), pp. 255–73.
5. Combrune, *Essay*, p. 59.
6. Ashworth, *Customs and Excise*, pp. 265–8.
7. B. Martin, *The Theory of the Genuine Hydrometer, Farther Illustrated* (London, [*c.* 1762]), p. 17.
8. [Richardson], *Observations*, pp. 54–5. For the topography of Fleet Street as the centre of the capital's instruments retail trade, see Millburn, *Benjamin Martin*.
9. Richardson, *Theoretic Hints*, p. 24.
10. Frost, *Address*, pp. 4, 48–9; Aldabella and Barnard, *Hull and East Riding Breweries*, p. 81.
11. A. W. Skempton and J. Brown, 'John and Edward Troughton, Mathematical Instrument Makers', *Notes and Records of the Royal Society*, 27:2 (1973), pp. 233–62; A. McConnell, *Instrument Makers to the World: A History of Cooke, Troughton and Simms* (York:

William Sessions, 1992); A. McConnell, 'From Craft Workshop to Big Business: The London Scientific Instrument Trade's Response to Increasing Demand, 1750–1820', *London Journal*, 19:1 (1994), pp. 36–53; J. Richardson, *The Philosophical Principles of the Science of Brewing*, 3rd edn (York: printed for the author, 1805), p. 457.

12. Richardson, *Statical Estimates*, p. 241.
13. A. Walters, 'Tools of Enlightenment: The Material Culture of Science in Eighteenth-Century England' (PhD dissertation, University of California at Berkeley, 1992), pp. 146–52.
14. Troughton's trade catalogue of this time is included in the end matter to J. Harris, *The Description and Use of the Globes, and the Orrery* (London, 1783).
15. A copy of the 'Tables and Directions for Using the Saccharometer' is held by Cambridge University Library, and a digital version is included in the Eighteenth Century Collections Online (ECCO) collection.
16. Richardson, *Statical Estimates*, pp. 2–6.
17. Ibid., pp. 8–9, 205–37; brewing reference on p. 226. The 'friend', who signs himself 'W. D.', is unidentified. He was evidently habituated to precision measurement; if he was involved in the instruments trade, no obvious candidate matches the initials given.
18. Richardson, *Statical Estimates*, p. iv; [Richardson], *Observations*, p. 7.
19. The National Library of Ireland holds a book of tables printed in 1795 for Thomas Saunders, who was contracted to produce the saccharometer in Dublin. The correction figures are identical to Troughton's.
20. Richardson, *Statical Estimates*, pp. [iii], xviii.
21. Ibid., pp. 89, 238–40.
22. Ibid., pp. viii–ix.
23. Hayman, *Practical Treatise*, p. iv; W. L. Tizard, *The Theory and Practice of Brewing Illustrated*, 2nd edn (London: sold by the author, 1846), p. vi–vii.
24. [Ellis], *London and Country Brewer*, 1st edn, part 1, pp. 12–13.
25. Richardson, *Statical Estimates*, pp. 66–7, 80–3, emphasis original; Richardson, *Philosophical Principles*, 3rd edn, p. 253, n.
26. Combrune, 'Historical Account' (1762 MS) (MS is unpaginated; images in the author's microfilm copy are numbered 0028 and 0032); TNA: CUST 48/16, fol. 277; CUST 48/19, fols 319–21, 334–5.
27. Richardson, *Statical Estimates*, p. xiii.
28. The articles usually cited in support of this ban are as given in the Statutes at Large (hereafter Statutes), 15 Car. II (1663), c. 11, s. 12; 1 Wm. & Mar. (1689), first session, c. 24, s. 11.
29. Statutes: 22 Geo. III (1782), c. 68; S. Dowell, *History of Taxation and Taxes in England from the Earliest Times to the Year 1885*, 4 vols (London: Longmans, 1888), vol. 4, p. 77.
30. Richardson, *Statical Estimates*, p. 148.
31. Ibid., pp. xiii–xv, 148. For the frustration of established practice by the mixing regulation, see TNA: CUST 48/34, fols 154–5.
32. Richardson, *Statical Estimates*, pp. 144–7.
33. Ibid., pp. 85–6.
34. Ibid., pp. 200–3.
35. [Booth], *Art of Brewing*, pp. 44–6.
36. Martin, *Theory of the Genuine Hydrometer*, pp. 18–22.
37. Reddington, *Practical Treatise*, pp. 1–3.
38. Richardson, *Statical Estimates*, pp. 24–5; J. Richardson, *Remarks on a Pamphlet Entitled Hydrometrical Observations and Experiments in the Brewery* (London, 1785), p. 47.

39. Martin certainly proposed the possibility, if not the mechanism, of applying the hydrometer to worts as distinct from beers (note 7 on p. 237), a fact which Richardson's recollection notably elides. A more plausible account, in a letter from Henry Thrale of 1770, suggests that Martin abandoned wort hydrometry because he did not think his instrument, calibrated in the distillery, would give sufficiently linear readings for practical use: see J. Baverstock, *Treatises on Brewing*, ed. J. H. Baverstock (London: Whittaker, 1824), pp. xvi–xvii.
40. Shaw, *Three Essays*, p. 62.
41. Clarke, *Notice*, pp. 26–45 and back matter, emphasis original.
42. Patent 1259, 27 June 1780.
43. [Richardson], *Observations*, p. 30.
44. B. Martin, *A Sure Guide to Distillers, and All Dealers in Spirituous Liquors* (London, 1759); Ashworth, *Customs and Excise*, pp. 263–72.
45. Richardson, *Statical Estimates*, p. 235, n.
46. Ibid., p. 21.
47. Ibid., pp. 3, 30–1.
48. C. Bensaude-Vincent, '"The Chemist's Balance for Fluids": Hydrometers and their Multiple Identities, 1770–1810', in F. Holmes and T. Levere (eds), *Instruments and Experimentation in the History of Chemistry* (London: MIT Press, 2000), pp. 153–83, on p. 166.
49. Richardson, *Statical Estimates*, p. 64.
50. Ibid., p. 65; Richardson, *Remarks*, p. 45.
51. J. Reynoldson, *Practical and Philosophical Principles of Making Malt*, 2nd edn (London: Gale and Curtis, 1808), p. 57, n.
52. H. Jackson, *An Essay on Bread, Wherein the Bakers and Millers are Vindicated* (London, 1758), p. 34.
53. Richardson, *Statical Estimates*, pp. 22–3.
54. Ibid., pp. 25–8.
55. Ibid., pp. 151–3.
56. Richardson, *Theoretic Hints*, pp. 57–8.
57. Richardson, *Statical Estimates*, p. 156, emphasis original.
58. S. Slyboots, *The Tavern Scuffle: Or, the Club in an Uproar* (London, 1726); Quincy, *Pharmacopœia*, pp. 224–5.
59. P. Clark, 'The "Mother Gin" Controversy in the Early Eighteenth Century', *Transactions of the Royal Historical Society*, 38 (1988), pp. 63–84.
60. [D. Defoe], *A Brief Case of the Distillers, and of the Distilling Trade in England* (London, 1726), p. 51.
61. A. Fourcroy, 'D'un memoire du cit. Fabroni [*sic*], sur les fermentations etc', *Annales de chimie*, 31 (1799), pp. 299–327, on p. 320 states that Rouelle advanced this view around 1760 or earlier.
62. J. Priestley, *Experiments and Observations on Different Kinds of Air*, vol. 2 (London, 1775), pp. 269–71.
63. T. Henry, *Experiments and Observations on Ferments and Fermentation* (Warrington, 1785).
64. M. Dobson, *A Medical Commentary on Fixed Air* (Chester, 1779), p. 69, n.
65. T. Garnett, *A Treatise on the Mineral Waters of Harrogate* ([Bradford], 1792), p. 94; G. Gregory, *The Economy of Nature Explained and Illustrated*, 3 vols (London, 1796), vol. 2, p. 524.

66. Richardson, *Statical Estimates*, pp. 172–3.
67. Ibid., pp. 162–6.
68. Ibid., p. 170; Richardson, *Philosophical Principles*, 2nd edn, pp. 381–401.
69. The only subsequent mention of Richardson's theory I have traced is the commentary in [Booth], *Art of Brewing*, p. 23, which accepts the reality of the effect but doubts Richardson's explanation.
70. J. Baverstock, *Hydrometrical Observations and Experiments in the Brewery* (London, 1785), pp. vi–viii, and see also Baverstock, *Treatises*, pp. xvi–xvii, 144.
71. J. Baverstock, *Practical Observations on the Prejudices against the Brewery* (London, 1811), p. 27; Baverstock, *Hydrometrical Observations*, p. ix.
72. *Transactions of the Society, Instituted at London for the Encouragement of Arts*, 2 (1784), p. 110.
73. Baverstock, *Practical Observations*, p. 27, emphasis original.
74. Baverstock, *Treatises*, p. [126].
75. Baverstock, *Hydrometrical Observations*, pp. 88–90.
76. Ibid., pp. 61–2.
77. Richardson, *Remarks*, pp. 2–5.
78. Baverstock, *Hydrometrical Observations*, pp. 65–6; Richardson, *Remarks*, p. 56.
79. Richardson, *Remarks*, pp. 3–4.
80. Ibid., p. 5.
81. Ibid., pp. 76–82.
82. Ibid., pp. 43–4, 52.
83. Ibid., pp. 81–2.
84. *Monthly Review*, 74 (1786), p. 397.
85. *Critical Review*, 61 (1786), p. 155; *Monthly Review*, 81 (1789), p. 273.
86. More to Baverstock, 3 December 1785, reproduced in Baverstock, *Treatises*, p. 302.
87. Baverstock to Robert Bogle, 24 May 1785, reproduced in Baverstock, *Treatises*, pp. 305–6. For Wright, see Millburn, *Benjamin Martin*, pp. 165–7.
88. RSA: index of fellowships. For the Ramsbottoms, see R. G. Thorne, *The House of Commons, 1790–1820*, 5 vols (London: Secker and Warburg, 1986), vol. 3, p. 8.
89. Baverstock, *Treatises*, pp. xviii–xix; Morrice, *Treatise on Brewing*, 2nd edn, p. 92; Shannon, *Practical Treatise*, pp. 248–55. For John Ramsbottom's role in co-ordinating petitions on taxation and materials, see for instance TNA: CUST 48/34, fols 323–4 (1802); CUST 48/40, fols 51–4 (1806).
90. Morrice, *Treatise on Brewing*, 2nd edn, p. 45; *Remarks and Experiments on Different Parts of the Process of Brewing* (Oxford: R. Bliss, 1807), p. 16.
91. *Transactions of the Society Instituted at London for the Encouragement of Arts*, 8 (1790), pp. 193–203.
92. Blake, *Theoretical and Practical Remarks*, title page, pp. 79–84; Baverstock, *Treatises*, p. 259; T. A. Glendinning, 'A Short Account of the Brewers' Saccharometer', *Journal of the Federated Institutes of Brewing*, 6 (1900), pp. 357–64, on p. 364.
93. Evidence of John Martineau: 'Report from the Committee to whom the Several Petitions, Relative to the Making of Malt, Were Referred', session 1806, paper 224 (hereafter 1806 Malt Petitions Report), p. 44.
94. Thomas Fowell Buxton's book of observations, LMA: B/THB/G/63, fol. 30, verso.
95. Morrice, *Treatise on Brewing*, 2nd edn, appendix; Hayman, *Practical Treatise*, pp. 34–5; Shore, *Practical Treatise*, p. 15; G. L. Worthington, *The Brewer's Guide: A New Work* (London, 1812), pp. 68–9.

96. G. S. Keith, 'Observations on the Papers Presented to the House of Commons, Relating to the Experiments Made for Ascertaining the Relative Quality of Malt Made from Barley and Scotch Bigg', *Farmer's Magazine*, 8 (1807), pp. 476–500, on pp. 487–8; Accum, *Art of Brewing*, p. 104. See also A. Morrice, *A Practical Treatise on Brewing the Various Sorts of Malt-Liquor*, 7th edn (London, 1827), pp. 42–3, 141–50, showing clumsy modifications to the text to endorse Dring and Fage, who were now the book's retailers.
97. Levesque, *Art of Brewing*, 2nd edn, front matter.
98. *Encyclopaedia Britannica*, 2nd edn (1778–83), vol. 2, pp. 1378–85; Shannon, *Practical Treatise*, pp. 156–233.
99. Black, *Practical Treatise*, 1st edn, p. 37.
100. M. Teich, *Bier, Wissenschaft und Wirtschaft in Deutschland, 1800–1914* (Vienna: Böhlau, 2000), pp. 60–1.
101. [Booth], *Art of Brewing*, pp. 11–15; F. Faulkner, *The Theory and Practice of Modern Brewing*, 2nd edn (London: F. W. Lyon, 1888), p. 67.
102. D. Knox, 'The Development of the London Brewing Industry, 1830–1914, with special reference to Whitbread and Company' (BLitt dissertation, University of Oxford, 1956), p. 149; *Gentleman's Magazine*, new series, 9 (1838), p. 554; Abraham Rees (ed.), *Cyclopaedia*, 28 (1819), s. v. 'Porter'. The *Cyclopaedia* entry outlived Philetus Richardson, propagating to the *Britannica* and other works.
103. 1806 Hope Report, p. 68; Baverstock, *Treatises*, p. 258; Tizard, *Theory and Practice*, 2nd edn, pp. 216–7.
104. Glendinning, 'Short Account', p. 363.
105. [Booth], *Art of Brewing*, p. 16.
106. *Vinum Britannicum*, p. 30; [Ellis], *London and Country Brewer*, 1st edn, part 1, p. 19; Reddington, *Practical Treatise*, p. 25.
107. Richardson, *Statical Estimates*, pp. 67–70.
108. Richardson, *Philosophical Principles*, 3rd edn, pp. 249–52. The figures presented in the 1784 edition, and those added for 1787–8, give some clear support, but after this date there is often little difference between Richardson's values for 'pale' and for 'brown and pale mixed'.
109. *Annals of Philosophy*, 10 (1817), p. 466.
110. The prices for brown and pale malts between 1784 and 1830 from Truman's rest books, reproduced in Mathias, *Brewing Industry*, p. 548, clearly show this effect at work. See also pp. 72–3 of the same work, and James Hinton Baverstock's comments in Baverstock, *Treatises*, pp. 189–90.

5 Chemists, Druggists and Beer Doctors

1. J. Malcolm, *A Compendium of Modern Husbandry*, 3 vols (London: Baldwin, 1805), vol. 1, p. 296.
2. 'Minutes Taken (in session 1818) before the Committee to whom the Petition of Several Inhabitants of London and its Vicinity, Complaining of the High Price and Inferior Quality of Beer, was Referred', session 1819, paper 220 (hereafter 1818 Beer Minutes), pp. 58–9.
3. R. Green, *The Porter Brewer Detected* (London, 1764).
4. The fluctuations in prices, and the strong upward trend of taxation, are summarized in Mathias, *Brewing Industry*, pp. 546–50; and see also p. 114 on the squeezing of publicans' profit margins.

5. M. Douglas, *Purity and Danger: An Analysis of Concepts of Pollution and Taboo* (1966; new edn Abingdon: Routledge, 2002), pp. 35–40.
6. As reproduced in *The Poetical Works of John Skelton, With Notes*, ed. A. Dyce, 2 vols (London: Thomas Rodd, 1843), vol. 1, pp. 101–2.
7. Bennett, *Ale, Beer and Brewsters*, pp. 123–4 considers Elynour Rummyng in the light of contemporary attitudes to women as proprietors of knowledge and trade.
8. Sumner, 'Status, Scale'.
9. [T. Smollett], *The Expedition of Humphry Clinker*, 3 vols (London, 1771), vol. 2, pp. 2–11.
10. *Read's Weekly Journal*, 29 January 1757; [P. Markham], *Poison Detected: or Frightful Truths; and Alarming to the British Metropolis* (London, 1757); P. Markham, *Syhoroc: or, Considerations on the Ten Ingredients Used in the Adulteration of Bread-Flour, and Bread*, 2nd edn (London, 1758), pp. 16–17. The controversy is covered in F. Filby, *A History of Food Adulteration and Analysis* (London: George Allen and Unwin, 1934), pp. 79–104.
11. For brewers in Parliament, see Mathias, *Brewing Industry*, pp. 330–8.
12. [H. Dalrymple], *Rodondo, or, the State Jugglers, Canto III* (London, 1770), pp. 17–18.
13. Child, *Every Man his Own Brewer*, pp. 5–6.
14. Threale, *The Complete Family-Brewer*, derived largely from a 1740s source, adds an appendix borrowed from Child, calling confusedly for 'Creolus Indian [*sic*] Berry'.
15. For instance, [Charles Williams?], 'Iohn Bull in the Cellar', *Scourge*, 5 (1813), pp. 262–4 and colour plate for March, at http://ia700407.us.archive.org/19/items/scourgeormonthly05crui/scourgeormonthly05crui.pdf [accessed 13 January 2013].
16. *Discourse on the Preparation*, pp. 66–7; Combrune, 'Historical Account', image 0050.
17. W. Randall, *The State of the Hop Plantations* (London, 1800), p. 5, n.
18. James Gillray, 'Election Candidates, or the Republican Goose at the Top of the Poll', 20 May 1807; 'Overthrow of the Republican Babel', 1 May 1809.
19. Satirical Prints Collection, British Museum (hereafter BM Satires), 10574. Useful references are provided in the 1947 catalogue entry for this item, prepared by M. Dorothy George and reproduced in the online catalogue at http://www.britishmuseum.org/research/search_the_collection_database.aspx [accessed 13 January 2013].
20. For Quassi or Kwasí, see R. Price, 'Kwasímukámba's Gambit', *Bijdragen tot de Taal-, Land- en Volkenkunde*, 135:1 (1979), pp. 151–69; L. Schiebinger, *Plants and Empire: Colonial Bioprospecting in the Atlantic World* (Cambridge, MA: Harvard University Press, 2007), pp. 211–4.
21. 'First Report from the Committee on the State of the Police of the Metropolis, with Minutes', session 1817, paper 233 (hereafter 1817 Police Minutes), pp. 119–26.
22. *The Trials at Large of Joseph Merceron, Esq. for Fraud* (London: printed for W. Wright, 1819).
23. 1817 Police Minutes; 1818 Beer Minutes; *The Alarm Bell, or, Hints to Travellers* (London: T. Harvey, [1817]), pp. 5–7.
24. For instance, 1818 Beer Minutes, pp. 35–6. For the artificial definition of the 'great eleven' see Sumner, 'Status, Scale', p. 296.
25. 'Report from the Committee on Public Breweries', Session 1818, paper 399 (hereafter 1818 Breweries Report).
26. Memo from [C.?] N. Buxton (of Truman's) to W. [C.?] [Higgins?] (Whitbread), 1812: 'We tried an experiment on a small scale with some bitter substitute a short time ago', LMA: LMA/4453/A/15/006.
27. 1818 Beer Minutes, p. 25.

28. Mathias, *Brewing Industry*, pp. 59–62, 82–98; Pearson, *British Breweries*, pp. 27–40.
29. Evidence of John Adams, 1818 Beer Minutes, p. 54.
30. TNA: CUST 48/30, fols 275–6, 318–9; CUST 48/32, fols 1–2.
31. Statutes: 42 Geo. III (1802), c. 38, ss. 20, 21.
32. W. H. R. Brown, *Golden Lane Brewery. The Attorney General versus Brown, Parry, and Others* (London: printed for J Johnson, [1808]), p. [ix]; TNA: CUST 103/66 (Excise Trials, 1807–10), fol. 952.
33. For instance: *Morning Post*, 2 September 1807 (Walsby); *Hampshire Telegraph and Sussex Chronicle*, 19 September 1808 (James Paffard, Portsmouth); *Morning Post*, 6 January 1809 (Bankside Brewery, Southwark).
34. Handbill regarding Hunt, TNA: HO 44/18, fol. 489, emphasis original; *An Impartial Narrative of the Late Melancholy Occurrences in Manchester* (Liverpool: Henry Fisher, 1819), p. 19.
35. 'A Practical Joke, or, a Trick upon Travellers', 1810, BM Satires 11657.
36. [Smollett], *Humphry Clinker*, vol. 2, p. 11.
37. Although [Ellis], *London and Country Brewer*, 2nd edn, p. 212, describes an enterprising gang who managed to pass off a hopped and fermented sugar extract as malt ale.
38. *Monthly Magazine*, 2 (1796), p. 596.
39. Malcolm, *Modern Husbandry*, vol. 1, pp. 294–6.
40. 1818 Beer Minutes, pp. 40–1.
41. *The Crying Frauds of the London Markets: Proving their Deadly Influence upon the Two Great Pillars of Life, Bread & Porter* (London, 1795), p. 5.
42. W. Cobbett, *Cottage Economy, Containing Information Relative to the Brewing of Beer* (London: Clement, 1822), p. 46, emphasis original.
43. *Morning Post*, 23 January 1807. The passage also appeared in the *Oracle*, *Dispatch*, *Bell's Weekly Messenger* and *Observer*. *The Times* declined to print the 'atrocious representations': 12 Feburary 1807.
44. *The Times*, 12 Feburary 1807; 'Imperial Parliament', *Morning Chronicle*, 14 March 1818.
45. Harrison, 'Description of England', pp. 169–70; M[arkham], *English Hus-wife*, p. 123; Yworth, *Cerevisiarii Comes*, p. 54; *Discourse on the Preparation*, p. 40; [Ellis], *London and Country Brewer*, 2nd edn, p. 285; [Watkins], *Compleat Brewer*, p. 123.
46. Statutes: 1 Wm. & Mar., session 1 (1688), c. 24, s. 17; 10 & 11 Will. III (1699), c. 21, s. 34; 13 Will. III (1701), c. 5; 9 Anne, c. 12 (1710), s. 24.
47. Sambrook, *Country House Brewing*, pp. 118, 131–2.
48. R. Flower, *Observations on Beer and Brewers* (Cambridge: B. Flower, 1802), p. 22.
49. [Ellis], *London and Country Brewer*, 2nd edn, pp. 291–3.
50. 'True Method of Brewing Strong Ales', annexed to J. Lightbody, *The Gauger and Measurer's Companion* (London, 1694); M[ortimer], *Whole Art of Husbandry*, pp. 595–6. For isinglass in the mid-eighteenth century, see chapter 3.
51. Best MSS: Henry Fanner to Best, 3 and 7 February 1776 and 5 and 7 March 1776, and 'Observations upon Brewing made at Lowestoft Novbr: 4th. 1775'.
52. Statutes: 39 and 40 Geo. III (1800), c. 62; 41 Geo. III (1800), c. 5.
53. [Booth], *Art of Brewing*, p. 19.
54. I discuss the complications raised by innovations in fining below. For a parallel case involving yeast, see Mathias, *Brewing Industry*, pp. 50–1.
55. Hayman, *Practical Treatise*, p. 102.
56. Statutes: 13 Will. III (1701), c. 5.
57. Patent 2625, 31 May 1802; *ODNB*.

58. TNA: CUST 48/38, fols 408–11; CUST 48/40, fols 51–6, 449–54.
59. 'The immaculate senator', *Morning Post*, 21 May 1810, p. [3]; 'The Brewer's Druggist', *Satirist*, 6 (1810), pp. 437–9; 'The Late Sheriff Wood', *Satirist*, 7 (1810), pp. 556–8; 'Convenient Arrangements', *Morning Post*, 29 July 1820, p. [3].
60. [C. F. Lawler], *The Lamentations of the Porter Vat Which Exploded of the Drug-Gripes, October 17th, 1814* (London: John Fairburn, [1814]).
61. Treasury to Excise, 23 April 1811, TNA: CUST 48/51, fol. 91.
62. Statutes: 51 Geo. III (1811), c. 87; 52 Geo. III (1812), c. 65.
63. S. Hartlib, *The Reformed Common-wealth of Bees, Presented in Severall Letters and Observations* (London, 1655), pp. 18–19; 33–5; S. Hartlib, *Samuel Hartlib his Legacy of Husbandry*, 3rd edn (London, 1655), pp. 20–1; 119; 145, emphasis original.
64. Birch, *History of the Royal Society*, vol. 1, p. 171, vol. 2, p. 104; Sprat, *History of the Royal-Society*, p. 226; J. Mortimer, *The Whole Art of Husbandry: or, the Way of Managing and Improving Land*, 4th edn, 2 vols (London, 1716), vol. 1, pp. 354–5.
65. W. Charleton, *Two Discourses ... [containing] the Mysterie of Vintners* (London, 1669), p. 149; Glauber, *Works*, pp. 159–60; [Shaw], *Juice of the Grape*, p. 53; Mills, *Practical Husbandry*, vol. 5, pp. 174–7; Boerhaave, *Elements of Chemistry* (1735 Dallowe edn), vol. 1, pp. 71–2; *The Chemical Works of Caspar Neumann, M. D.*, ed. W. Lewis (London, 1759), p. 449.
66. Combrune, *Theory and Practice*, pp. 272–8.
67. A detailed, if slightly hagiographic account of Jackson's role in this incident appears in Filby, *History of Food Adulteration*, pp. 79–104.
68. Jackson, *Essay on Bread*, pp. 32–42; Best MSS: 'Proposals', p. 2; Jackson, *Essay on British Isinglass*, pp. 37, 40, 71, 76.
69. Jackson, *Essay on British Isinglass*, p. 79, emphasis original.
70. [Richardson], *Observations*, pp. 26, 39, emphasis original.
71. [Booth], *Art of Brewing*, p. 46. Compare Combrune, *Theory and Practice*, p. 275: Combrune's 'as much pounded salt of steel as will lay upon a shilling' is upgraded by Richardson to a half-crown piece's worth.
72. Piozzi, *Letters*, vol. 1, p. 79, n.
73. Jackson, *Essay on British Isinglass*, pp. 60–1.
74. Piozzi, *Autobiography*, vol. 1, pp. 257–8.
75. *Crying Frauds*, p. 5; Malcolm, *Modern Husbandry*, vol. 1, p. 296.
76. C. Hutton, G. Shaw and R. Pearson (eds), *The Philosophical Transactions of the Royal Society of London ... Abridged, with Notes*, 18 vols (London: Baldwin, 1809), vol. 13, p. 362, n. The unsigned note mistakenly attributes the account to Piozzi's *Anecdotes*, rather than her *Letters* of Johnson.
77. For Jackson's evident (if not necessarily unusual) culpability in this practice see N. Landau, *Law, Crime and English Society, 1660–1830* (Cambridge: Cambridge University Press, 2002), p. 67.
78. T. Thomson, *History of the Royal Society, from its Institution to the End of the Eighteenth Century* (London: Baldwin, 1812), p. 499; [Thomson], 'Brewing', p. 485. Thomson must have collected considerable private information on the brewery through his Excise consultancy (chapter 6), but his characterization of Jackson seems to derive purely from Piozzi or from the 1809 account.
79. [Richardson], *Observations*, p. 44.
80. *Monthly Magazine*, 24 (1807), pp. 346–8.

81. Records of Attorney General v. Brown and Parry, 1809, TNA: CUST 103/66, fols 913–1048, on fol. 928, punctuation regularized.
82. *The Times*, 21 April 1806.
83. 'Cattle Insurance', *Satirist*, 4 (1809), pp. 566–75; 'Sleeping Partners', *Satirist*, 5 (1809), pp. 230–5; 'School of Projects', *Satirist*, 5 (1809), pp. 313–7.
84. Mathias, *Brewing Industry*, p. 250.
85. TNA: CUST 48/47, fols 257–64.
86. TNA: CUST 103/66, fols 988–9, 1000.
87. TNA: CUST 103/66, fol. 1020.
88. Solicitor-General's reply, TNA: CUST 103/66, fols 1023, 1028.
89. Summing up, TNA: CUST 103/66, fols 1040–1.
90. Statutes: 56 Geo. III (1816), c. 58.
91. 1818 Beer Minutes, pp. 16–22, 80–4.
92. A. Cooley, *A Cyclopædia of Practical Receipts*, 2nd edn (London: John Churchill, 1845), p. 278.
93. 1818 Beer Minutes, pp. 80–4; Patent 4112, 28 March 1817.
94. *London Gazette*, 21 November 1818, p. 2083 (dissolution of partnership); insurance document of 1812, LMA: CLC/B/192/F/001/MS11936/459/867077. For this largely obscure period in Imlay's career, see W. Verhoeven, *Gilbert Imlay: Citizen of the World* (London: Pickering & Chatto, 2008), pp. 208–9.
95. *Gentleman's Magazine*, 89 (1819), p. 648.
96. Faraday to Simon Cock, 12 August 1818, in F. A. J. L. James (ed.), *The Correspondence of Michael Faraday*, vol. 1 (London: Institution of Electrical Engineers, 1991), pp. 164–5.
97. Wheeler v. Malins, Court of Chancery, 1818, reported in *The Times*, 19 August 1818; Rex v. Wheeler, Court of King's Bench, 1819, reported in W. Carpmael, *Law Reports of Patent Cases*, vol. 1 (London: A Macintosh, 1843), pp. 394–400; *Post Office London Directory for 1820* (London: Critchett and Woods, 1820), p. 229. Malins was insured as a 'coffee and malt roaster' in 1818 and as a 'malt roaster' thereafter: records of the Sun Fire Office, LMA: CLC/B/192/F/001/MS11936.
98. Statutes: 5 Vic. (1842), c. 30, s. 2; W. Ford, *An Historical Account of the Malt Trade and Laws* (London, 1849), pp. 68–70.
99. E. Stieb, *Drug Adulteration: Detection and Control in Nineteenth-Century Britain* (Madison, WI: University of Wisconsin Press, 1966), pp. 164–5.
100. Golinski, *Science as Public Culture*, p. 260; R. J. Cole, 'Friedrich Accum (1769–1838). A Biographical Study', *Annals of Science*, 7:2 (1951), pp. 128–43, on pp. 134–5; G. Fisher, *Life of Benjamin Silliman*, 2 vols (New York: Charles Scribner, 1866), vol. 1, p. 142.
101. F. C. Accum, *A Treatise on Adulterations of Food, and Culinary Poisons*, 1st edn (London: Longman, 1820). An analysis of the quotation is given in Cole, 'Friedrich Accum', p. 137, n. 64. The price and format are as reported in *The Times*, 24 January 1820, p. [4].
102. Filby, *History of Food Adulteration*, especially p. 19, has guided subsequent accounts on this point.
103. Surveyed in C. A. Browne, 'Life and Chemical Services of Fredrick Accum', *Journal of Chemical Education*, 2 (1925), pp. 829–51, 1008–34, 1140–9, on 1027–34.
104. C. Mathews, *Mr. Mathews' Memorandum-book, of Peculiarities, Character, and Manners* (London: Duncombe, [*c.* 1825]), p. 24.
105. F. C. Accum, *A Treatise on Adulterations of Food, and Culinary Poisons*, 2nd edn (London: Longman, 1820), pp. x–xi; *Deadly Adulteration and Slow Poisoning, or, Disease and Death in the Pot and Bottle* (London: Sherwood, [*c.* 1830]), pp. 39–40; Browne,

'Life and Chemical Services', pp. 1140–4; C. A. Browne, 'Recently Acquired Information Concerning Fredrick Accum, 1769–1838', *Chymia*, 1 (1948), pp. 1–9, on p. 7. For the allegation, which seems to have been well founded, see Cole, 'Friedrich Accum', pp. 137–42; K. D. C. Vernon, 'The Library of the Royal Institution, 1799–1954', *Nature*, 174:4423 (1954), pp. 242–4, on p. 244.

106. Accum, *Treatise on Adulterations*, 1st edn, pp. 159–60, 206–9; compare 1818 Beer Minutes, pp. 29–31, 36–8.
107. Accum, *Treatise on Adulterations*, 1st edn, pp. 211–6.
108. Ibid., pp. 204–5.
109. E. Howe and J. Child, *The Society of London Bookbinders, 1780–1951* (London: Sylvan, 1952), p. 45, n. 20; E. Howe, *A List of London Bookbinders, 1648–1815* (London: Bibliographical Society, 1950), p. 94; [J. Tuck], *A Guide to Young Brewers, Particularly Adapted to the Families of the Nobility and Gentry* (London: F. Westley, 1820).
110. Tuck, *Private Brewer's Guide*, pp. [viii]–ix.
111. Accum, *Art of Brewing*, p. 55, for instance, borrows without attribution from Ploughman, *Oeconomy in Brewing*, p. 10.
112. Tuck, *Private Brewer's Guide*, pp. 63, n., 84, 92, 99, 106, 130.
113. Ibid., pp. 137–8.
114. Ibid., pp. 60–2.
115. Ibid., pp. xi–xii.
116. W. Smyth, *The Publican's Guide, or, Key to the Distill-House* (London, 1781), pp. 22–3, 64–6; J. Hardy, *The Retail Compounder, or, Publican's Friend* (London, [*c.* 1795]), pp. 78–9; P. Boyle, *The Publican and Spirit Dealer's Daily Companion* (London, [*c.* 1800]), p. 37.
117. Boyle, *Publican*, [title page], pp. 60–73, [146]; 1818 Beer Minutes, p. 35.
118. Morrice, *Treatise on Brewing*, 2nd edn, pp. 131–48; Shannon, *Practical Treatise*.
119. Moir, *Brewing Made Easy*, pp. 30–1.
120. Accum, *Treatise on Adulterations*, 2nd edn, pp. 150–3.
121. Tuck, *Private Brewer's Guide*, pp. 84, 182; compare Accum, *Art of Brewing*, p. 196; Child, *Every Man his Own Brewer*, p. 8.
122. *A New System of Practical Domestic Economy, Founded on Modern Discoveries*, 3rd edn (London: Henry Colburn, 1823), pp. 250–1.
123. W. J. Forbes, 'Death (a Dealer) to his London correspondent', in R. Dagley (ed.), *Death's Doings, Consisting of Numerous Original Compositions, in Prose and Verse* (London: Andrews and Cole, 1826), pp. 297–300.
124. Tizard, *Theory and Practice*, 2nd edn, pp. 491–3; [W. Loftus], *The Brewer: A Familiar Treatise on the Art of Brewing*, new edn (London: Loftus, [1877]), front matter.
125. Stieb, *Drug Adulteration*, pp. 16, 32, 166, 299, n. 4; H. Schuette, 'Death in the Pot', *Transactions of the Wisconsin Academy of Sciences, Arts and Letters*, 35 (1943), pp. 283–303, on p. 283. Accum's crusading pose recalls *Crying Frauds*, p. [2].
126. *Quarterly Review*, 24 (1820–1), pp. 341–52, on p. 347; *London Literary Gazette*, 15 January 1820, pp. 33–38, on p. 38.

6 Professors in the Brewhouse

1. [T. Thomson], 'Brewing', p. 460.
2. S. Schaffer, 'Scientific Discoveries and the End of Natural Philosophy', *Social Studies of Science*, 16:3 (1986), pp. 387–420; Golinski, *Science as Public Culture*.

3. J. Topham, 'Scientific Publishing and the Reading of Science in Nineteenth-Century Britain: A Historiographical Survey and Guide to Sources', *Studies in History and Philosophy of Science*, 31:4 (2000), pp. 559–612, on p. 560.
4. J. Morrell, 'The Chemist Breeders: The Research Schools of Liebig and Thomas Thomson', *Ambix*, 19:1 (1972), pp. 1–46.
5. 'Report from the Committee on the Re-Committed Report Respecting the Rate of Duty Payable on Malt made from Barley of the Growth of England, and from Barley and Bigg of the Growth of Scotland', session 1803–4, paper 129 (hereafter 1804 Malt Report), pp. 4–5.
6. 'Report Respecting the Distilleries in Scotland, &c. &c. Ordered to be Printed 12th July 1799' (hereafter 1799 Distilleries Report), pp. 471–80; Scottish Excise Board minute books, National Archives of Scotland, Register House, Edinburgh (hereafter NAS): CE2/7, fol. 30; 1804 Malt Report, pp. 23–4; G. S. Keith, *A General View of the Agriculture of Aberdeenshire* (Aberdeen: A. Brown, 1811), pp. 318–21.
7. 1804 Malt Report, pp. 34, 4–11.
8. NAS: CE2/8, fols 76–7.
9. For Hope's and Thomson's roles in teaching at Edinburgh, see J. Morrell, 'Practical Chemistry in the University of Edinburgh, 1799–1843', *Ambix*, 16:1 (1969), pp. 66–80, on pp. 65–70; for Coventry, *ODNB*.
10. NAS: CE2/13, fol. 216; CE2/25, fols 236–7.
11. 1806 Hope Report, p. 3.
12. NAS: CE2/8, fols 172, 177–8, 181–5, 253; CE2/9, fols 266–7; CE2/13, fols 241, 314–5.
13. 1806 Hope Report, pp. 52–7; NAS: CE2/9, fol. 238; CE2/10, fols 62, 115.
14. 1806 Hope Report, pp. 4–7; NAS: CE2/8, fols 120–2.
15. 1806 Hope Report, pp. 13–14.
16. Ibid., pp. 21–2, 32.
17. Ibid., pp. 104, 107.
18. Ibid., pp. 7, 12, 105.
19. NAS: CE2/8, fols 146–7.
20. Keith, 'Observations', p. 478.
21. Ibid., pp. 480–1.
22. Ibid., p. 489.
23. 'Report from the Select Committee on Petitions Complaining of the Additional Duty on Malt in Scotland', session 1821, paper 598 (hereafter 1821 Malt Petitions Report), pp. 3–5, 67.
24. Ashworth, *Customs and Excise*, pp. 267–8.
25. D. P. Miller, 'Between Hostile Camps: Sir Humphry Davy's Presidency of the Royal Society of London, 1820–1827', *British Journal for the History of Science*, 16:1 (1983), pp. 1–47; J. Gascoigne, *Science in the Service of Empire: Joseph Banks, The British State and the Uses of Science in the Age of Revolution* (Cambridge: Cambridge University Press, 1998), p. 26.
26. Ashworth, *Customs and Excise*, pp. 272–8; TNA: DSIR 26/88; DSIR 26/89. A detailed summary of the investigation appears in F. G. H. Tate, *Alcoholometry: An Account of the British Method of Alcoholic Strength Determination* (London: HMSO, 1930), pp. 4–13.
27. Keith, *Agriculture of Aberdeenshire*, pp. 318–21.
28. Unlike the other sections, 'Brewing' bears a specific statement of authorship: 1806 Hope Report, p. 59.

29. Lavoisier's struggles with the atomic basis of fermentation are charted in F. Holmes, *Lavoisier and the Chemistry of Life: An Exploration of Scientific Creativity* (Madison, WI: University of Wisconsin Press, 1985), pp. 372–84.
30. J. Dalton, *A New System of Chemical Philosophy*, part 1 (Manchester: Bickerstaff, 1808), p. 219. T. Thomson, *A System of Chemistry*, 3rd edn, 5 vols (Edinburgh: Bell and Bradfute, 1807), vol. 3, p. 481.
31. Thomson in 1806 Hope Report, pp. 46–7, and compare T. Thomson, *A System of Chemistry*, 4th edn, 5 vols (Edinburgh: Bell and Bradfute, 1810), vol. 5, pp. 412–17.
32. Thomson in 1806 Hope Report, pp. 45–8.
33. 1806 Hope Report, pp. 70–1.
34. Ibid., pp. 106–8.
35. Ibid., pp. 45–9, 109–18. For this episode in the context of Thomson's general character, see Morrell, 'Practical Chemistry', pp. 247–8.
36. Thomson in 1806 Hope Report, p. 114, emphasis original.
37. NAS: CE2/12, fol. 290; CE2/13, fol. 131.
38. 1806 Hope Report, p. 68.
39. For Allan, see D. J. Bryden, *Scottish Scientific Instrument-Makers, 1600–1900* (Edinburgh: [Royal Scottish Museum], 1972), p. 15.
40. Report from the Select Committee on the Use of Molasses in Breweries and Distilleries, with an Appendix and Index', session 1831, paper 297 (hereafter 1831 Molasses Report), p. 243.
41. 1806 Hope Report, p. 66; W. H. Roberts, *The Scottish Ale-Brewer and Practical Maltster* (Edinburgh: A. and C. Black, 1847), pp. 29–30.
42. Statutes: 56 Geo. III (1816), c. 106, ss. 20–1.
43. *ODNB*. Agnes's father, 'R. Colquhoun', was perhaps the Robert Colquhoun, Jr who in the 1780s was a partner in the large-scale Underwood Distillery, on the Forth and Clyde Canal outside Falkirk: *Scots Magazine*, 50 (1788), p. 206.
44. Letter from Fabbroni to Jean-Baptiste van Mons, reproduced in *Annales de chimie*, 30 (1800), pp. 220–4; Fourcroy, 'D'un memoire du cit. Fabroni'; [Nicholson's] *Journal of Natural Philosophy, Chemistry, and the Arts*, 4 (1801), pp. 45–6.
45. W.T. Brande, 'Experiments to Ascertain the State in which Spirit Exists in Fermented Liquors', *Philosophical Transactions*, 101 (1811), pp. 337–46; W. T. Brande, 'Additional Remarks on the State in which Spirit Exists in Fermented Liquors', *Philosophical Transactions*, 103 (1813), pp. 82–7; *Annales de chimie*, 86 (1813), pp. 175–7.
46. Brande, 'Experiments', p. 338.
47. *Quarterly Journal of Science and the Arts*, 4 (1818), pp. 289–91.
48. The same period saw values of 6.20 per cent for Edinburgh ale, 5.56 per cent for Dorchester and London porters averaging 4.20 per cent; small beer, in London, averaged 1.28 per cent. *Quarterly Journal of Science and the Arts*, 5 (1818), pp. 124, 152–4.
49. Accum, *Treatise on Adulterations*, 1st edn, pp. 219–22; Accum, *Art of Brewing*, p. 11. Some of the values Accum attributes to Brande's 1811 paper were in fact first published later in the *Quarterly Journal*.
50. Accum, *Treatise on Adulterations*, 1st edn, pp. 173–4.
51. Statutes: 1 Will. IV (1830), c. 51, c. 64. For the 1830 reforms, see Gourvish and Wilson, *British Brewing Industry*, pp. 3–22; and for the background to this development, J. Nicholls, *The Politics of Alcohol: A History of the Drink Question in England* (Manchester: Manchester University Press, 2009), pp. 80–95.
52. Yeo, *Encyclopaedic Visions*, pp. 251–76.

53. Thomson to Napier, 30 March 1816, Manuscripts and Archives, British Library (hereafter BL): Add MS 34611, letter 375.
54. [Thomson], 'Brewing', p. 461.
55. Ibid., pp. 460–1. Thomson was no doubt objecting to passages such as Shannon, *Practical Treatise*, pp. 16–25, evidently an attempt to paraphrase an oxygen-theory account of fermentation without clear understanding of the concepts involved.
56. *Edinburgh Encyclopaedia*, 1830, vol. 4, pp. 459–75, on p. 475.
57. *Encyclopaedia Edinensis*, 1827, vol. 2, pp. 4–25. Periodical notices place the appearance of the relevant part around the end of 1817: *Literary Panorama*, 7 (1818), p. 610.
58. Thomson to Napier, 9 November 1815, BL: Add MS 34611, letter 310; Thomson to Napier, 20 April 1817, BL: Add MS 34612, letter 72. The material referred to appears in [Thomson], 'Brewing', p. 488, prior to the 'Explanation of the Plates'. Details of this work for the English Excise have not come to light. Thomson's comments in the 1831 Molasses Report, pp. 235–47, refer briefly to London (on p. 246) but focus on his more extensive Edinburgh experience.
59. [Thomson], 'Brewing', pp. 480–1.
60. T. Thomson, *A System of Chemistry*, 2nd edn, 4 vols (Edinburgh: Bell and Bradfute, 1804), vol. 4, p. 450 (Westrumb); Thomson, *System of Chemistry*, 3rd edn, vol. 5, pp. 412–3 (Thénard); T. de Saussure, 'On the Conversion of Starch into Sugar', *Annals of Philosophy*, 6 (1815), pp. 424–31; T. Thomson, *A System of Chemistry*, 5th edn, 4 vols (London: Baldwin, Cradock and Joy, 1817), vol. 4: p. 24 (composition of sugar).
61. Thomson to Napier, 9 November 1815, BL: Add MS 34611, letter 310; Thomson to Napier, 18 March 1818, BL: Add MS 34612, letter 176.
62. *London Encylopedia* 1829, vol. 4, p. 523.
63. [Thomson], 'Brewing', p. 461. Advance publicity named Thomson among the contributors to the *Supplement*. Individual entries, as published, were identified by initials – 'J.' in Thomson's case – which were matched to their authors in lists bound into some releases of some parts of the work.
64. W. V. Farrar, 'Andrew Ure, F.R.S., and the Philosophy of Manufactures', *Notes and Records of the Royal Society*, 27 (1973), pp. 299–324, on p. 308.
65. Farrar, 'Andrew Ure', p. 322, n. 36.
66. [A. Ure], 'A System of Chemistry, in Four Volumes, 8vo. by Thomas Thomson', [*Quarterly*] *Journal of Science and the Arts*, 4 (1818), pp. 299–321; A. Ure, *A Dictionary of Chemistry, on the Basis of Mr Nicholson's*, 1st edn (London: Thomas and George Underwood, 1821), pp. xviii–xix, and s.v. 'Fermentation'.
67. T. Thomson, 'Answer to the Review of the Sixth Edition of Dr Thomson's System', *Annals of Philosophy*, 3 (1822), pp. 241–75.
68. *Dublin Philosophical Journal*, 1 (1825), pp. 85–94.
69. Baverstock, *Treatises*, p. ix.
70. [Thomson], 'Brewing', p. 485.
71. Ibid., pp. 473, 485–6, 464.
72. T. Thomson, 'On Pepper', *Journal of Natural Philosophy*, 9 (1804), pp. 68–78, on p. 69, emphasis original.
73. 1806 Hope Report, p. 47, compare Thomson, *System of Chemistry*, 4th edn, vol. 5, p. 404.
74. Reynoldson, *Making Malt*, pp. 67–8. For Bent, see also 1806 Malt Petitions Report, pp. 67–70.

75. The correspondence, which ran in vol. 14 of the *Political Register* between 12 November and 24 December 1808, is reproduced in full in Baverstock, *Treatises*, pp. 125–62.
76. Baverstock, *Treatises*, pp. 138–42.
77. 1806 Hope Report, p. 46. None of the figures which follow distinguish 'first mash', 'second mash', etc., as would have been the case in conventional brewery book-keeping.
78. Baverstock, *Practical Observations*, pp. 48–9.
79. Baverstock, *Treatises*, pp. 144–53, emphasis original; evidence of John Martineau, 1806 Malt Petitions Report, p. 44; 'Report from the Committee on the Distillation of Sugar and Molasses', session 1808, paper 178 (hereafter 1808 Sugar Distilling Report), pp. 235–6.
80. Baverstock, *Treatises*, pp. 154–60. Emphasis original.
81. Ham's diverse career, which later extended to chemical manufacturing, may be glimpsed in autobiographical memoirs left by his sister: E. Ham, *Elizabeth Ham by Herself, 1783–1820* (London: Faber, 1945). The account of mucilage in the 'I. H.' letter recurs in J. Ham, *The Theory and Practice of Brewing, from Malted and Unmalted Corn, and from Potatos* [*sic*] (London: Simpkin and Marshall, 1829), pp. 58–9.
82. Baverstock, *Treatises*, pp. 160–2.
83. *Annals of Philosophy*, 4 (1814), pp. 236–7; *Annals of Philosophy*, 5 (1815), pp. 156–7.
84. Baverstock, *Treatises*, pp. 112–3, notes; compare 209, n.
85. [Booth], *Art of Brewing*, pp. 4–5.
86. [Thomson], 'Brewing', p. 467.
87. G. A. Wigney, *A Philosophical Treatise on Malting and Brewing* (Worthing: Worthing Press, 1823), p. 27.
88. A useful near-contemporary survey of this literature is W. Procter, 'Observations on Dextrine and Diastase', *American Journal of Pharmacy*, 5:4 (1840), pp. 265–79.
89. J.-B. Biot and J.-F. Persoz, 'Mémoire sur les modifications que la fécule et la gomme', *Nouvelles annales du museum d'histoire naturelle*, 2 (1833), pp. 109–25.
90. A. Payen and J.-F. Persoz, 'Mémoire sur la diastase, les principaux produits de ses réactions et leurs applications aux arts industriels', *Annales de chimie et de physique*, 53 (1833), pp. 73–92.
91. W. Irvine, 'On Fermentation' (1785), in W. Irvine, *Essays, Chiefly on Chemical Subjects* (London: J. Mawman, 1805), pp. 307–26, on pp. 316–19.
92. Combrune, *Essay*, p. 159.
93. [Booth], *Art of Brewing*, p. 35; Hayman, *Practical Treatise*, pp. 85–7.
94. [Thomson], 'Brewing', p. 464; J. Ham, *Instructions for Brewing by a New Method, from Unmalted Corn*, 4th edn (London: Baldwin, Cradock and Joy, 1822) p. 4, emphasis original.
95. Ham, *Instructions for Brewing*, pp. 27, 35. No copy has been traced of the first edition of this pamphlet; the three following editions are all dated 1822.
96. Ham, *Elizabeth Ham*, p. 212; Ham, *Theory and Practice*, pp. 41–55, 73–94. For Ham's role in the reform movement in Bristol, see G. Amey, *City under Fire: The Bristol Riots and Aftermath* (Guildford: Lutterworth Press, 1979), p. 110.
97. Payen and Persoz, 'Mémoire sur la diastase', p. 73, translation mine.
98. A. Payen and J.-F. Persoz, 'Mémoire sur l'amidone (substance intérieure de la fécule) et suite de recherches sur la diastase', *Annales de chimie et de physique*, 56 (1834), pp. 337–71, on p. 361.
99. The only prior publication on diastase I have been able to find in English is a translation, without commentary, of a brief report from the *Annales des sciences naturelles*: H. Dutro-

chet, 'On Diastase and its Importance in Vegetable Physiology', *Magazine of Botany and Gardening*, 2 (1834), pp. 91–3.

100. Black, *Practical Treatise*, 1st edn, pp. 143–6.
101. T. Thomson, *Chemistry of Organic Bodies: Vegetables* (London: J. B. Baillière, 1838), pp. 666–8, 1011–8.
102. A. Ure, *A Dictionary of Arts, Manufactures, and Mines*, 1st edn (London: Longman, 1839), pp. 97–8.
103. T. Graham, *Elements of Chemistry* (London: Hippolyte Baillière, 1842), p. 745.
104. T. Thomson and W. Stewart, *Brewing and Distillation, with Practical Instructions* (Edinburgh: A. and C. Black, 1849), preface and p. 198; and for the marginal role of mucilage in this scheme, see p. 204.
105. *Encyclopaedia Britannica*, 8th edn (1854), vol. 5, pp. 319–20.

7 Treatises for the Trade

1. W. Black, *A Practical Treatise on Brewing, Based on Chemical and Economical Principles*, 2nd edn (London: Smith and Elder, 1840), p. 127.
2. See, for instance, the soda manufacturer James Muspratt's judgment: R. Bud and G. Roberts, *Science Versus Practice: Chemistry in Victorian Britain* (Manchester: Manchester University Press, 1984), p. 60.
3. D. S. L. Cardwell, *James Joule: a Biography* (Manchester: Manchester University Press, 1991); Sibum, 'Reworking the Mechanical Value of Heat'.
4. Ham, *Instructions for Brewing*, pp. 6–7; Ham, *Theory and Practice*.
5. Worthington, *Brewer's Guide*, title page.
6. J. Topham, 'Science and Popular Education in the 1830s: The Role of the Bridgewater Treatises', *British Journal for the History of Science*, 25:4 (1992), pp. 397–430; R. Kinraide, 'The Society for the Diffusion of Useful Knowledge and the Democratization of Learning in Early Nineteenth-Century Britain' (PhD dissertation, University of Wisconsin–Madison, 2006). Secord, *Victorian Sensation*, pp. 48–51 presents the Society's work in the context of the several contemporary cheap print series on natural knowledge.
7. 'Society for the Diffusion of Useful Knowledge' (prospectus, 1827), bound in with [H. Brougham], *A Discourse of the Objects, Advantages, and Pleasures of Science* (London: Baldwin and Cradock, 1827).
8. [H. Brougham], 'Library of Useful Knowledge', *Edinburgh Review*, 50 (1829), pp. 181–93, on pp. 183–4. Brougham's authorship is confirmed by W. Houghton, et al. (eds), *The Wellesley Index to Victorian Periodicals, 1824–1900*, 5 vols (Toronto: University of Toronto Press, 1966): vol. 1, p. 471.
9. *ODNB*; *London Gazette*, 7 November 1826, p. 2635; John Gilchrist to Booth, 16 January 1827, University College London Special Collections held at the National Archives, Kew: MS.Add 193.
10. [Booth], *Art of Brewing*, pp. 1–2, 9.
11. Ibid., p. 53.
12. Booth to Thomson, 25 June; Thomson to Booth, 6 July; compare Booth to Thomas Coates, 4 July 1829, in David Booth letters, Society for the Diffusion of Useful Knowledge correspondence (Box 15), University College London Special Collections held at the National Archives (hereafter Booth Letters), emphasis original.
13. Booth Letters: Booth to Coates, undated 'Friday eveng' [1830], emphasis original.
14. [Booth], *Art of Brewing*, back matter.

15. Booth Letters: Booth–Coates correspondence.
16. Booth Letters: Anthony Carlisle to Booth, 5 December 1838.
17. J. N. Hays, 'The Rise and Fall of Dionysius Lardner', *Annals of Science*, 38:5 (1981), pp. 527–42, on pp. 535–8.
18. M. Donovan, *Domestic Economy, Volume 1: Brewing, Distilling, Wine-making, Baking, &c.* (London: Longman, 1830), p. [v].
19. *Fraser's Magazine*, 1 (1830), pp. 209–15.
20. Donovan, *Domestic Economy*, p. vi.
21. Ibid., pp. 148–9.
22. Ibid., pp. 189–208.
23. Ibid., p. 211.
24. Ibid., p. 215.
25. D. Booth, *The Art of Brewing*, new edn (London: F. J. Mason, 1834), preface to the third part, p. iv.
26. Black, *Practical Treatise*, 1st edn, pp. 12, 40, 95, 106, 121.
27. Ibid., p. 18.
28. Ibid., pp., [iii]–vi, [1]–2; compare Tuck, *Private Brewer's Guide* (see above, p. 130); Ham, *Instructions for Brewing*, p. 46.
29. Black, *Practical Treatise*, 1st edn, pp. 74–8, emphasis original.
30. Ibid., pp. 135–6, 4–5.
31. Black, *Practical Treatise*, 2nd edn, p. 6.
32. Black, *Practical Treatise*, 1st edn, p. 95. An expanded version appears in Black, *Practical Treatise*, 2nd edn, pp. 126–8.
33. Black, *Practical Treatise*, 1st edn, pp. 11–4; compare, for instance, H. Davy, *A Syllabus of a Course of Lectures on Chemistry* (London: Royal Institution, 1802), p. 62.
34. Black, *Practical Treatise*, 1st edn, p. 30.
35. Graham's notebook for 1837–1838, Wellcome: MS 2554, contains a note of Black's home and business addresses.
36. For the Liverpool meeting in the context of institutional shifts in British chemistry, see R. Bud, 'The Discipline of Chemistry: The Origins and Early Years of the Chemical Society of London' (PhD dissertation, University of Pennsylvania, 1980), pp. 112–7; J. Morrell and A. Thackray, *Gentlemen of Science: Early Years of the British Association for the Advancement of Science* (Oxford: Clarendon, 1981), p. 489.
37. W. Brock, *Justus von Liebig: The Chemical Gatekeeper* (Cambridge: Cambridge University Press, 1997), p. 97; G. K. Judel, *Justus Liebig in Grossbritannien: Justus Liebigs Briefe aus Grossbritannien an seine Frau Henriette* (Giessen: Justus Liebig-Gesellschaft, 2003), p. 39.
38. *Report of the Seventh Meeting of the British Association for the Advancement of Science* (London: John Murray, 1838), pp. 38–41.
39. *Literary Gazette*, 16 September 1837, pp. 588–9.
40. Black, *Practical Treatise*, 2nd edn, p. iv.
41. W. Black, *A Practical Treatise on Brewing, Based on Chemical and Economical Principles*, 3rd edn (London: Longman, 1844), p. [v].
42. Black, *Practical Treatise*, 2nd edn, pp. 127, 160–1.
43. Ibid., p. 157.
44. Black, *Practical Treatise*, 3rd edn, pp. 83–4.
45. For Wigney as herald of a national change in brewing culture, see Sibum, 'Reworking the Mechanical Value of Heat', pp. 83–91.

46. Wigney, *Theoretical and Practical Treatise*, pp. 252–61.
47. Wigney, *Elementary Dictionary*, pp. 68–9.
48. Wigney, *Philosophical Treatise*, p. 91.
49. Ibid., p. xi. For Sir Richard Phillips's guerrilla authorship tactics, see M. Kinnell, 'Sceptreless, Free, Uncircumscribed? Radicalism, Dissent and Early Children's Books', *British Journal of Educational Studies*, 36:1 (1988), pp. 49–71.
50. Wigney, *Philosophical Treatise*, pp. viii–ix, 114–5.
51. Wigney, *Theoretical and Practical Treatise*, pp. iv–v; Wigney, *Elementary Dictionary*, pp. ii–v.
52. Wigney, *Philosophical Treatise*, pp. 98–100.
53. Wigney, *Theoretical and Practical Treatise*, p. 143.
54. *Monthly Magazine*, 37 (1814), p. 577; Wigney, *Theoretical and Practical Treatise*, p. iv; Wigney, *Elementary Dictionary*, p. ii; *London Gazette*, 21 October 1817, p. 2159; *London Gazette*, 13 January 1818, p. 112. For the Wigney family, see C. Fleet, *Glimpses of our Ancestors in Sussex*, second series (Lewes: Farncombe, 1883), pp. 47–53; 'George Wigney', Barcombe and Hamsey website, n. d., at http://www.bandhpast.co.uk/barcombe/b1200other.php [accessed 13 January 2013].
55. G. A. Wigney, *An Introductory Treatise on the Theory and Practice of Malting and Brewing* (Brighton: Frederick Wigney, 1850), p. i; *London Gazette*, 3 December, 1830.
56. Wigney to Cobb and Company of Margate, 3 February 1835; 5 March 1836; 25 February 1837, Cobb of Margate family and business papers, Kent History and Library Centre, Maidstone (hereafter Cobb papers): EK-U1453/B2/40/588.
57. Wigney, *Theoretical and Practical Treatise*, pp. 167, 190.
58. Ibid., pp. [i]–vii.
59. Wigney to Cobb, 5 March 1836, Cobb papers: EK-U1453/B2/40/588.
60. Wigney, *Introductory Treatise*, pp. 9–10.
61. Wigney, *Elementary Dictionary*, p. vi.
62. Wigney, *Introductory Treatise*, pp. 11–12.
63. Wigney, *Philosophical Treatise*, p. ix.
64. Wigney, *Introductory Treatise*, pp. 57–62.
65. Ibid., pp. 191–6.
66. Ibid., *Introductory Treatise*, p. 65.
67. Patent 8921, 5 October 1841.
68. The price is confirmed by Tizard, *Theory and Practice*, 2nd edn, 'Testimonials' (end matter), p. 7.
69. W. L. Tizard, *The Theory and Practice of Brewing Illustrated*, 1st edn (London: sold by the author, 1843), pp. 151, 348–50, 405–7.
70. Tizard's known publications are listed in the bibliography. See also Patent 14119, 8 May 1852; Patent 1647, 26 July 1854; Patent 227, 26 January 1857.
71. Tizard, *Theory and Practice*, 2nd edn, p. 106.
72. Tizard, *Theory and Practice*, 1st edn, pp. 171–95.
73. Tizard, *Theory and Practice*, 2nd edn, pp. 154–5.
74. Tizard, *Theory and Practice*, 1st edn, p. 3.
75. Tizard, *Theory and Practice*, 2nd edn, pp. 205–9; compare T. Hitchcock, *A Practical Treatise on Brewing, Showing the Cause and Prevention of Acidity in Malt Liquors* (London: printed by R. Boyd, 1842), pp. 17, 26.
76. Tizard, *Theory and Practice*, 1st edn, pp. 4, 119; Tizard, *Theory and Practice*, 2nd edn, p. 151.

77. Tizard, *Theory and Practice*, 1st edn, pp. 110, 169; Tizard, *Theory and Practice*, 2nd edn, pp. 149, 478.
78. Tizard, *Theory and Practice*, 1st edn, pp. 94–5, 101–2.
79. For Ure's status in this period as a private consultant on chemical and industrial questions – in some ways providing a pattern for specialists such as Tizard himself – see Farrar, 'Andrew Ure', p. 314.
80. A. Ure, *Recent Improvements in Arts, Manufactures, and Mines: Being the Second Edition of a Supplement to the Third Edition of his Dictionary* (London: Longman, 1845), pp. x, 13–24, italics original. For earlier borrowings from Black see Ure, *Dictionary of Arts*, 1st edn, pp. 118, 460–1.
81. For the character of the *Mechanics'*, see B. Marsden, 'Carriages, Coffee-Cups and Dynamometers: Representing French Technical Cultures in the London Mechanics' Magazine, 1823–1848', in P. Bret, I. Gouzévitch and L. Pérez (eds), *Les techniques et la technologie entre la France et la Grande-Bretagne (XVIIe-XIXe siècles)* (Paris: Centre de documentation d'histoire des techniques, 2011), pp. 243–54.
82. *Mechanics' Magazine*, 42 (1845), pp. 70–2.
83. A. Ure, *A Dictionary of Arts, Manufactures, and Mines*, 3rd edn (London: Longman, 1843), p. 100; Tizard, *Theory and Practice*, 2nd edn, pp. 152, 202–4, 505–8.
84. Patent 3651, 20 February 1813 (John Roberts); Patent 5091, 10 February 1825 (George Augustus Lamb), and see also *Register of the Arts and Sciences*, 16 September 1826, p. 179.
85. Patent 10593, 7 April 1845 (Thomas Jarvis); Patent 10952, 18 November 1845 (Dircks); [*Newton's*] *London Journal of Arts*, 29 (1846), pp. 249–51. The original scheme under Jarvis's name had been announced, apparently in expectation of a patent, several years earlier: *Morning Chronicle*, 13 August 1840, p. [1]. For the retail scheme, see, for instance, *Mechanics' Magazine*, 45 (1846), p. 47; *Derby Mercury*, 16 September 1846, p. [2].
86. Tizard, *Theory and Practice*, 2nd edn, p. 274; contrast *Economist*, 14 June 1845, p. 572.
87. *Mechanics' Magazine*, 43 (1845), pp. 437–40; *Reynolds's Miscellany*, 1 (1846), p. 13.
88. Tizard, *Theory and Practice*, 3rd edn, part 3, p. 6 includes a testimonial from Crowson.
89. *Mechanics' Magazine*, 45 (1846), pp. 389–91.
90. Ibid., 45 (1846), pp. 442–3.
91. *Mechanics' Magazine*, 45 (1846), pp. 494–5. The spat rumbled on, inconclusively, with another missive on each side: *Mechanics' Magazine*, 45 (1846), pp. 537–40, 575; *Mechanics' Magazine*, 46 (1847), p. 38.
92. Tizard, *Theory and Practice*, 3rd edn, pp. 455–60.
93. Tizard, *Theory and Practice*, 2nd edn, pp. 246–8, 491–3, 501.
94. W. L. Tizard, *The Theory and Practice of Brewing Illustrated*, 4th edn (London: sold by the author, 1857), back matter.
95. Tizard, *Theory and Practice*, 1st edn, pp. [vii]–viii.
96. Tizard, *Theory and Practice*, 1st edn, pp. 320–5; see also Tizard, *Theory and Practice*, 4th edn, p. 371.
97. My general account of the Chemical Society is drawn from Bud, 'Discipline of Chemistry', supplemented by Bud and Roberts, *Science Versus Practice*.
98. Bud, 'Discipline of Chemistry', p. 25.
99. *Pharmaceutical Journal*, 9 (1867–8), pp. 493–4; Bud, 'Discipline of Chemistry', pp. 134–5. Warington was in fact hired in November 1830, according to Truman's records (cited below, note 102).

100. Bud, 'Discipline of Chemistry', p. 138, and see also R. Bud, *The Uses of Life: A History of Biotechnology* (Cambridge: Cambridge University Press, 1993), pp. 17–8; Anderson, 'Pattern of Brewing Research', p. 86.
101. R. Warington, 'Observations on the Action of Animal Charcoal', *Memoirs and Proceedings of the Chemical Society of London* 2 (1845), pp. 326–9, on p. 327.
102. Truman's clerks' salary book, 1819–1845, LMA: B/THB/F/1, fols 65 (Warington) and 95 (Henry); clerks' rest book, 1840–1854, LMA: B/THB/B/115 (Henry's 'Experiments').
103. Quoted in Bud and Roberts, *Science Versus Practice*, p. 48.
104. 'Anniversary Meeting', *Journal of the Chemical Society, Transactions*, 43 (1883), pp. 224–66, on p. 255.
105. S. Piesse, 'Some Observations on Brewing', *Proceedings of the Chemical Society of London* [vol. 2] (1841–2), pp. 23–4.
106. B. Stevenson, 'John Nokes Furze, 1817–1859', *Micscape*, 181 (November 2010), at http://www.microscopy-uk.org.uk/mag//artnov10/bs-FurzeJN.html [accessed 13 January 2013]. Furze's partnership is confirmed by a note of creditorship in the *Law Journal*'s list of 'Bankrupts, Certificates and Dividends' for 1841, p. 15. For the background to the firm, see Ure, *Dictionary of Arts*, 1st edn, p. 97.
107. [D. P. Deurbroucq, and (?) Nichols], *Observations on the Vinous Fermentation; with a Description of a Patent Apparatus to Improve the Same* (London, 1823); Ham, *Theory and Practice*, pp. 21–3; [Booth], *Art of Brewing*, p. 39; Black, *Practical Treatise*, 3rd edn, pp. 97–8.
108. J. Furze, 'Observations on Fermentation', *Memoirs and Proceedings of the Chemical Society of London*, 2 (1845), pp. 21–5.
109. Bud and Roberts, *Science Versus Practice*, pp. 48–51, 65, 108; Bud, 'Discipline of Chemistry', pp. 332–5.
110. *Journal of the Chemical Society*, 21 (1868), pp. xxxi–xxxiv.
111. Warington, 'Observations'.

8 Analysis and Synthesis

1. Faulkner, *The Theory and Practice of Modern Brewing*, 2nd edn, p. xi.
2. G. Roberts, 'The Establishment of the Royal College of Chemistry: An Investigation of the Social Context of Early-Victorian Chemistry', *Historical Studies in the Physical Sciences*, 7 (1976), pp. 437–85; *ODNB*.
3. C. Russell, *Edward Frankland: Chemistry, Controversy and Conspiracy in Victorian England* (Cambridge: Cambridge University Press, 2003), pp. 166–7.
4. Anderson, 'Pattern of Brewing Research', pp. 89–94 surveys this generation. See also A. von Hofmann, E. Fischer and H. Caro, 'Zur Erinnerung an Peter Griess', *Berichte der Deutschen chemischen Gesellschaft*, 24 (1891), pp. 1006–78, on pp. 1040–50; E. R. Ward, 'Peter Griess (1829–1888) and the Burton Breweries', *Journal of the Royal Institute of Chemistry*, 82 (1958), pp. 458–63; H. D. O'Sullivan, *The Life and Work of Cornelius O'Sullivan, F.R.S.* (Guernsey: Guernsey Star and Gazette, [1934]); H. T. Brown, 'Reminiscences of Fifty Years' Experience of the Application of Scientific Methods to Brewing', *Journal of the Institute of Brewing*, 22 (1916), pp. 267–354, on pp. 268–70; H. E. A[rmstrong], 'Adrian Brown', *Journal of the Chemical Society, Transactions*, 121 (1922), pp. 2899–907.
5. Ward, 'Peter Griess', pp. 458–60.

6. H. Gay, *The History of Imperial College London, 1907–2007* (London: Imperial College Press, 2007), p. 19.
7. For the Alliance, see B. Harrison, *Drink and the Victorians: The Temperance Question in England, 1815–1872* (Pittsburgh, PA: University of Pittsburgh Press, 1971); and for the later emergence of 'Trade Defence', D. W. Gutzke, *Protecting the Pub: Brewers and Publicans against Temperance* (Woodbridge: Royal Historical Society, 1989).
8. *Deadly Adulteration*; *Lancet*, 16 (1831), pp. 485–7.
9. For Hassall and the Commission see C. Hamlin, *A Science of Impurity: Water Analysis in Nineteenth-Century Britain* (Bristol: Adam Hilger, 1990); S. D. Smith, 'Coffee, Microscopy, and the *Lancet*'s Analytical Sanitary Commission', *Social History of Medicine*, 14:2 (2001), pp. 171–97; B. Charnley, 'Arguing over Adulteration: The Success of the Analytical Sanitary Commission', *Endeavour*, 32 (2008), pp. 129–33.
10. Brown, 'Reminiscences', pp. 270–1; compare Anderson, 'Pattern of Brewing Research', pp. 94, 106, n. 62.
11. *Medical Times and Gazette*, 25 (1852), p. 294.
12. *Bristol Temperance Herald*, 15 (1851), p. 88.
13. *Medical Times and Gazette*, 25 (1852), p. 346.
14. Ibid., p. 402; J. S. Bushnan, *Burton and its Bitter Beer* (London: William S. Orr, 1853), p. [iii]. Bushnan resigned his editorship part-way through the controversy. His 1853 volume, mysteriously, contains an almost fawning vindication of Allsopp: it is hard to escape the suspicion that pressure had been applied.
15. *Medical Times and Gazette*, 25 (1852), pp. 376–7.
16. Hofmann, Fischer and Caro, 'Zur Erinnerung an Peter Griess', pp. 1040–2.
17. For the water investigations, see Hamlin, *Science of Impurity*, p. 104.
18. Statutes: 10 Vic. (1847), c. 5.
19. P. Hammond and H. Egan, *Weighed in the Balance: A History of the Laboratory of the Government Chemist* (London: HMSO, 1992), pp. 49–59 places this episode in the context of the development of the Laboratory of the Government Chemist. See also G. Fownes, 'On the Value in Absolute Alcohol of Spirits of Different Specific Gravities', *Philosophical Transactions*, 137 (1847), pp. 249–51; report of Thomas Dobson and George Phillips, 2 August 1847, TNA: DSIR 26/979; W. H. Johnston, *Loftus's Inland Revenue Officer's Manual* (London: Loftus, 1865), pp. 304–15.
20. For the text of the report, see for instance *Quarterly Journal of Chemistry*, 5 (1853), pp. 229–56, in which the acknowledgements to brewers appear on p. 253.
21. Ward, 'Peter Griess', p. 387; entries for 10 and 12 April 1852 in Graham's notebook, Wellcome: MS 2567.
22. T. Graham and A. Hofmann, 'Report upon the Alleged Adulteration of Pale Ales by Strychnine', 26 April 1852; published in, for instance, *Edinburgh Medical and Surgical Journal*, 78 (1852), pp. 246–9.
23. *Lancet*, 59 (1852), pp. 473–9.
24. Brock, *Justus von Liebig*, p. 140.
25. For instance: *Lancet*, 59 (1852), pp. 551–2.
26. *Medical Times and Gazette*, 25 (1852), pp. 469–71.
27. *Monthly Journal of Medical Science*, 15 (1852), pp. 414–6.
28. For instance: *Quarterly Review*, 181 (1852), number for June, end matter. For the advertisement's appearance in the 'Bleak House Advertiser' see D. Hack, *The Material Interests of the Victorian Novel* (Charlottesville, VA: University of Virginia Press, 2005), pp. 57–61.

29. Roberts, 'Establishment of the Royal College of Chemistry', p. 464.
30. *Monthly Journal of Medical Science*, 15 (1852), pp. 138–42; *Punch*, 4 November 1854, p. 183. The *Punch* piece further suggested that 'Mr. Justus Liebig ought to be a jolly good *judge* of beer'.
31. *Medical Times and Gazette*, 26 (1852), pp. 439–40.
32. *The Times*, 28 December 1852, 7C. For humuline see Tizard, *Theory and Practice*, 2nd edn, pp. 268–74.
33. Hofmann, Fischer and Caro, 'Zur Erinnerung an Peter Griess', p. 1042 ('*Die Strychnin-Episode hatte der Chemie in Burton on Trent zu einem Ansehen verholfen.*')
34. S. Muspratt (ed.), *Chemistry, Theoretical, Practical and Analytical, as Applied and Relating to the Arts and Manufactures*, 2 vols (Glasgow: W. Mackenzie, 1860).
35. Brown, 'Reminiscences', p. 269; Anderson, 'Pattern of Brewing Research', p. 89. For Muspratt see M. Stephens and G. Roderick, 'The Muspratts of Liverpool', *Annals of Science*, 29:3 (1972), pp. 287–311.
36. Muspratt, *Chemistry*, vol. 1, p. 248; compare Donovan, *Domestic Economy*, p. 211.
37. For the contemporary patent abolition movement, see C. MacLeod, 'Concepts of Invention and the Patent Controversy in Victorian Britain', in R. Fox (ed.) *Technological Change: Methods and Themes in the History of Technology* (Amsterdam: Harwood Academic, 1996), pp. 137–53.
38. Muspratt, *Chemistry*, vol. 1, p. 264.
39. *Brewers' Journal*, 61 (1920), p. 387.
40. For Pontifex, see *Brewers' Journal*, 2 (1866), pp. 99–100.
41. *Brewers' Journal*, 8 (1872), pp. 259–60.
42. For the *Grocer*'s stance, see I. Paulus, *The Search for Pure Food: A Sociology of Legislation in Britain* (London: Martin Robertson, 1974), p. 92.
43. *Brewers' Journal and Hop and Malt Trades Review 1865–1965: 100 Years* (London: William Reed, 1965), pp. 93, 112.
44. *Brewers' Journal*, 2 (1866), p. 117.
45. *Brewers' Journal*, 3 (1867), p. 114
46. *Brewers' Journal*, 8 (1872), pp. 96–7.
47. L. Richmond and A. Turton, *The Brewing Industry: A Guide to Historical Records* (Manchester: Manchester University Press, 1990), p. 389.
48. *Brewers' Journal*, 3 (1867), p. 1; for Campbell, see *Journal of the Chemical Society, Transactions*, 43 (1883), pp. 252–3.
49. For the public analysts see R. G. Anderson, *Brewers and Distillers by Profession: A History of the Institute of Brewing and Distilling* (London: IBD, 2012), pp. 25–6.
50. A. Williamson, *A Plea for Pure Science, Being the Inaugural Lecture at the Opening of the Faculty of Science* (London: Taylor and Francis, 1870), p. 25; see also Bud and Roberts, *Science Versus Practice*, pp. 85–6, 146–7.
51. A. C. Chapman, 'Charles Graham' [obituary], *Journal of the Chemical Society, Transactions* 97 (1910), pp. 677–80; Bud and Roberts, *Science Versus Practice*, pp. 156–7.
52. For the Cantor series, see Hudson and Luckhurst, *Royal Society of Arts*, pp. 293–4.
53. 'Practice with science' was also the motto of the Royal Agricultural Society, founded in 1838: the allusion was presumably deliberate, given Graham's background. Agricultural engagement with science offered a number of useful precedents, for which see Bud and Roberts, *Science Versus Practice*, pp. 55–9.
54. *Journal of the Society of Arts*, 22 (1874), pp. 188–90. For British brewers' absence from the global trade, see Anderson, *Brewers and Distillers*, pp. 7–8.

55. *Journal of the Society of Arts*, 22 (1874), pp. 221–4.
56. Ibid., p. 338.
57. Anderson, *Brewers and Distillers*, pp. 18–20.
58. L. Pasteur, *Studies on Fermentation: The Diseases of Beer, their Causes, and the Means of Preventing them* (originally *Études sur la Bière*, 1876), trans. F. Faulkner and D. C. Robb (London: Macmillan, 1879), pp. 22–4.
59. *Journal of the Society of Arts*, 22 (1874), pp. 298–302.
60. Ibid., p. 338.
61. C. O'Sullivan, 'On the Transformation-Products of Starch', *Journal of the Chemical Society*, 25 (1872), pp. 579–88, on p. 580.
62. A.-P. Dubrunfaut, 'Note sur le glucose', *Annales de chimie et de physique*, 21 (1847), pp. 178–80, refers to the substance as *glucose de malt*. O'Sullivan was perhaps misled by later sources in believing that Dubrunfaut had coined 'maltose': the *OED* traces its first use in English to an 1862 translation of Leopold Gmelin's work. It was not widely used in English until O'Sullivan's publication.
63. C. O'Sullivan, 'On Maltose', *Journal of the Chemical Society*, 29 (1876), pp. 478–82; C. O'Sullivan, 'On the Action of Malt-Extract on Starch', *Journal of the Chemical Society*, 30 (1876), pp. 125–44. For Booth's account of temperature-dependence, see p. 152.
64. Patent 2225, 24 August 1871; R. G. Anderson, 'Louis Pasteur (1822–1895): An Assessment of his Impact on the Brewing Industry', paper presented at the European Brewery Convention, Brussels, 1995.
65. F. Faulkner, *The Art of Brewing Practical and Theoretical*, 2nd edn (London: F. W. Lyon, 1878), pp. 13–14.
66. Patent 3909, 12 November 1874.
67. *Journal of the Chemical Society, Transactions*, 37 (1880), p. 260.
68. *Brewers' Journal*, 52 (1916), p. 161.
69. W. G. Valentin, 'On the Preparation of Dextrine-Maltose (Malt-Sugar) and its Use in Brewing', *Journal of the Society of Arts*, 24 (1875), pp. 404–14, on pp. 404–8.
70. R. Warington [Jr], 'Notes on the Chemistry of Tartaric and Citric Acid', *Journal of the Chemical Society*, 28 (1875), pp. 925–94, on p. 994.
71. *Journal of the Society of Arts*, 24 (1876), pp. 484, 501, 516–17, 573.
72. *Brewers' Journal*, 41 (1925), p. 484; Patent 1612, 3 May 1873.
73. *Journal of the Chemical Society, Transactions*, 51 (1887), pp. 476–7; *Brewers' Journal*, 22 (1886), pp. 130–1.
74. A detailed bibliography of the first decade of Faulkner's contributions appears in H. Stopes, *Malt and Malting* (London: F. W. Lyon, 1885), pp. 441–3.
75. *Brewers' Journal*, 11 (1875), pp. 210–11; *Brewers' Journal*, 12 (1876), pp. 103–6.
76. *Brewers' Guardian*, 6 (1876), pp. 1–2.
77. Faulkner, *Art of Brewing*, preface.
78. E. R. Southby, *Brewing Practically and Scientifically Considered* (London: Unwin, 1877).
79. Faulkner, *Art of Brewing*, preface, pp. 1, 150.
80. Ibid., pp. 2, 128, 198.
81. Southby, *Brewing*, p. 140.
82. *Brewers' Journal*, 2 (1866), pp. 114–15; *Brewers' Journal*, 8 (1872), pp. 98–100, 261, 265–6.
83. *Brewers' Journal*, 12 (1876), pp. 187–90.
84. Southby, *Brewing*, p. 138; *Brewers' Guardian*, 7 (1877), p. 33.
85. *Brewers' Journal*, 12 (1876), p. 238; Pasteur, *Studies on Fermentation*.

86. Faulkner, *Art of Brewing*, p. 152.
87. H. T. Brown, 'The Nitrogen Question in Brewing', *Journal of the Institute of Brewing*, 13 (1907), pp. 394–457, on p. 397; Faulkner, *Art of Brewing*, pp. 187–96.
88. Faulkner, *Theory and Practice*, 2nd edn, p. 106.
89. *Brewers' Journal*, 41 (1925), p. 484.
90. F. Faulkner, *The Theory and Practice of Modern Brewing*, 1st edn (London: F. W. Lyon, 1884), pp. 63–4; W. S. Squire, 'On the Processes Concerned in the Conversion of Starch into Alcohol, and their Relation to Brewing and Distilling', *Brewers' Guardian*, 14 (1884), pp. 285–6, 303–5, 316–19, on p. 285, n.; *Brewers' Journal*, 20 (1884), p. 300. For Squire, see *Journal of the Society of Chemical Industry*, 25 (1906), p. 523; for Veley, J. A. Gardner, 'Victor Herbert Veley, 1856–1933', *Obituary Notices of Fellows of the Royal Society*, 1 (1934), pp. 228–35.
91. *Brewers' Journal*, 20 (1884), pp. 440–1; *Brewers' Journal*, 21 (1885), pp. 73–9.
92. *Brewers' Journal*, 21 (1885), pp. 80–1, 112; *Brewers' Guardian*, 15 (1885), pp. 92–3, 107, 123.
93. Brown, 'Reminiscences', pp. 274–5.
94. *Brewers' Journal*, 12 (1876), p. 190.
95. Faulkner, *Art of Brewing*, p. 161.
96. *Brewers' Journal*, 12 (1876), pp. 236–7.
97. Ibid., pp. 280–1.
98. Muspratt, *Chemistry*, vol. 1, p. 281.
99. *Journal of the Society of Arts*, 22 (1874), p. 338.
100. *Brewers' Journal*, 11 (1875), pp. 211, 228.
101. Statutes: 25 Vic., c. 22, ss. 20, 22; 37 and 38 Vic., c. 16, ss. 14–17.
102. *Good and Cheap Beer for the Million, by the Use of Sugar and Molasses in Public Breweries* (London: Pelham Richardson, 1846).
103. Faulkner, *Art of Brewing*, pp. 6–15, 27.
104. *Brewers' Journal*, 2 (1866), pp. 137–8.
105. Faulkner, *Art of Brewing*, p. 129.
106. Ham, *Theory and Practice*, pp. 11–12.
107. Southby, *Brewing*, p. [1].
108. *Journal of the Society of Arts*, 22 (1874), pp. 371–2.
109. For the origins and operation of the Act, see C. Clark, *The British Malting Industry Since 1830* (London: Hambledon Press, 1998), pp. 65–88.
110. Hansard: HC Deb., vol. 252, col. 1646 (10 June 1880).
111. Statutes: 43 and 44 Vic., c. 20.
112. *The Times*, 10 September 1880.
113. *Brewers' Journal*, 16 (1880), pp. 329–30.
114. For which see D. W. Gutzke, *Pubs and Progressives: Reinventing the Public House in England, 1896–1960* (DeKalb, IL.: Northern Illinois University Press, 2005).
115. *Brewers' Journal*, 16 (1880), pp. 327–9.
116. Ibid., 16 (1880), pp. 399–401.
117. For teetotalism and the Liberals after 1872, see A. E. Dingle, *The Campaign for Prohibition in Victorian England: the United Kingdom Alliance, 1872–1895* (London: Croom Helm, 1980).
118. Faulkner, *Theory and Practice*, 2nd edn, p. xi.
119. Anderson, *Brewers and Distillers*, pp. 33–58.

Conclusion

1. The earliest instance yet identified of 'bung' used as a slang term for 'brewer' (from the bung of a cask). It was commonly applied to both brewers and publicans in the later nineteenth century.
2. Tizard, *Theory and Practice*, 1st edn, p. [v]. Emphasis original.
3. *Punch* 26 (1854), p. 4. The Paterfamilias story is discussed in R. Noakes, 'Representing "A Century of Inventions": Nineteenth-Century Technology and Victorian *Punch*', in L. Henson, et al. (eds), *Culture and Science in the Nineteenth-Century Media* (Aldershot: Ashgate, 2004), pp. 151–63.
4. *Punch*, 26 (1854), pp. 42–3, 58–9, 70–1.
5. A. Brown, *Laboratory Studies for Brewing Students* (London: Longman, 1904), pp. [v]–vi.
6. E. G. Hooper, *The Manual of Brewing, Scientific and Technical*, 1st edn (London: Sheppard and St John, 1882), p. [iii]. For Hooper, see Anderson, *Brewers and Distillers*, p. 26.
7. A. C. Chapman, *Brewing* (Cambridge: Cambridge University Press, 1912), p. [vii]. For Chapman, see *Biochemical Journal*, 26 (1932), pp. 1715–17, and for the Cambridge Manuals of Science and Literature series, see P. Bowler, *Science for All: the Popularization of Science in Early Twentieth-Century Britain* (Chicago, IL: University of Chicago Press, 2009), pp. 126–8.
8. Tyndall, 'Fermentation', p. 131.
9. Anderson, 'Pattern of Brewing Research', pp. 96–8.
10. Brown, 'Nitrogen Question', p. 448.
11. J. L. Baker, 'Horace Tabberer Brown', *Biochemical Journal*, 19 (1925), pp. 165–7.
12. Anderson, *Brewers and Distillers*, pp. 196–213.
13. Squire, 'Processes', p. 285.
14. *Brewers' Journal*, 20 (1884), pp. 380–2.
15. *Brewers' Journal*, 21 (1885), p. 35.
16. *Brewers' Journal*, 20 (1884), p. 327.

WORKS CITED

Archives

Best MSS: James Best collection, U480/B874 (not itemized), Medway Archives, Strood, Kent

Booth Letters: Letters of David Booth, Society for the Diffusion of Useful Knowledge correspondence (Box 15), University College London Special Collections (as below)

BL: Manuscripts and Archives, British Library

BM Satires: Satirical Prints Collection, British Museum. Largely accessible online via www.britishmuseum.org/research.aspx [accessed 13 January 2013]

Brakspear MSS: Brakspear of Henley collection, Oxfordshire History Centre, Oxford

Cobb Papers: Cobb of Margate family and business papers, EK-U1453, Kent History and Library Centre, Maidstone

Cotton MSS: Cotton family of Madingley and Landwade collection, 588/DR, Cambridgeshire Archives, Cambridge

LMA: London Metropolitan Archives, Clerkenwell, London

NAS: National Archives of Scotland, Register House, Edinburgh

RDS Minutes: Minute Books of the Proceedings of Council, Royal Dublin Society Archives

RS: Archives of the Royal Society, London

RSA: Archives of the Royal Society of Arts, London

Rylands: University of Manchester Special Collections, John Rylands Library, Manchester

UCL Spec Coll: University College London Special Collections (held at the National Archives)

TNA: The National Archives, Kew

Wellcome: Archives and Manuscripts, Wellcome Library, Euston Road, London

Abbreviations for Standard Works

DNB: L. Stephen, et al. (eds), *Dictionary of National Biography* (London: Smith, Elder, 1885–1900).

DSB: C. C. Gillispie, F. L. Holmes and N. Koertge (eds), *The Complete Dictionary of Scientific Biography*, 1970–2007. All entries taken from the online edition within the Gale Virtual Reference Library [accessed 13 January 2013].

Hansard: Official Report of Parliamentary Debates, sourced from the transcripts at hansard.millbanksystems.com/ [accessed 13 January 2013].

OED: Oxford English Dictionary, online edition at www.oed.com [accessed 13 January 2013].

ODNB: H. C. G. Matthew and B. Harrison (eds), *Oxford Dictionary of National Biography*, Oxford: Oxford University Press, 2004 onwards, online edition at www.oxforddnb.com [accessed 13 January 2013].

Statutes: Information on wording of Acts of Parliament is taken from Ruffhead's 'Statutes at Large' and its successor volumes, and the nineteenth-century 'Public General Statutes' series.

Short Titles Used for Parliamentary Reports and Records of Minutes

Page references given in the notes are from the page numbering of the reports themselves. Page ranges according to the distinct running pagination used for the nineteenth-century *Parliamentary Papers* series are also recorded below for ease of finding.

1799 Distilleries Report: 'Report Respecting the Distilleries in Scotland, &c. &c. Ordered to be Printed 12th July 1799'.

1804 Malt Report: 'Report from the Committee on the Re-Committed Report Respecting the Rate of Duty Payable on Malt Made from Barley of the Growth of England, and from Barley and Bigg of the Growth of Scotland', session 1803–4, paper 129; *Parliamentary Papers* 5, pp. 787–875.

1806 Hope Report: 'Papers ... Relating to Experiments Made by Order of the Commissioners of Excise for Scotland, to Ascertain the Relative Qualities of Malt Made from Barley and Scotch Bigg, &c.', session 1806, paper 202 (Includes Thomas Hope, Andrew Coventry and Thomas Thomson's 'Report of the Experiments Made'); *Parliamentary Papers* 2, pp. 425–543.

1806 Malt Petitions Report: 'Report from the Committee to whom the Several Petitions, Relative to the Making of Malt, Were Referred', session 1806, paper 224; *Parliamentary Papers* 2, pp. 335–424.

1808 Sugar Distilling Report: 'Report from the Committee on the Distillation of Sugar and Molasses', session 1808, paper 178; *Parliamentary Papers* 4, pp. 1–317.

1817 Police Minutes: 'First Report from the Committee on the State of the Police of the Metropolis, with Minutes', session 1817, paper 233; *Parliamentary Papers* 7, pp. 1–319.

1818 Beer Minutes: 'Minutes Taken (in session 1818) before the Committee to whom the Petition of Several Inhabitants of London and its Vicinity, Complaining of the High Price and Inferior Quality of Beer, was Referred', session 1819, paper 220; *Parliamentary Papers* 5, pp. 453–565.

1818 Breweries Report: 'Report from the Committee on Public Breweries', session 1818, paper 399; *Parliamentary Papers* 3, pp. 295–300.

1821 Malt Petitions Report: 'Report from the Select Committee on Petitions Complaining of the Additional Duty on Malt in Scotland', session 1821, paper 598; *Parliamentary Papers* 8, pp. 211–301.

1831 Molasses Report: 'Report from the Select Committee on the Use of Molasses in Breweries and Distilleries, with an Appendix and Index', session 1831, paper 297; *Parliamentary Papers* 7, pp. 109–353.

Published Sources

Accum, F. C., *A Treatise on Adulterations of Food, and Culinary Poisons*, 1st edn (London: Longman, 1820).

—, *A Treatise on Adulterations of Food, and Culinary Poisons*, 2nd edn (London: Longman, 1820).

—, *A Treatise on the Art of Brewing*, 1st edn (London: Longman, 1820).

Adams, G., *A Catalogue of Optical, Philosophical, and Mathematical Instruments* ([London], [*c.* 1765]).

The Alarm Bell, or, Hints to Travellers (London: T. Harvey, [1817]).

Aldabella, P. and R. Barnard, *Hull and East Riding Breweries: From the Eighteenth Century to the Present* ([Beverley]: East Yorkshire Local History Society, 1997).

Allan, D. G. C., 'The Society for the Encouragement of Arts, Manufactures and Commerce: Organisation, Membership and Objectives in the First Three Decades (1755–84)' (PhD dissertation, University of London (external), 1979).

Amey, G., *City Under Fire: The Bristol Riots and Aftermath* (Guildford: Lutterworth Press, 1979).

Anderson, J., *The Constitutions of the Antient and Honourable Fraternity of Free and Accepted Masons* (London, 1756).

Anderson, R. G., 'The Pattern of Brewing Research: A Personal View of the History of Brewing Chemistry in the British Isles', *Journal of the Institute of Brewing*, 98:2 (1992), pp. 85–109.

—, 'Louis Pasteur (1822–1895): An Assessment of his Impact on the Brewing Industry', Paper Presented at the European Brewery Convention, Brussels, 1995.

—, *Brewers and Distillers by Profession: A History of the Institute of Brewing and Distilling* (London: IBD, 2012).

An Answer to a Late Scurrilous Pamphlet, Published by one Baker and his Accomplices Respecting Dr. James's Powder (London, [*c.* 1754]).

Appleby, J. H., 'Humphrey Jackson, FRS, 1717–1801: A Pioneering Chemist', *Notes and Records of the Royal Society of London*, 40:2 (1986), pp. 147–68.

—, 'Erasmus King: Eighteenth-Century Experimental Philosopher', *Annals of Science*, 47:4 (1990), pp. 375–92.

The Application of Eight Thousand Pounds Granted by Parliament to the Dublin Society (Dublin, 1764).

A[rmstrong], H. E., 'Adrian Brown', *Journal of the Chemical Society, Transactions*, 121 (1922), pp. 2899–907.

Ashworth, W., *Customs and Excise: Trade, Production and Consumption in England 1640–1845* (Oxford: Oxford University Press, 2003).

—, 'The Intersection of Industry and the State in Eighteenth-Century Britain', in L. Roberts, S. Schaffer and P. Dear (eds), *The Mindful Hand: Inquiry and Invention from the Late Renaissance to Early Industrialisation* (Amsterdam: KNAvW, 2007), pp. 349–77.

Bacon, F., *The Philosophical Works of Francis Bacon*, ed., trans. and rev. P. Shaw, 3 vols (London, 1733).

Bailey, N., *Dictionarium Domesticum, Being a New and Compleat Houshold Dictionary* (London, 1736).

Baillon, J.-F., 'Early Eighteenth-Century Newtonianism: The Huguenot Contribution', *Studies in the History and Philosophy of Science*, 35:3 (2004), pp. 533–48.

Baker, J. L., 'Horace Tabberer Brown', *Biochemical Journal*, 19 (1925), pp. 165–7.

Barnard, H. C., *The British Isles in Pictures* (London: Black, 1910).

Baron, S., *Brewed in America: A History of Beer and Ale in the United States* (Boston, MA: Little, Brown, 1962).

[Barrow, J.], *A Supplement to the New and Universal Dictionary* (London, 1754).

Baverstock, J., *Hydrometrical Observations and Experiments in the Brewery* (London, 1785).

—, *Practical Observations on the Prejudices against the Brewery* (London, 1811).

—, *Treatises on Brewing*, ed. J. H. Baverstock (London: Whittaker, 1824).

[Beale, J.], *Herefordshire Orchards, a Pattern for all England* (London, 1657).

Beckett, J., 'Land Tax or Excise: The Levying of Taxation in Seventeenth- and Eighteenth-Century England', *English Historical Review*, 100:2 (1985), pp. 285–308.

Bennett, J. M., *Ale, Beer and Brewsters in England: Women's Work in a Changing World, 1300–1600* (New York: Oxford University Press, 1996).

Bensaude-Vincent, B., '"The Chemist's Balance for Fluids": Hydrometers and their Multiple Identities, 1770–1810', in F. Holmes and T. Levere (eds), *Instruments and Experimentation in the History of Chemistry* (London: MIT Press, 2000), pp. 153–83.

[Beresford, J.], *Observations on the Brewing Trade of Ireland, Submitted to the Publick* ([Dublin?], [*c.* 1774]).

Best, M. R., 'The Mystery of Vintners', *Agricultural History*, 50:2 (1976), pp. 363–76.

Biot, J.-B. and J.-F. Persoz, 'Mémoire sur les modifications que la fécule et la gomme', *Nouvelles annales du museum d'histoire naturelle*, 2 (1833), pp. 109–25.

Birch, T., *The History of the Royal Society of London for Improving of Natural Knowledge*, 4 vols (London, 1757).

Black, W., *A Practical Treatise on Brewing, and on Storing of Beer*, 1st edn (London: Smith, Elder, 1835).

—, *A Practical Treatise on Brewing, Based on Chemical and Economical Principles*, 2nd edn (London: Smith and Elder, 1840).

—, *A Practical Treatise on Brewing, Based on Chemical and Economical Principles*, 3rd edn (London: Longman, 1844).

Blake, G., *Strictures on a New Mode of Brewing, &c.* (London, [1791]).

—, *Theoretical and Practical Remarks on G. Blake's System of Malting and Brewing* (London, 1817).

Boerhaave, H., *A New Method of Chemistry*, trans. P. Shaw and E. Chambers (London, 1727).

—, *Elements of Chemistry: Being the Annual Lectures of Herman Boerhaave, M. D.*, trans. T. Dallowe, 2 vols (London, 1735).

Booth, D., *The Art of Brewing*, new edn (London: F. J. Mason, 1834).

[Booth, D.], *The Art of Brewing*, Society for the Diffusion of Useful Knowledge partwork (London: Baldwin and Cradock, 1829).

Bowler, P., *Science for All: The Popularization of Science in Early Twentieth-Century Britain* (Chicago, IL: University of Chicago Press, 2009).

Boyle, P., *The Publican and Spirit Dealer's Daily Companion* (London, [*c.* 1800]).

Boyle, R., *The Philosophical Works of the Honourable Robert Boyle*, ed. P. Shaw, 3 vols (London, 1725).

—, *The Works of the Honourable Robert Boyle*, ed. T. Birch, 6 vols (London, 1772).

[Boys, J.], *Directions for Brewing Malt Liquors* (London, 1700).

Brande, W., *The Town and Country Brewery Book* (London: Dean and Munday, [*c.* 1830]).

Brande, W. T., 'Experiments to Ascertain the State in which Spirit Exists in Fermented Liquors', *Philosophical Transactions*, 101 (1811), pp. 337–46.

—, 'Additional Remarks on the State in which Spirit Exists in Fermented Liquors', *Philosophical Transactions*, 103 (1813), pp. 82–7.

Brewers' Journal and Hop and Malt Trades Review 1865–1965: 100 Years (London: William Reed, 1965).

Brock, W., *Justus von Liebig: The Chemical Gatekeeper* (Cambridge: Cambridge University Press, 1997).

[Brougham, H.], 'Library of Useful Knowledge', *Edinburgh Review*, 50 (1829), pp. 181–93.

Brown, A., *Laboratory Studies for Brewing Students* (London: Longman, 1904).

Brown, H. T., 'The Nitrogen Question in Brewing', *Journal of the Institute of Brewing*, 13 (1907), pp. 394–457.

—, 'Reminiscences of Fifty Years' Experience of the Application of Scientific Methods to Brewing', *Journal of the Institute of Brewing*, 22 (1916), pp. 267–354.

Brown, W. H. R., *Golden Lane Brewery. The Attorney General versus Brown, Parry, and Others* (London: printed for J. Johnson [1808]).

Browne, C. A., 'Life and Chemical Services of Fredrick Accum', *Journal of Chemical Education*, 2 (1925), pp. 829–51, 1008–34, 1140–9.

—, 'Recently Acquired Information Concerning Fredrick Accum, 1769–1838', *Chymia*, 1 (1948), pp. 1–9.

Bryden, D. J., *Scottish Scientific Instrument-Makers, 1600–1900* (Edinburgh: [Royal Scottish Museum], 1972).

Bud, R., 'The Discipline of Chemistry: The Origins and Early Years of the Chemical Society of London' (PhD dissertation, University of Pennsylvania, 1980).

—, *The Uses of Life: A History of Biotechnology* (Cambridge: Cambridge University Press, 1993).

— and G. Roberts, *Science Versus Practice: Chemistry in Victorian Britain* (Manchester: Manchester University Press, 1984).

Burnaby, A., *An Essay upon the Excising of Malt* (London, 1696).

Bushnan, J. S., *Burton and its Bitter Beer* (London: William S. Orr, 1853).

Cardwell, D. S. L., *James Joule: A Biography* (Manchester: Manchester University Press, 1991).

Carpmael, W., *Law Reports of Patent Cases*, vol. 1 (London: A Macintosh, 1843).

Carter, H. B., *Sir Joseph Banks, 1743–1820* (London: British Museum (Natural History), 1988).

Castle, T., *The Female Thermometer: Eighteenth-Century Culture and the Invention of the Uncanny* (New York: Oxford University Press, 1995).

Chadwick, W., *A Practical Treatise on Brewing, with the Mode of Using the Thermometer and Saccharometer* (London: Whittaker, 1835).

Chambers, N., 'The Society of Arts and Joseph Banks: A First Step in London Learned Society', *Notes and Records of the Royal Society*, 61:3 (2007), pp. 313–25.

Chapman, A., 'From Alchemy to Airpumps: The Foundations of Oxford Chemistry to 1700', in R. Williams, A. Chapman and J. Rowlinson (eds), *Chemistry at Oxford: A History from 1600 to 2005* (London: Royal Society of Chemistry, 2009), pp. 17–51.

Chapman, A. C., 'Charles Graham' [obituary], *Journal of the Chemical Society, Transactions*, 97 (1910), pp. 677–80.

—, *Brewing* (Cambridge: Cambridge University Press, 1912).

Charleton, W., *Two Discourses ... [containing] the Mysterie of Vintners* (London, 1669).

Charnley, B., 'Arguing over Adulteration: The Success of the Analytical Sanitary Commission', *Endeavour*, 32 (2008), pp. 129–33.

Child, S., *Every Man his Own Brewer* (London, [*c.* 1790]).

Chomel, N., *Dictionaire Oeconomique: or, the Family Dictionary*, trans. R. Bradley (London, 1727).

Christie, I., 'Samuel Bentham's Library in Russia', *Slavonic and East European Review*, 65:1 (1987), pp. 26–37.

Christie, J. R. R., 'Historiography of Chemistry in the Eighteenth Century: Hermann Boerhaave and William Cullen', *Ambix*, 41:1 (1994), pp. 4–19.

Clark, C., *The British Malting Industry Since 1830* (London: Hambledon Press, 1998).

Clark, P., 'The "Mother Gin" Controversy in the Early Eighteenth Century', *Transactions of the Royal Historical Society*, 38 (1988), pp. 63–84.

Clark, W., J. Golinski and S. Schaffer (eds), *The Sciences in Enlightened Europe* (Chicago, IL: University of Chicago Press, 1999).

Clarke, R., *Notice is Hereby Given, to all Dealers ... that the Hydrometer, or Brandy-Prover ... is now Brought to its Utmost Perfection* (London, [1746]).

Clifford, J. L. (ed.), *Dr. Campbell's Diary of a Visit to England in 1775* (Cambridge: Cambridge University Press, 1947).

Clow, A. and N. L. Clow, *The Chemical Revolution: A Contribution to Social Technology* (London: Batchworth, 1952).

Cobbett, W., *Cottage Economy, Containing Information Relative to the Brewing of Beer* (London: Clement, 1822).

de Coetlogon, D., *An Universal History of Arts and Sciences*, 2 vols (London, 1745).

Cole, R. J., 'Friedrich Accum (1769–1838). A Biographical Study', *Annals of Science*, 7:2 (1951), pp. 128–43.

Combrune, M., *An Essay on Brewing, with a View of Establishing the Principles of the Art* (London, 1758).

—, 'An Historical Account of the English Brewery' (unpublished manuscript, 1762: Beinecke Library, Yale University, Osborn c602).

—, 'An Historical Accompt of the English Brewery' (unpublished manuscript, 1767: Beinecke Library, Yale University, Osborn c603).

—, *The Theory and Practice of Brewing* (London, 1762).

—, *Theorie und Praxis des Bierbrauens*, trans. C. H. Reichel (Leipzig, 1796).

—, *L'art de brasser, traduit de l'anglais*, translator unidentified (Paris: le Normant, 1802).

—, *The Theory and Practice of Brewing*, new edn (London: Vernor and Hood, 1804).

[Combrune, M.], *Some Account of the Rise, Progress and Present State of the Brewery ... in a Letter to a Friend* (London, 1757).

[—], *An Enquiry into the Prices of Wheat, Malt, and Occasionally of Other Provisions* (London, 1768).

The Complete Family Brewer (London, 1789).

The Complete Family-Piece, and Country Gentleman and Farmer's Best Guide, 2nd edn (London, 1737).

The Complete Maltster and Brewer: Being a Brief Dissertation in Defence of Long Grown Malts (London, 1765).

Cook, J., 'The Method Taken for Preserving the Health of the Crew of His Majesty's Ship the Resolution during her Late Voyage round the World', *Philosophical Transactions*, 66 (1776), pp. 402–6.

Cooley, A., *A Cyclopædia of Practical Receipts*, 2nd edn (London: John Churchill, 1845).

Cornell, M., *Beer: the Story of the Pint* (London: Headline, 2003).

Corran, H. S., *A History of Brewing* (Newton Abbot: David and Charles, 1975).

Crawford, R., *Poetry, Enclosure, and the Vernacular Landscape, 1700–1830* (Cambridge: Cambridge University Press, 2002).

Crellin, J. K., 'A Note on Dr James's Fever Powder', *Transactions of the British Society for the History of Pharmacy*, 1:3 (1974), pp. 136–43.

Curteis, T., *Essays on the Preservation and Recovery of Health* (London, 1704).

The Crying Frauds of the London Markets: Proving their Deadly Influence upon the Two Great Pillars of Life, Bread & Porter (London, 1795).

Dagley, R. (ed.), *Death's Doings* (London: Andrews and Cole, 1826).

[Dalrymple, H.], *Rodondo, or, the State Jugglers, Canto III* (London, 1770).

Dalton, J., *A New System of Chemical Philosophy*, part 1 (Manchester: Bickerstaff, 1808).

Davy, H., *A Syllabus of a Course of Lectures on Chemistry* (London: Royal Institution, 1802).

Deadly Adulteration and Slow Poisoning, or, Disease and Death in the Pot and Bottle (London: Sherwood, [*c.* 1830]).

[Defoe, D.], *A Brief Case of the Distillers, and of the Distilling Trade in England* (London, 1726).

[Deurbroucq, D. P. and (?) Nichols], *Observations on the Vinous Fermentation; with a Description of a Patent Apparatus to Improve the Same* (London, 1823).

Dictionarium Rusticum & Urbanicum (London, 1704).

Digby, K., *The Closet of the Eminently Learned Sir Kenelme Digbie Kt. Opened* (London, 1669).

Dingle, A. E., *The Campaign for Prohibition in Victorian England: The United Kingdom Alliance, 1872–1895* (London: Croom Helm, 1980).

A Discourse on the Preparation, Preservation, and Restoration of Malt-Liquors (London, 1733).

The Distiller of London, Compiled and Set Forth by the Speciall Licence and Command of the Kings most Excellent Majesty (London, 1639).

The Distiller of London, with the Clavis to Unlock the Deepest Secrets of that Mysterious Art (London, 1652).

Dobson, M., *A Medical Commentary on Fixed Air* (Chester, 1779).

Donovan, M., *Domestic Economy, Volume 1: Brewing, Distilling, Wine-making, Baking, &c.* (London: Longman, 1830).

Dossie, R., *Memoirs of Agriculture*, 3 vols (London, 1768).

[Dossie, R.], *The Elaboratory Laid Open, or, the Secrets of Chemistry and Pharmacy Revealed* (London, 1758).

Douglas, M., *Purity and Danger: An Analysis of Concepts of Pollution and Taboo* (1966; new edn Abingdon: Routledge, 2002).

Dowell, S., *History of Taxation and Taxes in England from the Earliest Times to the Year 1885*, 4 vols (London: Longmans, 1888).

Dublin Society, *Instructions for Planting and Managing Hops* (Dublin, 1733).

Dubrunfaut, A.-P., 'Note sur le glucose', *Annales de chimie et de physique*, 21 (1847), pp. 178–80.

Dutrochet, H., 'On Diastase and its Importance in Vegetable Physiology', *Magazine of Botany and Gardening*, 2 (1834), pp. 91–3.

Eamon, W., *Science and the Secrets of Nature: Books of Secrets in Medieval and Early Modern Culture* (Princeton, NJ: Princeton University Press, 1994).

Edgeworth, R. L. and M. Edgeworth, *Memoirs of Richard Lovell Edgeworth*, 3rd edn (London: Bentley, 1844).

Ellis, W., *The Country Housewife's Family Companion* (London, 1750).

—, *The New Art of Brewing and Improving Malt Liquors* (London, 1761).

—, *Ellis's Husbandry, Abridged and Methodized*, 2 vols (London, 1772).

[Ellis, W.,], *The London and Country Brewer*, [1st edn of the first part] (London, [*c.* 1735]).

[—], *The London and Country Brewer* [first part] (Dublin, 1735).

[—], *The London and Country Brewer... third and last Part*, 1st edn (London, 1738).

[—], *The London and Country Brewer*, 2nd edn (London: Thomas Astley, 1743).

[—], *The London and Country Brewer*, 5th edn (London, 1744).

Entick, J., *A New and Accurate History and Survey of London*, 4 vols (London, 1766).

An Essay upon Excisi[n]g Several Branches that have Hitherto Escaped the Duty of the Brewing Trade (n.p., [1699]).

Every Man his Own Brewer: or, a Compendium of the English Brewery (London, 1768).

The Farmers' Real Friend, Shewing How to Brew with Three-Fourths Barley to One-Fourth Malt (London: Groombridge, 1847).

Farrar, W. V., 'Andrew Ure, F.R.S., and the Philosophy of Manufactures', *Notes and Records of the Royal Society*, 27 (1973), pp. 299–324.

Faulkner, F., *The Art of Brewing Practical and Theoretical*, 2nd edn (London: F. W. Lyon, 1878).

—, *The Theory and Practice of Modern Brewing*, 1st edn (London: F. W. Lyon, 1884).

—, *The Theory and Practice of Modern Brewing*, 2nd edn (London: F. W. Lyon, 1888).

Figala, K. and U. Petzold, 'Alchemy in the Newtonian Circle: Personal Acquaintances and the Problem of the Late Phase of Isaac Newton's Alchemy', in J. V. Field and F. James (eds), *Renaissance and Revolution: Humanists, Scholars, Craftsmen and Natural Philosophers in Early Modern Europe* (Cambridge: Cambridge University Press, 1993), pp. 173–92.

Filby, F., *A History of Food Adulteration and Analysis* (London: George Allen and Unwin, 1934).

Fisher, G., *Life of Benjamin Silliman*, 2 vols (New York: Charles Scribner, 1866).

Fleet, C., *Glimpses of our Ancestors in Sussex*, second series (Lewes: Farncombe, 1883).

Flower, R., *Observations on Beer and Brewers* (Cambridge: B. Flower, 1802).

Forbes, W. J., 'Death (a Dealer) to his London Correspondent', in R. Dagley (ed.), *Death's Doings, Consisting of Numerous Original Compositions, in Prose and Verse* (London: Andrews and Cole, 1826), pp. 297–300.

Ford, W., *An Historical Account of the Malt Trade and Laws* (London, 1849).

Forsyth, J. S., *The Farmer, Maltster, Distiller, & Brewer's Practical Memorandum Book* (London: D. Cox, [*c.* 1823]).

Fourcroy, A., 'D'un memoire du cit. Fabroni [*sic*], sur les fermentations etc', *Annales de chimie*, 31 (1799), pp. 299–327.

Fownes, G., 'On the Value in Absolute Alcohol of Spirits of Different Specific Gravities', *Philosophical Transactions*, 137 (1847), pp. 249–51.

Frost, C., *An Address, Delivered to the Literary and Philosophical Society at Kingston upon Hull* (Hull: I. Wilson, 1831).

Fuller, T., *Pharmacopoia Extemporanea* (London, 1710).

Furze, J., 'Observations on Fermentation', *Memoirs and Proceedings of the Chemical Society of London*, 2 (1845).

Gardner, J. A., 'Victor Herbert Veley, 1856–1933', *Obituary Notices of Fellows of the Royal Society*, 1 (1934), pp. 228–35.

Garnett, T., *A Treatise on the Mineral Waters of Harrogate* ([Bradford], 1792).

Gascoigne, J., *Science in the Service of Empire: Joseph Banks, The British State and the Uses of Science in the Age of Revolution* (Cambridge: Cambridge University Press, 1998).

Gay, H., *The History of Imperial College London, 1907–2007* (London: Imperial College Press, 2007).

A General Description of all Trades (London, 1747).

'George Wigney', Barcombe and Hamsey website, n. d., at http://www.bandhpast.co.uk/barcombe/b1200other.php [accessed 13 January 2013].

Gibbs, F. W., 'Peter Shaw and the Revival of Chemistry', *Annals of Science*, 7:3 (1951), pp. 211–37.

Glauber, J. R., *The Works of the Highly Experienced and Famous Chymist, John Rudolph Glauber*, trans. C. Packe (London, 1689).

Glendinning, T. A., 'A Short Account of the Brewers' Saccharometer', *Journal of the Federated Institutes of Brewing*, 6 (1900), pp. 357–64.

Golinski, J., 'Peter Shaw: Chemistry and Communication in Augustan England', *Ambix*, 30:1 (1983), pp. 19–29.

—, *Science as Public Culture* (Cambridge: Cambridge University Press, 1992).

—, 'Barometers of Change: Meterological Instruments as Machines of Enlightenment', in W. Clark, J. Golinski and S. Schaffer (eds), *The Sciences in Enlightened Europe* (Chicago, IL: University of Chicago Press, 1999), pp. 69–93.

—, '"Fit Instruments": Thermometers in Eighteenth-Century Chemistry', in F. L. Holmes and T. H. Levere (eds), *Instruments and Experimentation in the History of Chemistry* (Cambridge, MA: MIT Press, 2000), pp. 185–210.

—, 'Chemistry', in R. Porter (ed.) *The Cambridge History of Science, vol. 4: Eighteenth Century Science* (Cambridge: Cambridge University Press, 2003), pp. 375–96.

—, *British Weather and the Climate of Enlightenment* (Chicago, IL: University of Chicago Press, 2007).

Good and Cheap Beer for the Million, by the Use of Sugar and Molasses in Public Breweries (London: Pelham Richardson, 1846).

Gourvish, T. and R. Wilson, *The British Brewing Industry, 1830–1980* (Cambridge: Cambridge University Press, 1994).

Grabiner, J., '"Some Disputes of Consequence": Maclaurin among the Molasses Barrels', *Social Studies of Science*, 28:1 (1998), pp. 139–68.

Graham, T., *Elements of Chemistry* (London: Hippolyte Baillière, 1842).

— and A. Hofmann, 'Report upon the Alleged Adulteration of Pale Ales by Strychnine', *Edinburgh Medical and Surgical Journal*, 78 (1852), pp. 246–9.

Green, R., *The Porter Brewer Detected* (London, 1764).

Gregory, G., *The Economy of Nature Explained and Illustrated*, 3 vols (London, 1796).

Günther, A., 'Matthew Maty MD, FRS (1718–76) and Science at the Foundation of the British Museum, 1753–80', *Bulletin of the British Museum (Natural History), Historical Series*, 15:1 (1987), pp. 1–58.

Gutzke, D. W., *Protecting the Pub: Brewers and Publicans against Temperance* (Woodbridge: Royal Historical Society, 1989).

—, *Pubs and Progressives: Reinventing the Public House in England, 1896–1960* (DeKalb, IL: Northern Illinois University Press, 2005).

Hack, D., *The Material Interests of the Victorian Novel* (Charlottesville, VA: University of Virginia Press, 2005).

Hale, T., *A Compleat Body of Husbandry*, 4 vols (London, 1756).

Ham, E., *Elizabeth Ham by Herself, 1783–1820* (London: Faber, 1945).

Ham, J., *Instructions for Brewing by a New Method, from Unmalted Corn*, 4th edn (London: Baldwin, Cradock and Joy, 1822).

—, *The Theory and Practice of Brewing, from Malted and Unmalted Corn, and from Potatos* [*sic*] (London: Simpkin and Marshall, 1829).

Hamlin, C., *A Science of Impurity: Water Analysis in Nineteenth-Century Britain* (Bristol: Adam Hilger, 1990).

Hammond, P. and H. Egan, *Weighed in the Balance: A History of the Laboratory of the Government Chemist* (London: HMSO, 1992).

Hardy, J., *The Retail Compounder, or, Publican's Friend* (London, [*c.* 1795]).

Harley, J. A. G., 'Samuel Whitbread's First Enterprise: A Survey of the Organisation and Methods of an Eighteenth Century London Brewery, 1742–50', *Guildhall Miscellany*, 1:9 (1958), pp. 3–26.

Harris, J., *The Description and Use of the Globes, and the Orrery* (London, 1783).

Harrison, B., *Drink and the Victorians: The Temperance Question in England, 1815–1872* (Pittsburgh, PA: University of Pittsburgh Press, 1971).

Harrison, W., 'The Description of England', in R. Holinshed, et al., *The First and Second Volumes of Chronicles* ([London], 1587), pp. 131–250.

Hartlib, S., *The Reformed Common-wealth of Bees, Presented in Severall Letters and Observations* (London, 1655).

—, *Samuel Hartlib his Legacy of Husbandry*, 3rd edn (London, 1655).

Hasted, E., *The History and Topographical Survey of the County of Kent*, 12 vols (Canterbury, 1797–1801).

Hayman, E. N., *A Practical Treatise to Render the Art of Brewing More Easy ... Inculcating the Use of the Saccharometer* (London: Longman, 1819).

Hays, J. N., 'The Rise and Fall of Dionysius Lardner', *Annals of Science*, 38:5 (1981), pp. 527–42, on pp. 535–8.

Henry, T., *Experiments and Observations on Ferments and Fermentation* (Warrington, 1785).

Heun, F. W., *Versuch der Kunst, alle Arten Biere nach englischen Grundsätzen zu brauen* (Leipzig, 1777).

Higgs, H., *Bibliography of Economics, 1751–1775* (Cambridge: Cambridge University Press, 1935).

Hitchcock, T., *A Practical Treatise on Brewing, Showing the Cause and Prevention of Acidity in Malt Liquors* (London: printed by R. Boyd, 1842).

von Hofmann, A., E. Fischer and H. Caro, 'Zur Erinnerung an Peter Griess', *Berichte der Deutschen chemischen Gesellschaft*, 24 (1891), pp. 1006–78.

Holmes, F., *Lavoisier and the Chemistry of Life: An Exploration of Scientific Creativity* (Madison, WI: University of Wisconsin Press, 1985).

Hooke, R., *The Posthumous Works of Robert Hooke* (London: Royal Society, 1705).

Hooper, E. G., *The Manual of Brewing, Scientific and Technical*, 1st edn (London: Sheppard and St John, 1882).

Houghton, W., et al. (eds), *The Wellesley Index to Victorian Periodicals, 1824–1900*, 5 vols (Toronto: University of Toronto Press, 1966), vol. 1.

Houghton, W. B., Jr, 'The History of Trades: Its Relation to Seventeenth-Century Thought', *Journal of the History of Ideas*, 2:1 (1941), pp. 33–60.

Howard-Jones, N., 'John Quincy, M. D. [d. 1722], Apothecary and Iatrophysical Writer', *Journal of the History of Medicine*, 6:2 (1951), pp. 149–75.

Howe, E., *A List of London Bookbinders, 1648–1815* (London: Bibliographical Society, 1950).

— and J. Child, *The Society of London Bookbinders, 1780–1951* (London: Sylvan, 1952).

Hudson, D. and K. W. Luckhurst, *The Royal Society of Arts, 1754–1954* (London: John Murray, 1954).

Hughes, E., *A Saving of Twenty Per Cent. A Treatise on the Brewing of Beer* (Uxbridge, 1796).

Hutton, C., G. Shaw and R. Pearson (eds), *The Philosophical Transactions of the Royal Society of London ... Abridged, with Notes*, 18 vols (London: Baldwin, 1809).

An Impartial Enquiry into the Present State of the British Distillery (London, 1736).

An Impartial Narrative of the Late Melancholy Occurrences in Manchester (Liverpool: Henry Fisher, 1819).

Irvine, W., 'On Fermentation' (1785), in W. Irvine, *Essays, Chiefly on Chemical Subjects* (London: J. Mawman, 1805), pp. 307–26.

Jackson, H., *Reflexions Concerning the Virtues of Tar Water* (London, 1744).

—, *An Essay on Bread, Wherein the Bakers and Millers are Vindicated* (London, 1758).

—, *An Essay on British Isinglass* (London, 1765).

—, *New Art of Hardening and Preserving Wood* (London, 1770).

—, 'An Account of the Discovery of the Manner of Making Isinglass in Russia; with a Particular Description of its Manufacture in England', *Philosophical Transactions*, 63 (1774), pp. 1–15.

James, F. A. J. L. (ed.), *The Correspondence of Michael Faraday*, vol. 1 (London: Institution of Electrical Engineers, 1991).

Janes, H., *The Red Barrel: A History of Watney Mann* (London: John Murray, 1963).

Janssens, U., *Matthieu Maty and the Journal Britannique, 1750–1755* (Amsterdam: Holland University Press, 1975).

Jefcoate, G., 'German Immigrants and the London Book Trade, 1700–1770', in R. Vigne and C. Littleton (eds), *From Strangers to Citizens: The Integration of Immigrant Communities in Britain, Ireland, and Colonial America, 1550–1750* (Brighton: Sussex Academic Press, 2001), pp. 503–10.

Johns, A., *The Nature of the Book: Print and Knowledge in the Making* (Chicago, IL: University of Chicago Press, 1998).

Johnston, W. H., *Loftus's Inland Revenue Officer's Manual* (London: Loftus, 1865).

Judel, G. K., *Justus Liebig in Grossbritannien: Justus Liebigs Briefe aus Grossbritannien an seine Frau Henriette* (Giessen: Justus Liebig-Gesellschaft, 2003).

Kalm, P., *Kalm's Account of his Visit to England on his Way to America in 1748*, trans. J. Lucas (London: Macmillan, 1892 [1748]).

Keith, G. S., 'Observations on the Papers Presented to the House of Commons, Relating to the Experiments Made for Ascertaining the Relative Quality of Malt Made from Barley and Scotch Bigg', *Farmer's Magazine*, 8 (1807), pp. 476–500.

—, *A General View of the Agriculture of Aberdeenshire* (Aberdeen: A. Brown, 1811).

Kinnell, M., 'Sceptreless, Free, Uncircumscribed? Radicalism, Dissent and Early Children's Books', *British Journal of Educational Studies*, 36:1 (1988), pp. 49–71.

Kinraide, R., 'The Society for the Diffusion of Useful Knowledge and the Democratization of Learning in Early Nineteenth-Century Britain' (PhD dissertation, University of Wisconsin–Madison, 2006).

Klein, U., 'Experiments at the Intersection of Experimental History, Technological Inquiry, and Conceptually Driven Analysis: A Case Study from Early Nineteenth-Century France', *Perspectives on Science*, 13:1 (2005), pp. 1–48.

— and W. Lefèvre, *Materials in Eighteenth-Century Science: A Historical Ontology* (Cambridge, MA: MIT Press, 2007).

Knox, D., 'The Development of the London Brewing Industry, 1830–1914, with special reference to Whitbread and Company' (BLitt dissertation, University of Oxford, 1956).

Landau, N., *Law, Crime and English Society, 1660–1830* (Cambridge: Cambridge University Press, 2002).

[Lawler, C. F.], *The Lamentations of the Porter Vat Which Exploded of the Drug-Gripes, October 17th, 1814* (London: John Fairburn, [1814]).

Lawrence, C., 'Disciplining Disease: Scurvy, the Navy, and Imperial Expansion, 1750–1825', in D. P. Miller and P. H. Reill (eds), *Visions of Empire: Voyages, Botany, and Representations of Nature* (Cambridge: Cambridge University Press, 1996), pp. 80–106.

[Leigh, P.], 'The Chemistry of a Pint of Beer', *Household Words*, 2 (15 February 1851), pp. 498–502.

Levesque, J., *The Art of Brewing and Fermenting, in the Summer, and all other Seasons*, 2nd edn (London: Thomas Hurst, 1836).

[Lewis, W.], *The New Dispensatory* (London, 1753).

Lightbody, J., *The Gauger and Measurer's Companion* (London, 1694).

—, *Every Man His Own Gauger ... To Which is Added, the True Art of Brewing Beer, Ale, Mum ...* (London, 1695).

Lillywhite, B., *London Coffee Houses: A Reference Book of Coffee Houses of the Seventeenth, Eighteenth and Nineteenth Centuries* (London: G. Allen and Unwin, 1963).

Lindeboom, G. A., *Herman Boerhaave: The Man and his Work* (London: Methuen, 1968).

Loch, D., *Letters Concerning the Trade and Manufactures of Scotland*, 2nd edn (Edinburgh, 1774).

—, *Essays on the Trade, Commerce, Manufactures, and Fisheries of Scotland*, 3 vols (Edinburgh, 1778).

[Loftus ,W.], *The Brewer: A Familiar Treatise on the Art of Brewing*, new stereotyped edn (London: Loftus, 1863).

[—], *Loftus's Almanack for Brewers, Distillers, and the Wine and Spirit Trades* (London: Loftus, 1869).

[—], *The Brewer: a Familiar Treatise on the Art of Brewing*, new edn (London: Loftus, [1877]).

Long, J., *A Treatise on Malting* (Dublin, 1800).

Luu, L. B., *Immigrants and the Industries of London, 1500–1700* (Aldershot: Ashgate, 2005).

Lynch, P. and J. Vaizey, *Guinness's Brewery in the Irish Economy, 1759–1876* (Cambridge: Cambridge University Press, 1960).

MacBride, D., *Experimental Essays on the Following Subjects: I. On the Fermentation of Alimentary Mixtures ...* 1st edn (London, 1764).

McConnell, A., *Instrument Makers to the World: A History of Cooke, Troughton and Simms* (York: William Sessions, 1992).

—, 'From Craft Workshop to Big Business: The London Scientific Instrument Trade's Response to Increasing Demand, 1750–1820', *London Journal*, 19:1 (1994), pp. 36–53.

McCulloch, J. R., *The Literature of Political Economy: A Classified Catalogue* (London: Longman, 1845).

Macdonagh, O., 'The Origins of Porter', *Economic History Review*, 16:3 (1964), pp. 530–5.

[Macky, J.], *A Journey through England, in Familiar Letters from a Gentleman*, 2 vols (London, 1722).

MacLeod, C., 'Concepts of Invention and the Patent Controversy in Victorian Britain', in R. Fox (ed.) *Technological Change: Methods and Themes in the History of Technology* (Amsterdam: Harwood Academic, 1996).

Malcolm, J., *A Compendium of Modern Husbandry*, 3 vols (London: Baldwin, 1805).

M[arkham], G., *The English Hus-wife, Contayning, The Inward and Outward Vertues which Ought to be in a Compleat Woman* (London, 1615).

Markham, P., *Syhoroc: or, Considerations on the Ten Ingredients Used in the Adulteration of Bread-Flour, and Bread*, 2nd edn (London, 1758).

[Markham, P.], *Poison Detected: or Frightful Truths; and Alarming to the British Metropolis* (London, 1757).

Marsden, B., 'Carriages, Coffee-cups and Dynamometers: Representing French Technical Cultures in the London Mechanics' Magazine, 1823–1848', in P. Bret, I. Gouzévitch and L. Pérez (eds), *Les techniques et la technologie entre la France et la Grande-Bretagne (XVIIe-XIXe siècles)* (Paris: Centre de documentation d'histoire des techniques, 2011), pp. 243–54.

Martin, B., *A Sure Guide to Distillers, and All Dealers in Spirituous Liquors* (London, 1759).

—, *The Theory of the Genuine Hydrometer, Farther Illustrated* (London, [*c.* 1762]).

—, *A Description of the Nature, Construction, and Use of the Torricellian, or Simple Barometer* (London, 1766).

Mathews, C., *Mr. Mathews' Memorandum-book, of Peculiarities, Character, and Manners* (London: Duncombe, [*c.* 1825]).

Mathias, P., *The Brewing Industry in England, 1700–1830* (Cambridge: Cambridge University Press, 1959).

Middleton, W. E. K., *A History of the Thermometer and its Use in Meteorology* (Baltimore, MD: Johns Hopkins University Press, 1966).

Millburn, J. R., *Benjamin Martin: Author, Instrument-maker, and 'Country Showman'* (Leyden: Noordhoff, 1976).

—, *Benjamin Martin: Supplement* (London: Vade-Mecum, 1986).

Miller, D. P., 'Between Hostile Camps: Sir Humphry Davy's Presidency of the Royal Society of London, 1820–1827', *British Journal for the History of Science*, 16:1 (1983), pp. 1–47.

—, 'The "Hardwicke Circle": The Whig Supremacy and its Demise in the 18th-Century Royal Society', *Notes and Records of the Royal Society*, 52:1 (1998), pp. 73–92.

—, 'The Usefulness of Natural Philosophy: The Royal Society and the Culture of Practical Utility in the Later Eighteenth Century', *British Journal for the History of Science*, 32:2 (1999), pp. 185–201.

—, 'Seeing the Chemical Steam through the Historical Fog: Watt's Steam Engine as Chemistry', *Annals of Science*, 65:1 (2007), pp. 47–72.

Miller, P., *The Gardeners Dictionary*, 2nd edn (London, 1733).

— *The Gardeners Dictionary*, 6th edn (London, 1752).

Mills, J., *A New System of Practical Husbandry*, 5 vols (London, 1767).

Mirowski, P., *The Effortless Economy of Science?* (Durham, NC: Duke University Press, 2004).

Moir, W., *Brewing Made Easy, Being a Compendium of all the Directions that have Hitherto been Published* (London: Scott, 1802).

Moray, R., 'An Account of the Manner of Making Malt in Scotland', *Philosophical Transactions*, 12 (1678), pp. 1069–71.

Morrell, J., 'Practical Chemistry in the University of Edinburgh, 1799–1843', *Ambix*, 16:1 (1969), pp. 66–80.

—, 'The Chemist Breeders: The Research Schools of Liebig and Thomas Thomson', *Ambix*, 19:1 (1972), pp. 1–46.

— and A. Thackray, *Gentlemen of Science: Early Years of the British Association for the Advancement of Science* (Oxford: Clarendon, 1981).

Morrice, A., *A Treatise on Brewing; Wherein is Exhibited the Whole Process of the Art and Mystery of Brewing*, 2nd edn (London, 1802).

—, *A Practical Treatise on Brewing the Various Sorts of Malt-Liquor*, 7th edn (London, 1827).

Mortimer, C., 'A Discourse Concerning the Usefulness of Thermometers in Chemical Experiments', *Philosophical Transactions*, 44 (1746–7), pp. 672–95.

M[ortimer], J., *The Whole Art of Husbandry, or, the Way of Managing and Improving of Land*, 1st edn (London, 1707).

Mortimer, J., *The Whole Art of Husbandry, or, the Way of Managing and Improving of Land*, 4th edn, 2 vols (London, 1716).

Muspratt, S. (ed.), *Chemistry, Theoretical, Practical and Analytical, as Applied and Relating to the Arts and Manufactures*, 2 vols (Glasgow: W. Mackenzie, 1860).

Nangle, B., *The Monthly Review, First Series, 1749–1789: Indexes of Contributors and Articles* (Oxford: Clarendon, 1934).

—, *The Monthly Review, Second Series, 1790–1815: Indexes of Contributors and Articles* (Oxford: Clarendon, 1955).

Neumann, C., *The Chemical Works of Caspar Neumann, M. D.*, ed. W. Lewis (London, 1759).

A New System of Practical Domestic Economy, Founded on Modern Discoveries, 3rd edn (London: Henry Colburn, 1823).

Nicholls, J., *The Politics of Alcohol: A History of the Drink Question in England* (Manchester: Manchester University Press, 2009).

Noakes, R., 'Representing "A Century of Inventions": Nineteenth-Century Technology and Victorian *Punch*', in L. Henson, et al. (eds), *Culture and Science in the Nineteenth-Century Media* (Aldershot: Ashgate, 2004), pp. 151–63.

O'Sullivan, C., 'On the Transformation-Products of Starch', *Journal of the Chemical Society*, 25 (1872), pp. 579–88.

—, 'On the Action of Malt-Extract on Starch', *Journal of the Chemical Society*, 30 (1876), pp. 125–44.

—, 'On Maltose', *Journal of the Chemical Society*, 29 (1876), pp. 478–82.

O'Sullivan, H. D., *The Life and Work of Cornelius O'Sullivan, F.R.S.* (Guernsey: Guernsey Star and Gazette, [1934]).

Ochs, K., 'The Royal Society of London's History of Trades Programme: An Early Episode in Applied Science', *Notes and Records of the Royal Society of London*, 39:2 (1985), pp. 129–58.

Pasteur, L., *Studies on Fermentation: The Diseases of Beer, their Causes, and the Means of Preventing Them* (originally *Études sur la Bière*, 1876), trans. F. Faulkner and D. C. Robb (London: Macmillan, 1879).

Paulus, I., *The Search for Pure Food: A Sociology of Legislation in Britain* (London: Martin Robertson, 1974).

Payen, A. and J.-F. Persoz, 'Mémoire sur la diastase, les principaux produits de ses réactions et leurs applications aux arts industriels', *Annales de chimie et de physique*, 53 (1833), pp. 73–92.

— and J.-F. Persoz, 'Mémoire sur l'amidone (substance intérieure de la fécule) et suite de recherches sur la diastase', *Annales de chimie et de physique*, 56 (1834), pp. 337–71.

Pearson, L., *British Breweries: An Architectural History* (London: Hambledon, 1999).

Piesse, S., 'Some Observations on Brewing', *Proceedings of the Chemical Society of London* [vol. 2] (1841–2), pp. 23–4.

Piozzi, H. L. (ed.), *Letters To and From the Late Samuel Johnson*, 2 vols (London, 1788).

—, *Autobiography, Letters and Literary Remains*, ed. A. Hayward, 1st edn, 2 vols (London: Longman, 1861).

—, *Thraliana: The Diary of Mrs. Hester Lynch Thrale (later Mrs. Piozzi), 1776–1809*, ed. K. C. Balderston, 1st edn, 2 vols (Oxford: Clarendon, 1951).

Platt, H. L., *Shock Cities: The Environmental Transformation and Reform of Manchester and Chicago* (Chicago, IL: University of Chicago Press, 2005).

Ploughman, W., *Oeconomy in Brewing* (Romsey, Hants, 1797).

Porter, R., *Quacks: Fakers and Charlatans in English Medicine* (Stroud: Tempus, 2000).

Post Office London Directory for 1820 (London: Critchett and Woods, 1820).

Premiums Offered by the Society Instituted at London for the Encouragement of Arts, Manufactures and Commerce (London, 1766).

The Present State of the London Brewery: Recommended to the Perusal of Those Concerned in, or with the Trade (London, 1760).

Price, R., 'Kwasímukámba's Gambit', *Bijdragen tot de Taal-, Land- en Volkenkunde*, 135:1 (1979), pp. 151–69.

Priestley, J., *Experiments and Observations on Different Kinds of Air*, vol. 2 (London, 1775).

Pringle, J., *A Discourse upon some Late Improvements of the Means for Preserving the Health of Mariners* (London, 1776).

Procter, W., 'Observations on Dextrine and Diastase', *American Journal of Pharmacy*, 5:4 (1840), pp. 265–79.

Quincy, J., *Pharmacopœia Officinalis & Extemporanea: or, a Compleat English Dispensatory* (London, 1718).

Randall, W., *The State of the Hop Plantations* (London, 1800).

Rapp, D., 'Social Mobility in the Eighteenth Century: The Whitbreads of Bedfordshire, 1720–1815', *Economic History Review*, 27:3 (1974), pp. 380–94.

Reddington, W., *A Practical Treatise on Brewing* (London, 1760).

Reinarz, J., 'Fit for Management: Apprenticeship and the English Brewing Industry, 1870–1914', *Business History*, 43:3 (2001), pp. 33–53.

Remarks and Experiments on Different Parts of the Process of Brewing (Oxford: R. Bliss, 1807).

Report of the Seventh Meeting of the British Association for the Advancement of Science (London: John Murray, 1838).

Reynoldson, J., *Practical and Philosophical Principles of Making Malt*, 2nd edn (London: Gale and Curtis, 1808).

Richardson, J., *Theoretic Hints on an Improved Practice of Brewing Malt-Liquors* (London, 1777).

—, *Statical Estimates of the Materials for Brewing, or a Treatise on ... the Saccharometer* (London, 1784).

—, *Remarks on a Pamphlet Entitled Hydrometrical Observations and Experiments in the Brewery* (London, 1785).

—, *The Philosophical Principles of the Science of Brewing*, 1st edn (York, 1788).

—, *The Philosophical Principles of the Science of Brewing*, 2nd edn (Hull, 1798).

—, *The Philosophical Principles of the Science of Brewing*, 3rd edn (York: printed for the author, 1805).

[Richardson, J.], *Observations on the Art of Brewing Malt Liquors, in a Series of Strictures on a Secret System Inculcated in a Private course of Lectures ... by a Practical Brewer* (London, 1775).

Richmond, L. and A. Turton, *The Brewing Industry: A Guide to Historical Records* (Manchester: Manchester University Press, 1990).

Rigby, E., *Chemical Observations on Sugar* (London, 1788).

Roberts, G., 'The Establishment of the Royal College of Chemistry: An Investigation of the Social Context of Early-Victorian Chemistry', *Historical Studies in the Physical Sciences*, 7 (1976), pp. 437–85.

Roberts, L., 'The Death of the Sensuous Chemist: The "New" Chemistry and the Transformation of Sensuous Technology', *Studies in the History and Philosophy of Science Part A*, 26:4 (1995), pp. 503–29.

— and S. Schaffer, 'Preface', in L. Roberts, S. Schaffer and P. Dear (eds), *The Mindful Hand: Inquiry and Invention from the Late Renaissance to Early Industrialisation* (Amsterdam: KNAvW, 2007), pp. xiii–xxvii.

—, S. Schaffer and P. Dear (eds), *The Mindful Hand: Inquiry and Invention from the Late Renaissance to Early Industrialisation* (Amsterdam: KNAvW, 2007).

Roberts, W. H., *The Scottish Ale-Brewer and Practical Maltster* (Edinburgh: A. and C. Black, 1847).

Robinson, N., *A Compleat Treatise of the Gravel and Stone, with all their Causes* (London, 1721).

le Rougetel, H., *The Chelsea Gardener: Philip Miller, 1691–1771* (London: Natural History Museum, 1990).

Rusnock, A., 'Correspondence Networks and the Royal Society, 1700–1750', *British Journal for the History of Science*, 32:2 (1999), pp. 155–69.

Russell, C., *Edward Frankland: Chemistry, Controversy and Conspiracy in Victorian England* (Cambridge: Cambridge University Press, 2003).

Sambrook, P., *Country House Brewing in England, 1500–1900* (London: Hambledon, 1996).

de Saussure, T., 'On the Conversion of Starch into Sugar', *Annals of Philosophy*, 6 (1815), pp. 424–31.

Schaffer, S., 'Scientific Discoveries and the End of Natural Philosophy', *Social Studies of Science*, 16:3 (1986), pp. 387–420.

—, '"The Charter'd Thames": Naval Architecture and Experimental Spaces in Georgian Britain', in L. Roberts, S. Schaffer and P. Dear (eds), *The Mindful Hand: Inquiry and Invention from the Late Renaissance to Early Industrialisation* (Amsterdam: KNAvW, 2007), pp. 279–305.

Schiebinger, L., *Plants and Empire: Colonial Bioprospecting in the Atlantic World* (Cambridge, MA: Harvard University Press, 2007).

Schuette, H., 'Death in the Pot', *Transactions of the Wisconsin Academy of Sciences, Arts and Letters*, 35 (1943), pp. 283–303.

Schwartz, R. B., *Samuel Johnson and the New Science* (Madison, WI: University of Wisconsin Press, 1971).

Secord, J., *Victorian Sensation: The Extraordinary Publication, Reception, and Secret Authorship of Vestiges of the Natural History of Creation* (Chicago, IL: University of Chicago Press, 2000).

—, 'Knowledge in Transit', *Isis*, 95 (2004), pp. 654–72.

Sedgwick, J., *A New Treatise on Liquors* (London, 1725).

Shannon, R., *A Practical Treatise on Brewing, Distilling, and Rectification* (London, 1805).

Shapin, S., *A Social History of Truth: Civility and Science in Seventeenth-Century England* (Chicago, IL: University of Chicago Press, 1994).

Shaw, P., *Philosophical Principles of Universal Chemistry ... Drawn from the Collegium Jenense of Dr George Ernest Stahl* (London, 1730).

—, *Three Essays in Artificial Philosophy, or Universal Chemistry* (London, 1731).

—, *Chemical Lectures, Publickly Read at London, in the Years 1731, and 1732*, 1st edn (London, [1734]).

—, *Chemical Lectures, Publickly Read at London*, 2nd edn (London, 1755).

[Shaw, P.], *The Juice of the Grape: or, Wine preferable to Water* (London, 1724).

Shore, A., *A Practical Treatise on Brewing, Founded on Experiments Made with the Thermometer and Hydrometer*, 2nd edn (Nantwich, 1809).

Sibum, H. O., 'Reworking the Mechanical Value of Heat: Instruments of Precision and Gestures of Accuracy in Early Victorian England', *Studies in History and Philosophy of Science*, 26:1 (1995), pp. 73–106.

—, 'An Old Hand in a New System', in J.-P. Gaudillière and I. Löwy (eds), *The Invisible Industrialist: Manufactures and the Production of Scientific Knowledge* (Basingstoke: Macmillan, 1998), pp. 23–57.

—, 'Les gestes de la mesure: Joule, les pratiques de la brasserie et la science', *Annales: Histoire, Sciences Sociales*, 53:4–5 (1998), pp. 745–74.

Skelton, J., *The Poetical Works of John Skelton, With Notes*, ed. A. Dyce, 2 vols (London: Thomas Rodd, 1843).

Skempton, A. W. and J. Brown, 'John and Edward Troughton, Mathematical Instrument Makers', *Notes and Records of the Royal Society*, 27:2 (1973), pp. 233–62.

Slyboots, S., *The Tavern Scuffle: Or, the Club in an Uproar* (London, 1726).

Smith, G., *The Nature of Fermentation Explain'd* (London, 1729).

Smith, S. D., 'Coffee, Microscopy, and the *Lancet*'s Analytical Sanitary Commission', *Social History of Medicine*, 14:2 (2001), pp. 171–97.

[Smollett, T.], *The Expedition of Humphry Clinker*, 3 vols (London, 1771).

Smyth, W., *The Publican's Guide, or, Key to the Distill-House* (London, 1781).

'Society for the Diffusion of Useful Knowledge' (prospectus, 1827), bound in with [H. Brougham], *A Discourse of the Objects, Advantages, and Pleasures of Science* (London: Baldwin and Cradock, 1827).

Sorrenson, R., 'George Graham, Visible Technician', *British Journal for the History of Science*, 32:2 (1999), pp. 203–21.

Southby, E. R., *Brewing Practically and Scientifically Considered* (London: Unwin, 1877).

Sprat, T., *The History of the Royal-Society of London for the Improving of Natural Knowledge* (London, 1667).

Squire, W. S., 'On the Processes Concerned in the Conversion of Starch into Alcohol, and their Relation to Brewing and Distilling', *Brewers' Guardian*, 14 (1884), pp. 285–6, 303–5, 316–19.

Stephens, M. and G. Roderick, 'The Muspratts of Liverpool', *Annals of Science*, 29:3 (1972), pp. 287–311.

Steuens, C. C. E. and J. Liebault, *Maison Rustique, or the Covntrie Farme*, trans. R. Surflet (London, 1600).

Stevenson, B., 'John Nokes Furze, 1817–1859', *Micscape*, 181 (November 2010), at http://www.microscopy-uk.org.uk/mag//artnov10/bs-FurzeJN.html [accessed 13 January 2013].

Stewart, L., *The Rise of Public Science: Rhetoric, Technology and Natural Philosophy in Newtonian Britain, 1660–1750* (Cambridge: Cambridge University Press, 1992).

Stieb, E., *Drug Adulteration: Detection and Control in Nineteenth-Century Britain* (Madison, WI: University of Wisconsin Press, 1966).

Stopes, H., *Malt and Malting* (London: F. W. Lyon, 1885).

Stubbs, M., 'John Beale, Philosophical Gardener of Herefordshire. Part 1: Prelude to the Royal Society (1608–1663)', *Annals of Science*, 39:5 (1982), pp. 463–89.

—, 'John Beale, Philosophical Gardener of Herefordshire. Part 2: The Improvement of Agriculture and Trade in the Royal Society (1663–1683)', *Annals of Science*, 46:4 (1989), pp. 323–63.

Sumner, J., 'John Richardson, Saccharometry and the Pounds-Per-Barrel Extract: The Construction of a Quantity', *British Journal for the History of Science*, 34:3 (2001), pp. 255–73.

—, 'Early Heat Determination in the Brewery', *Brewery History*, 121 (2005), pp. 66–80.

—, 'Michael Combrune, Peter Shaw and Commercial Chemistry: The Chemical Origins of Brewing Thermometry', *Ambix*, 54:1 (2007), pp. 5–29.

—, 'Status, Scale and Secret Ingredients: The Retrospective Invention of London Porter', *History and Technology*, 24:3 (2008), pp. 289–306.

Tate, F. G. H., *Alcoholometry: An Account of the British Method of Alcoholic Strength Determination* (London: HMSO, 1930).

Teich, M., *Bier, Wissenschaft und Wirtschaft in Deutschland, 1800–1914* (Vienna: Böhlau, 2000).

The Theory and Practice of Malting and Brewing (Edinburgh, 1793).

Thomson, T., 'On Pepper', *Journal of Natural Philosophy*, 9 (1804), pp. 68–78.

—, *A System of Chemistry*, 2nd edn, 4 vols (Edinburgh: Bell and Bradfute, 1804).

—, *A System of Chemistry*, 3rd edn, 5 vols (Edinburgh: Bell and Bradfute, 1807).

—, *A System of Chemistry*, 4th edn, 5 vols (Edinburgh: Bell and Bradfute, 1810).

—, *History of the Royal Society, from its Institution to the End of the Eighteenth Century* (London: Baldwin, 1812).

—, *A System of Chemistry*, 5th edn, 4 vols (London: Baldwin, Cradock and Joy, 1817).

—, 'Answer to the Review of the Sixth Edition of Dr Thomson's System', *Annals of Philosophy*, 3 (1822), pp. 241–75.

—, *Chemistry of Organic Bodies: Vegetables* (London: J. B. Baillière, 1838).

— and W. Stewart, *Brewing and Distillation, with Practical Instructions* (Edinburgh: A. and C. Black, 1849).

[Thomson, T.], 'Brewing', *Encyclopaedia Britannica*, supplement to the 4th, 5th and 6th editions (1824).

Thorne, R. G., *The House of Commons, 1790–1820*, 5 vols (London: Secker and Warburg, 1986).

Threale, T., *The Complete Family-Brewer: or the Best Method of Brewing* (London, 1802).

Tizard, W. L., *The Theory and Practice of Brewing Illustrated*, 1st edn (London: sold by the author, 1843), p. 151.

—, *The Theory and Practice of Brewing Illustrated*, 2nd edn (London: sold by the author, 1846).

—, *The Theory and Practice of Brewing Illustrated*, 3rd edn (London: sold by the author, 1850).

—, *The Theory and Practice of Brewing Illustrated*, 4th edn (London: sold by the author, 1857).

Topham, J., 'Science and Popular Education in the 1830s: The Role of the Bridgewater Treatises', *British Journal for the History of Science*, 25:4 (1992), pp. 397–430.

—, 'Scientific Publishing and the Reading of Science in Nineteenth-Century Britain: A Historiographical Survey and Guide to Sources', *Studies in History and Philosophy of Science*, 31:4 (2000), pp. 559–612.

Townsend, J. R., *Trade and Plumb-Cake Forever, Huzza! The Life and Work of John Newbery, 1713–1767* (Cambridge: Colt, 1994).

The Trials at Large of Joseph Merceron, Esq. for Fraud (London: printed for W. Wright, 1819).

[Tryon, T.], *A New Art of Brewing Beer, Ale and Other Sorts of Liquors* (London, 1690).

Tuck, J., *The Private Brewer's Guide to the Art of Brewing Ale and Porter*, 2nd edn (London, 1822).

[Tuck, J.], *A Guide to Young Brewers, Particularly Adapted to the Families of the Nobility and Gentry* (London: F. Westley, 1820).

Tyndall, J., 'Fermentation and its Bearings on the Phenomena of Disease', *Popular Science Monthly*, 10:9 (1876), pp. 129–54.

Underwood, E. A., *Boerhaave's Men, at Leyden and After* (Edinburgh: Edinburgh University Press, 1977).

Unger, R., *Beer in the Middle Ages and the Renaissance* (Philadelphia, PA: University of Pennsylvania Press, 2004).

Ure, A., *A Dictionary of Chemistry, on the Basis of Mr Nicholson's*, 1st edn (London: Thomas and George Underwood, 1821).

—, *A Dictionary of Arts, Manufactures, and Mines*, 1st edn (London: Longman, 1839).

—, *A Dictionary of Arts, Manufactures, and Mines*, 3rd edn (London: Longman, 1843).

—, *Recent Improvements in Arts, Manufactures, and Mines: Being the Second Edition of a Supplement to the Third Edition of his Dictionary* (London: Longman, 1845).

[Ure, A.], 'A System of Chemistry, in Four Volumes, 8vo. by Thomas Thomson', [*Quarterly*] *Journal of Science and the Arts*, 4 (1818), pp. 299–321.

Valentin, W. G., 'On the Preparation of Dextrine-Maltose (Malt-Sugar) and its Use in Brewing', *Journal of the Society of Arts*, 24 (1875), pp. 404–14.

Verhoeven, W., *Gilbert Imlay: Citizen of the World* (London: Pickering & Chatto, 2008).

Vernon, K. D. C., 'The Library of the Royal Institution, 1799–1954', *Nature*, 174:4423 (1954), pp. 242–4.

Vinum Britannicum: or, an Essay on the Properties and Effects of Malt Liquors (Lo[n]don, 1727).

Votes of the [*Irish*] *House of Commons* ([Dublin], [1792]).

Walters, A., 'Tools of Enlightenment: The Material Culture of Science in Eighteenth-Century England' (PhD dissertation, University of California at Berkeley, 1992).

Ward, E. R., 'Peter Griess (1829–1888) and the Burton Breweries', *Journal of the Royal Institute of Chemistry*, 82 (1958), pp. 458–63.

Warington, R. [Jr], 'Notes on the Chemistry of Tartaric and Citric Acid', *Journal of the Chemical Society*, 28 (1875), pp. 925–94.

Warington, R., 'Observations on the Action of Animal Charcoal', *Memoirs and Proceedings of the Chemical Society of London*, 2 (1845), pp. 326–9.

Watkins, G., *The Compleat English Brewer, or, the Whole Art and Mystery of Brewing* (London, 1767).

[Watkins, G.], *The Compleat Brewer; or, the Art and Mystery of Brewing Explained* (London, 1760).

Webb, C., *London Livery Company Apprenticeship Registers, vol. 1: Brewers' Company 1685–1800* (London: Society of Genealogists, 1996).

Wheeler, G., *Home Brewing* (St Albans: CAMRA Books, 1997).

Whitaker, K., 'The Culture of Curiosity', in N. Jardine, J. Secord and E. Spary (eds), *Cultures of Natural History* (Cambridge: Cambridge University Press, 1996), pp. 75–90.

White, J., *Art's Treasury of Rareties, and Curious Inventions,* 4th edn (London, [*c.* 1690]).

Wigney, G. A., *A Philosophical Treatise on Malting and Brewing* (Worthing: Worthing Press, 1823).

—, *A Theoretical and Practical Treatise on Malting and Brewing* (Brighton: printed for the author, 1835).

—, *An Elementary Dictionary, or Cyclopædiæ, for the Use of Maltsters, Brewers, Distillers, Rectifiers, Vinegar Manufacturers, and Others* (Brighton: printed for the author, 1838).

—, *An Introductory Treatise on the Theory and Practice of Malting and Brewing* (Brighton: Frederick Wigney, 1850).

Williamson, A., *A Plea for Pure Science, Being the Inaugural Lecture at the Opening of the Faculty of Science* (London: Taylor and Francis, 1870).

Willmoth, F., *Sir Jonas Moore: Practical Mathematics and Restoration Science* (Woodbridge: Boydell, 1993).

Wiltshire, J., *Samuel Johnson in the Medical World* (Cambridge: Cambridge University Press, 1991).

Wise, M. N. (ed.), *The Values of Precision* (Princeton, NJ: Princeton University Press, 1995).

Worthington, G. L., *The Brewer's Guide: A New Work* (London, 1812).

Yeats, J., *The Technical History of Commerce* (London: Cassell, Petter and Galpin, 1871).

Yeo, R., *Encyclopaedic Visions* (Cambridge: Cambridge University Press, 2001).

The Young Brewer's Monitor, Comprising a Luminous and Scientific Summary of that Very Ancient and Important Art (London: Baldwin, 1824).

Yworth, W., *Cerevisiarii Comes, or, the New and True Art of Brewing* (London, 1692).

—, *The Compleat Distiller: or the Whole Art of Distillation Practically Stated,* 2nd edn (London, 1705).

INDEX

www.ingramcontent.com/pod-product-compliance
Lightning Source LLC
Chambersburg PA
CBHW020119200626
46936CB00024B/414
* 9 7 8 0 8 2 2 9 6 5 3 1 2 *